Crystallization of High Performance Metallic Materials

Crystallization of High Performance Metallic Materials

Guest Editors

Wangzhong Mu
Chao Chen

Basel • Beijing • Wuhan • Barcelona • Belgrade • Novi Sad • Cluj • Manchester

Guest Editors

Wangzhong Mu
Key Laboratory of
Electromagnetic Processing
of Materials (EPM)
Northeastern University
Shenyang
China

Chao Chen
Department of Metallurgy
Taiyuan University of Technology
Taiyuan
China

Editorial Office
MDPI AG
Grosspeteranlage 5
4052 Basel, Switzerland

This is a reprint of the Special Issue, published open access by the journal *Crystals* (ISSN 2073-4352), freely accessible at: www.mdpi.com/journal/crystals/special_issues/G43656XLVO.

For citation purposes, cite each article independently as indicated on the article page online and using the guide below:

Lastname, A.A.; Lastname, B.B. Article Title. *Journal Name* **Year**, *Volume Number*, Page Range.

ISBN 978-3-7258-3274-3 (Hbk)
ISBN 978-3-7258-3273-6 (PDF)
https://doi.org/10.3390/books978-3-7258-3273-6

© 2025 by the authors. Articles in this book are Open Access and distributed under the Creative Commons Attribution (CC BY) license. The book as a whole is distributed by MDPI under the terms and conditions of the Creative Commons Attribution-NonCommercial-NoDerivs (CC BY-NC-ND) license (https://creativecommons.org/licenses/by-nc-nd/4.0/).

Contents

About the Editors vii

Wangzhong Mu and Chao Chen
Crystallization of High-Performance Metallic Materials
Reprinted from: *Crystals* 2025, 15, 147, https://doi.org/10.3390/cryst15020147 1

Nashmi H. Alrasheedi, Mohamed M. El-Sayed Seleman, Mohamed M. Z. Ahmed and Sabbah Ataya
Fatigue and Fracture Behaviors of Short Carbon Fiber Reinforced Squeeze Cast AZ91 at 20 °C and 250 °C
Reprinted from: *Crystals* 2023, 13, 1469, https://doi.org/10.3390/cryst13101469 7

Chengzhi Yang, Bin Wu, Wenmin Deng, Shuzhen Li, Jianfeng Jin and Qing Peng
Assessment of the Interatomic Potentials of Beryllium for Mechanical Properties
Reprinted from: *Crystals* 2023, 13, 1330, https://doi.org/10.3390/cryst13091330 25

Qingdong Li, Shuai Liu, Binbin Liao, Baohua Nie, Binqing Shi and Haiying Qi et al.
Effect of Pore Defects on Very High Cycle Fatigue Behavior of TC21 Titanium Alloy Additively Manufactured by Electron Beam Melting
Reprinted from: *Crystals* 2023, 13, 1327, https://doi.org/10.3390/cryst13091327 44

Shuai Liu, Fangjun Liu, Zhanhao Yan, Baohua Nie, Touwen Fan and Dongchu Chen et al.
Nucleation of $L1_2$-Al_3M (M = Sc, Er, Y, Zr) Nanophases in Aluminum Alloys: A First-Principles ThermodynamicsStudy
Reprinted from: *Crystals* 2023, 13, 1228, https://doi.org/10.3390/cryst13081228 55

Yong Wang and Wangzhong Mu
Effect of Cooling Rate on Crystallization Behavior during Solidification of Hyper Duplex Stainless Steel S33207: An In Situ Confocal Microscopy Study
Reprinted from: *Crystals* 2023, 13, 1114, https://doi.org/10.3390/cryst13071114 65

Ying Wang, Ge Zhou, Xin Che, Feng Li and Lijia Chen
Effects of Ag on High-Temperature Creep Behaviors of Peak-Aged Al-5Cu-0.8Mg-0.15Zr-0.2Sc (-0.5Ag)
Reprinted from: *Crystals* 2023, 13, 1096, https://doi.org/10.3390/cryst13071096 80

Dengyu Liu, Qingqing Ding, Qian Zhou, Dingxin Zhou, Xiao Wei and Xinbao Zhao et al.
Microstructure, Mechanical Properties and Thermal Stability of Ni-Based Single Crystal Superalloys with Low Specific Weight
Reprinted from: *Crystals* 2023, 13, 610, https://doi.org/10.3390/cryst13040610 94

Xinmao Qin, Yilong Liang, Jiabao Gu and Guigui Peng
The Effect of Interatomic Potentials on the Nature of Nanohole Propagation in Single-Crystal Nickel: A Molecular Dynamics Simulation Study
Reprinted from: *Crystals* 2023, 13, 585, https://doi.org/10.3390/cryst13040585 107

Haoyu Zhang, Shuo Zhang, Shuai Zhang, Xuejia Liu, Xiaoxi Wu and Siqian Zhang et al.
High Temperature Deformation Behavior of Near-β Titanium Alloy Ti-3Al-6Cr-5V-5Mo at $\alpha + \beta$ and β Phase Fields
Reprinted from: *Crystals* 2023, 13, 371, https://doi.org/10.3390/cryst13030371 119

Kanwen Hou, Guohao Liu, Jia Yang, Wei Wang, Lixin Xia and Jun Zhang et al.
Vacuum Electrodeposition of Cu(In, Ga)Se$_2$ Thin Films and Controlling the Ga Incorporation Route
Reprinted from: *Crystals* **2023**, *13*, 319, https://doi.org/10.3390/cryst13020319 . . . . **136**

Zhiwen Lan, Hanjie Shao, Lei Zhang, Hong Yan, Mojia Huang and Tengfei Zhao
Elastic Constitutive Relationship of Metallic Materials Containing Grain Shape
Reprinted from: *Crystals* **2022**, *12*, 1768, https://doi.org/10.3390/cryst12121768 . . . . **150**

Xiufan Yang, Xinmao Qin, Wanjun Yan, Chunhong Zhang, Dianxi Zhang and Benhua Guo
Electronic Structure and Optical Properties of Cu$_2$ZnSnS$_4$ under Stress Effect
Reprinted from: *Crystals* **2022**, *12*, 1454, https://doi.org/10.3390/cryst12101454 . . . . **172**

About the Editors

Wangzhong Mu

Wangzhong Mu is an associate professor at the Royal Institute of Technology (KTH), Department of Materials Science and Engineering, and a visiting professor at Key Lab EPM, Northeastern University. He received both a Docent (Assoc. Prof./Reader) qualification and a Ph.D. degree from KTH. Moreover, he has worked at McMaster University (Canada), Tohoku University (Japan), and Ferritico AB (Sweden). He has close collaboration with the Max Planck Institute of Sustainable Materials (Düsseldorf, Germany), Hanyang University (Korea), Montan University Leoben (Austria), etc. His research interests include (i) particle behaviors in metals, (ii) sustainable metallurgy, (iii) artificial intelligence (AI)-based material design, etc. He has served as a PI for over 15 national-level and EU-level research grants, published over 70 peer-reviewed journal articles, and contributed over 20 plenary/keynote/invited lectures at international conferences.

Chao Chen

Dr. Chao Chen is an associate professor at the College of Materials Science and Engineering, Taiyuan University of Technology. He completed his Ph.D. degree in Material Science and Engineering at the KTH Royal Institute of Technology in 2015. He is a member of the Academic Committee on Metallurgical Reaction Engineering of the Chinese Society for Metals and a member of the National Academic Committee on Metallurgical Processes and Theories of the Metallurgical Physical Chemistry Division of the Chinese Society for Metals. His research interests include research on key technologies for producing special stainless steel; computational fluid dynamics (CFD) simulations of complex fluid flows in industrial processes; physical simulation and mathematical modeling of metallurgical processes; and preparation and application of catalysts for hydrogen production from electrolyzed water.

Editorial

Crystallization of High-Performance Metallic Materials

Wangzhong Mu [1,2,*] and Chao Chen [3,*]

[1] Key Laboratory of Electromagnetic Processing of Materials (Ministry of Education), School of Metallurgy, Northeastern University, Shenyang 110819, China
[2] Department of Materials Science and Engineering, KTH Royal Institute of Technology, Brinellvägen 23, SE-100 44 Stockholm, Sweden
[3] College of Materials Science and Engineering, Taiyuan University of Technology, Taiyuan 030024, China
* Correspondence: wmu@kth.se (W.M.); chenchao@tyut.edu.cn (C.C.)

Abstract: Crystallization includes liquid/solid and solid/solid phase transitions, important processes for improving engineering material performance, which have attracted significant attention in the community. The current Special Issue (SI) entitled '*Crystallization of High-Performance Metallic Materials*' has collected twelve research papers focusing on different aspects of the crystallization of metallic materials, e.g., the solidification of steel, fatigue and fracture behaviors of magnesium composites, nucleation of intermetallic compounds in aluminum alloys, microstructure evolution in nickel-based super-alloys, etc. The summary of crystallization behaviors at different temperature ranges in different metallic materials contributes to the state of the art of engineering material development.

Keywords: crystallization; high-performance metallic materials; solidification; steels; alloys

Received: 9 January 2025
Revised: 26 January 2025
Accepted: 27 January 2025
Published: 30 January 2025

Citation: Mu, W.; Chen, C. Crystallization of High-Performance Metallic Materials. *Crystals* **2025**, *15*, 147. https://doi.org/10.3390/cryst15020147

Copyright: © 2025 by the authors. Licensee MDPI, Basel, Switzerland. This article is an open access article distributed under the terms and conditions of the Creative Commons Attribution (CC BY) license (https://creativecommons.org/licenses/by/4.0/).

1. Introduction

Crystallization describes the general material process where a solid phase nucleates in a liquid or solid matrix. The atoms or molecules are highly organized into a structure known as a crystalline cluster. In the manufacturing field, the crystallization of metallic materials includes the formation and growth of a new solid phase during solidification [1], as well as the subsequent phase transformations [2]. Other processes, e.g., pyrolysis [3], could also be included in the broad concept of crystallization; however, they are not always mentioned within the scope. Regarding crystallization mechanisms, several fundamental aspects of thermodynamics and kinetics should be included [1–3].

The solidification process includes heat and mass transfer, and various reactions and morphology evolutions occur, e.g., dendrite growth, control of macro-segregation, and the columnar-to-equiaxed transition (CET) [4–6]. Understanding solidification will benefit the understanding of the casting process in the metallurgical industry, contributing to preventing defect formation, e.g., porosity, shrinkage [7–9], and non-metallic inclusions [10–12]. Subsequently, crystallization behaviors can also include the structural evolution of mold flux during continuous casting [13,14], post-microstructure evolutions, e.g., different types of ferrite formation [15], acicular ferrite nucleation from non-metallic inclusions [16,17], and bainite and martensitic transformation in solid-state metallic materials [18]. Last but not least, additive manufacturing (AM), as a novel and short-process technology, enables increased creativity and faster development. It has attracted much attention in the metallurgical community; crystallization in AM [19,20] is the focus for current and future research.

In the current SI, we intend to emphasize the crystallization behaviors in various high-performance metallic materials. Both liquid/solid and solid/solid transitions are considered. Furthermore, we include conventional engineering materials, e.g., steels, Ni-based superalloys, and Al alloys, as well as novel metallic materials, e.g., light-weight magnesium metal matrix composites, AM-built Ti alloys, and multicomponent alloys. State-of-the-art characterization methods (e.g., high-temperature confocal laser scanning microscopy, high-resolution microscopies) as well as modeling work (e.g., first-principles simulation, CALPHAD) regarding crystallization of metallic materials are included in this SI. Specifically, we discuss defect formations during the crystallization of different metallic materials, e.g., δ-ferrite formation and growth during solidification and segregation influenced by the cooling rate in duplex stainless steel, pore defect formation in additively manufactured TC21 Ti alloy and its influence on high-cycle fatigue behavior, etc. Additionally, the behaviors of intermetallic precipitates in solid-state high-performance alloys, e.g., the nucleation of $L1_2$-Al_3M (M = Sc, Er, Y, Zr) nanophase in advanced al alloys, are included. Finally, metallic alloys' mechanical properties associated with their crystallization behaviors, e.g., high-temperature creep behaviors of peak-aged Al-5Cu-0.8Mg-0.15Zr-0.2Sc(-0.5Ag) alloy, are highlighted. The current SI collects comprehensive research on crystallization behaviors in high-performance metallic materials, aiming to pave the way to understanding the correlation between process, structure, and properties in engineering materials.

2. An Overview of Published Articles

Manuscripts on various subjects related to crystallization behaviors were submitted for consideration for the current Special Issue (SI). After the peer-review process, twelve papers were finally accepted for publication. The contributions and their descriptions are listed in Table 1.

Contribution 1 investigated the fatigue and fracture behaviors of short carbon fiber-reinforced squeeze-cast AZ91 at different temperatures between 20 and 250 °C. In this work, mechanical properties were examined by tensile tests at the abovementioned test temperatures to find suitable fatigue testing stress and strain for stress- and strain-controlled tests. The obtained fatigue curves of stress against the number of cycles revealed that the fatigue strength of composite AZ91–carbon was approximately 55 MPa under high-cycle fatigue; additionally, the fatigue strength of the matrix alloy AZ91 was 37 MPa at 250 °C. This work finally revealed that the fracture types were mixed ductile/brittle contained fatigue serration, fiber fracture, and separation in the reinforced AZ91–carbon materials.

Contribution 2 assessed the interatomic potential of Beryllium to determine its mechanical properties. In this work, molecular dynamics simulations were used to calculate the mechanical properties of imperfect hexagonal close-packed (HCP)-type Beryllium. Through the simulation, three types of potentials, i.e., MEAM, Finnis–Sinclair, and Tersoff, were assessed. Furthermore, the volumetric change (VC) with pressure according to MEAM and Tersoff and the VC with temperature according to MEAM were consistent with the obtained experimental data. Finally, MEAM-type potential was found to deliver the most reasonable predictions of the targeted properties.

Contribution 3 investigated the effect of pore defects on the high-cycle fatigue behavior of TC21 Titanium alloy prepared by electron beam melting-type additive manufacturing (AM). The obtained stress–life cycle (S-N) curve of non-HIP specimens clearly showed a tendency to decrease in very-high-cycle regimes, and HIP treatment obviously improved fatigue properties. Finally, a fatigue indicator parameter model according to the pore defect characteristics investigated was established to predict the fatigue life of HIP and non-HIP samples.

Table 1. Summary of the contributions published in this Special Issue.

No. of Contribution	Research Area	Focus	Type of Research
1	Mechanical properties of AZ91 Mg alloy	Fatigue and fracture behaviors	Experimental study
2	Calculation of mechanical properties of Beryllium	Assessment of the interatomic potentials	Molecular dynamics simulation
3	Pore defects on the AM-built TC21 Titanium alloy	High-cycle fatigue behavior of Ti alloy	Experimental study
4	Nucleation of $L1_2$-Al_3M nanoparticles in Al alloys	Critical radius and nucleation energy of different intermetallic compounds	First-principles simulation
5	Solidification of hyper-duplex stainless steels	Crystallization kinetics and microstructure characterization	In situ characterization
6	Creep behaviors of multicomponent Al alloy	Effect of Ag addition on creep behavior and the mechanisms	Experimental study
7	Microstructure and properties of Ni-based single-crystal superalloy	Mechanical properties and thermal stability of Ni alloy	Experimental study
8	Nanohole propagation in single-crystal nickel	Effect of interatomic potentials on the properties of Ni	Molecular dynamics simulation study
9	Mechanical property evolution of near-β Titanium alloy	High-temperature deformation behavior of Ti alloy	Experimental study
10	Thin-film fabrication using vacuum electrodeposition	Preparation and characterization of a $Cu(In,Ga)Se_2$ thin film	Surface treatment study
11	Elastic constitutive relationship of metallic materials	Grain shape effect on the elastic properties of metals	Theoretical study
12	Effect of stress on properties of thin-film solar cell materials	Electronic structure and optical properties of Cu_2ZnSnS_4	First-principles and DFT calculations

Contribution 4 provided a first-principles and thermodynamic study to investigate the nucleation of $L1_2$-Al_3M (M = Sc, Er, Y, Zr) nanophases in Al alloys. The calculation results showed that the critical radius and nucleation energy of the $L1_2$-Al_3M phases decreased in the order Al_3Er, Al_3Y, Al_3Sc, and Al_3Zr.

Contribution 5 investigated the effect of cooling rate on the crystallization behavior of hyper-duplex stainless steel SAF™ 3207 HD (also named UNS S33207) during solidification. A combination of in situ observation using high-temperature confocal laser scanning microscopy (HT-CLSM) and differential scanning calorimetry (DSC) was used. The effect of the cooling rate, i.e., 4 and 150 °C/min, on the nucleation and growth behavior of δ-ferrite in S33207 during the solidification was investigated in situ. The results showed that S33207 steel's solidification mode was a ferrite–austenite type. Liquid to δ-ferrite transformation occurred at a certain degree of undercooling, and merging occurred during the growth of the δ-ferrite-phase dendrites.

Contribution 6 reported Ag's effect on the high-temperature creep behaviors of peak-aged Al-5Cu-0.8Mg-0.15Zr-0.2Sc(-0.5Ag) multicomponent alloys. The high-temperature creep performances of the proposed alloy were significantly improved with Ag addition.

Subsequently, constitutive relational models of the multicomponent alloy during high-temperature creep were built, and the activation energy could be calculated. The creep mechanism after Ag addition transitioned from lattice diffusion control to grain boundary diffusion control.

Contribution 7 provided a comprehensive study of the microstructure, mechanical properties, and thermal stability of Ni-based single-crystal superalloys with low specific weight (LSW). A multicomponent Ni-Co-Cr-Mo-Ta-Re-Al-Ti system was investigated. The alloys' mechanical properties were examined by tensile tests, elongation tests, and thermal exposure tests. The results of this work provided scientific insights for developing Ni-based single-crystal superalloys with LSW properties.

Contribution 8 performed a molecular dynamics simulation on the effect of interatomic potential on the nature of nanohole propagation in single-crystal nickel. It showed the difference between the different styles of interatomic potentials characterizing nanohole propagation in single-crystal Ni. Furthermore, it provided a theoretical basis for the selection of interatomic potentials using the molecular dynamics simulation methodology.

Contribution 9 studied the high-temperature deformation behavior of near-β Titanium alloy (Ti-3Al-6Cr-5V-5Mo); the flow stress behavior and processing maps in an α + β-phase field and β-phase field were investigated. The experimental data obtained from hot compressing simulations at 700 to 820 °C were used to establish constitutive models. After the deformation test, the maximum number of dynamic recrystallization grains and the minimum average grain size could be obtained. The current results are consistent with the high-power dissipation coefficient region, which is predicted by the processing map.

Contribution 10 studied the vacuum electrodeposition of $Cu(In,Ga)Se_2$ (CIGS) thin films prepared under a 3 kPa vacuum. Furthermore, the vacuum electrodeposition mechanism of CIGS was investigated. Meanwhile, the route of Ga incorporation into the thin films could be controlled in a vacuum environment by inhibiting pH changes at the cathode region. A higher current density and a lower diffusion impedance and charge transfer impedance were used in the abovementioned preparation process.

Contribution 11 investigated the elastic constitutive relationship of metallic materials containing different grain shapes; a new expression of the elastic constitutive relationship of polycrystalline materials containing different grain shape effects was established. The experimental results showed that the grain shape parameter was consistent with the theoretical results of the material's macroscopic mechanical properties.

Contribution 12 studied the electronic structure and optical properties of Cu_2ZnSnS_4; first-principles calculations based on density functional theory were applied for this study. Through this method, the band structure, density of states, and optical properties of Cu_2ZnSnS_4 under isotropic stress were calculated and analyzed. The results showed that Cu_2ZnSnS_4 is a direct band gap semiconductor under isotropic stress, and the lattice is tetragonal.

3. Summary

The current Special Issue (SI), *Crystallization of High-Performance Metallic Materials*, collects research contributions about solidification, casting, recrystallization during deformation, and mechanical property evolution of different engineering materials, e.g., stainless steels, Ni-based superalloys, Ti alloys, etc. Both experimental and simulation studies on crystallization topics were reported in different papers. Some research work on topics like the crystallization of slags (silicates) and mold flux has not been collected yet; we will continue to organize a Volume II SI on the same topic to collect more contributions on different topics of material crystallization.

Author Contributions: Conceptualization, W.M. and C.C.; methodology, W.M. and C.C.; investigation, W.M. and C.C.; resources, W.M. and C.C.; writing—original draft preparation, W.M.; writing—review and editing, W.M. and C.C.; project administration, W.M. and C.C.; funding acquisition, W.M. All authors have read and agreed to the published version of the manuscript.

Acknowledgments: As Guest Editors of the Special Issue (SI) *"Crystallization of High-Performance Metallic Materials"*, we (W.M. and C.C.) would like to express our deep appreciation to all the authors who contributed valuable work to publish in the current SI, and all the anonymous reviewers who provided their professional opinions to support the peer evaluation process.

Conflicts of Interest: The authors declare no conflicts of interest.

List of Contributions

1. Alrasheedi, N.H.; El-Sayed Seleman, M.M.; Ahmed, M.M.Z.; Ataya, S. Fatigue and Fracture Behaviors of Short Carbon Fiber Reinforced Squeeze Cast AZ91 at 20 °C and 250 °C. *Crystals* **2023**, *13*, 1469. https://doi.org/10.3390/cryst13101469.
2. Yang, C.; Wu, B.; Deng, W.; Li, S.; Jin, J.; Peng, Q. Assessment of the Interatomic Potentials of Beryllium for Mechanical Properties. *Crystals* **2023**, *13*, 1330. https://doi.org/10.3390/cryst13091330.
3. Li, Q.; Liu, S.; Liao, B.; Nie, B.; Shi, B.; Qi, H.; Chen, D.; Liu, F. Effect of Pore Defects on Very High Cycle Fatigue Behavior of TC21 Titanium Alloy Additively Manufactured by Electron Beam Melting. *Crystals* **2023**, *13*, 1327. https://doi.org/10.3390/cryst13091327.
4. Liu, S.; Liu, F.; Yan, Z.; Nie, B.; Fan, T.; Chen, D.; Song, Y. Nucleation of L12-Al3M (M = Sc, Er, Y, Zr) Nanophases in Aluminum Alloys: A First-Principles Thermodynamics Study. *Crystals* **2023**, *13*, 1228. https://doi.org/10.3390/cryst13081228.
5. Wang, Y.; Mu, W. Effect of Cooling Rate on Crystallization Behavior during Solidification of Hyper Duplex Stainless Steel S33207: An In Situ Confocal Microscopy Study. *Crystals* **2023**, *13*, 1114. https://doi.org/10.3390/cryst13071114.
6. Wang, Y.; Zhou, G.; Che, X.; Li, F.; Chen, L. Effects of Ag on High-Temperature Creep Behaviors of Peak-Aged Al-5Cu-0.8Mg-0.15Zr-0.2Sc(-0.5Ag). *Crystals* **2023**, *13*, 1096. https://doi.org/10.3390/cryst13071096.
7. Liu, D.; Ding, Q.; Zhou, Q.; Zhou, D.; Wei, X.; Zhao, X.; Zhang, Z.; Bei, H. Microstructure, Mechanical Properties and Thermal Stability of Ni-Based Single Crystal Superalloys with Low Specific Weight. *Crystals* **2023**, *13*, 610. https://doi.org/10.3390/cryst13040610.
8. Qin, X.; Liang, Y.; Gu, J.; Peng, G. The Effect of Interatomic Potentials on the Nature of Nanohole Propagation in Single-Crystal Nickel: A Molecular Dynamics Simulation Study. *Crystals* **2023**, *13*, 585. https://doi.org/10.3390/cryst13040585.
9. Zhang, H.; Zhang, S.; Zhang, S.; Liu, X.; Wu, X.; Zhang, S.; Zhou, G. High Temperature Deformation Behavior of Near-β Titanium Alloy Ti-3Al-6Cr-5V-5Mo at α + β and β Phase Fields. *Crystals* **2023**, *13*, 371. https://doi.org/10.3390/cryst13030371.
10. Hou, K.; Liu, G.; Yang, J.; Wang, W.; Xia, L.; Zhang, J.; Xu, B.; Yang, B. Vacuum Electrodeposition of Cu(In, Ga)Se$_2$ Thin Films and Controlling the Ga Incorporation Route. *Crystals* **2023**, *13*, 319. https://doi.org/10.3390/cryst13020319.
11. Lan, Z.; Shao, H.; Zhang, L.; Yan, H.; Huang, M.; Zhao, T. Elastic Constitutive Relationship of Metallic Materials Containing Grain Shape. *Crystals* **2022**, *12*, 1768. https://doi.org/10.3390/cryst12121768.
12. Yang, X.; Qin, X.; Yan, W.; Zhang, C.; Zhang, D.; Guo, B. Electronic Structure and Optical Properties of Cu$_2$ZnSnS$_4$ under Stress Effect. *Crystals* **2022**, *12*, 1454. https://doi.org/10.3390/cryst12101454.

References

1. Wang, Y.; Wang, Q.; Mu, W. In Situ Observation of Solidification and Crystallization of Low-Alloy Steels: A Review. *Metals* **2023**, *13*, 517. [CrossRef]
2. Wang, Y.; Chen, C.; Ren, R.J.; Xue, Z.X.; Wang, H.Z.; Zhang, Y.Z.; Wang, J.X.; Wang, J.; Chen, L.; Mu, W. Ferrite Formation and Decomposition in 316H Austenitic Stainless Steel Electro Slag Remelting Ingot for Nuclear Power Applications. *Mater. Charact.* **2024**, *218*, 114581. [CrossRef]
3. Wen, Y.; Wang, S.; Mu, W.; Yang, W.; Jönsson, P.G. Pyrolysis Performance of Peat Moss: A Simultaneous In-situ Thermal Analysis and Bench-scale Experimental Study. *Fuel* **2020**, *277*, 118173. [CrossRef]
4. Stawarz, M. Crystallization of Intermetallic Phases Fe_2Si, Fe_5Si_3 for High Alloyed Cast Irons. *Crystals* **2023**, *13*, 1033. [CrossRef]
5. Wang, Y.; Zhang, L. Study on the Positive Segregation in Columnar-to-equiaxed Transition Zone. *Metall. Res. Technol.* **2023**, *120*, 104. [CrossRef]
6. Lekakh, S.N.; O'malley, R.; Emmendorfer, M.; Hrebec, B. Control of Columnar to Equiaxed Transition in Solidification Macrostructure of Austenitic Stainless Steel Castings. *ISIJ Int.* **2017**, *57*, 824–832. [CrossRef]
7. Hardin, R.A.; Beckermann, C. Effect of porosity on deformation, damage, and fracture of cast steel. *Metall. Mater. Trans. A* **2013**, *44*, 5316–5332. [CrossRef]
8. Li, J.; Xu, X.W.; Ren, N.; Xia, M.X.; Li, J.G. A Review on Prediction of Casting Defects in Steel Ingots: From Macrosegregation to Multi-defect Model. *J. Iron Steel Res. Int.* **2022**, *29*, 1901–1914. [CrossRef]
9. Fan, J.; Li, Y.; Chen, C.; Ouyang, X.; Wang, T.; Lin, W. Effect of Uniform and Non-Uniform Increasing Casting Flow Rate on Dispersion and Outflow Percentage of Tracers in Four Strand Tundishes under Strand Blockage Conditions. *Metals* **2022**, *12*, 1016. [CrossRef]
10. Park, J.H.; Kang, Y. Inclusions in stainless steels—A review. *Steel Res. Inter.* **2017**, *88*, 1700130. [CrossRef]
11. Wang, Y.; Karasev, A.; Park, J.H.; Jönsson, P.G. Non-metallic inclusions in different ferroalloys and their effect on the steel quality: A review. *Metall. Mater. Trans. B* **2021**, *52*, 2892–2925. [CrossRef]
12. Ren, Y.; Wang, Y.; Li, S.; Zhang, L.; Zuo, X.; Lekakh, S.N.; Peaslee, K. Detection of Non-metallic Inclusions in Steel Continuous Casting Billets. *Metall. Mater. Trans. B* **2014**, *45*, 1291–1303. [CrossRef]
13. Wang, W.; Xu, H.; Zhai, B.; Zhang, L. A Review of the Melt Structure and Crystallization Behavior of Non-reactive Mold Flux for the Casting of Advanced High-strength Steels. *Steel Res. Int.* **2022**, *93*, 2100073. [CrossRef]
14. Park, J.Y.; Ko, E.Y.; Choi, J.; Sohn, I. Characteristics of Medium Carbon Steel Solidification and Mold Flux Crystallization Using the Multi-mold Simulator. *Metals Mater. Int.* **2014**, *20*, 1103–1114. [CrossRef]
15. Wu, K.M.; Inagawa, Y.; Enomoto, M. Three-dimensional Morphology of Ferrite Formed in Association with Inclusions in Low-Carbon Steel. *Mater. Charact.* **2004**, *52*, 121–127. [CrossRef]
16. Loder, D.; Michelic, S.K.; Mayerhofer, A.; Bernhard, C. On the Capability of Non-metallic Inclusions to Act as Nuclei for Acicular Ferrite in Different Steel Grades. *Metall. Mater. Trans. B* **2017**, *48*, 1992–2006. [CrossRef]
17. Jovanović, G.; Glišić, D.; Dikić, S.; Međo, B.; Marković, B.; Vuković, N.; Radović, N. Determining the Role of Acicular Ferrite Carbides in Cleavage Fracture Crack Initiation for Two Medium Carbon Microalloyed Steels. *Materials* **2023**, *16*, 7192. [CrossRef] [PubMed]
18. Lin, S.; Borgenstam, A.; Stark, A.; Hedström, P. Effect of Si on Bainitic Transformation Kinetics in Steels Explained by Carbon Partitioning, Carbide Formation, Dislocation Densities, and Thermodynamic Conditions. *Mater. Charact.* **2022**, *185*, 111774. [CrossRef]
19. Liu, H.; Jiang, Q.; Huo, J.; Zhang, Y.; Yang, W.; Li, X. Crystallization in Additive Manufacturing of Metallic Glasses: A review. *Addit. Manufact.* **2020**, *36*, 101568. [CrossRef]
20. Tobah, M.; Andani, M.T.; Sahu, B.P.; Misra, A. Microstructural and Hall-Petch Analysis of Additively Manufactured Ferritic Alloy Using 2507 Duplex Stainless Steel Powder. *Crystals* **2024**, *14*, 81. [CrossRef]

Disclaimer/Publisher's Note: The statements, opinions and data contained in all publications are solely those of the individual author(s) and contributor(s) and not of MDPI and/or the editor(s). MDPI and/or the editor(s) disclaim responsibility for any injury to people or property resulting from any ideas, methods, instructions or products referred to in the content.

Article

Fatigue and Fracture Behaviors of Short Carbon Fiber Reinforced Squeeze Cast AZ91 at 20 °C and 250 °C

Nashmi H. Alrasheedi [1], Mohamed M. El-Sayed Seleman [2], Mohamed M. Z. Ahmed [3,*] and Sabbah Ataya [1,*]

1. Department of Mechanical Engineering, College of Engineering, Imam Mohammad Ibn Saud Islamic University, Riyadh 11432, Saudi Arabia; nhrasheedi@imamu.edu.sa
2. Department of Metallurgical and Materials Engineering, Faculty of Petroleum and Mining Engineering, Suez University, Suez 43512, Egypt; mohamed.elnagar@suezuniv.edu.eg
3. Mechanical Engineering Department, College of Engineering at Al Kharj, Prince Sattam bin Abdulaziz University, Al Kharj 11942, Saudi Arabia
* Correspondence: moh.ahmed@psau.edu.sa (M.M.Z.A.); smataya@imamu.edu.sa (S.A.)

Abstract: AZ91 is one of the most broadly used Mg alloys because of its good castability and reasonable mechanical properties. Strengthening AZ91 with carbon short fibers aims to increase tensile and fatigue strength, creep, and wear resistance. One of the proposed applications of reinforced AZ91 is the production of pistons for trucks. Such reciprocating parts are subjected to alternating fatigue loads which can lead to fatigue failure. In this respect, studying the tensile and fatigue behavior of materials subjected to such loading conditions is of great interest. The alternating low-cycle fatigue (LCF) and high-cycle fatigue (HCF) of unreinforced AZ91 and carbon fiber-reinforced AZ91 (AZ91-C) were investigated at 20 °C and 250 °C. Tensile tests were carried out at the same testing temperature to find the appropriate fatigue testing stress and strain for stress-controlled and strain-controlled tests, respectively. The fatigue curves of stress against the number of cycles (S–N) revealed that the composite AZ91-C's fatigue strength was 55 MPa under HCF, while that of the matrix alloy AZ91 was only 37 MPa at 250 °C. Fracture investigations were conducted on the broken test samples. The fracture approach in the matrix material (AZ91) is mixed ductile/brittle containing fatigue serration, fiber fracture, and separation in the reinforced material (AZ91-C).

Keywords: AZ91; composites; carbon short fibers; tensile strength; fatigue life; LCF; HCF

Citation: Alrasheedi, N.H.; El-Sayed Seleman, M.M.; Ahmed, M.M.Z.; Ataya, S. Fatigue and Fracture Behaviors of Short Carbon Fiber Reinforced Squeeze Cast AZ91 at 20 °C and 250 °C. *Crystals* **2023**, *13*, 1469. https://doi.org/10.3390/cryst13101469

Academic Editors: Wangzhong Mu and Chao Chen

Received: 19 September 2023
Revised: 2 October 2023
Accepted: 6 October 2023
Published: 9 October 2023

Copyright: © 2023 by the authors. Licensee MDPI, Basel, Switzerland. This article is an open access article distributed under the terms and conditions of the Creative Commons Attribution (CC BY) license (https://creativecommons.org/licenses/by/4.0/).

1. Introduction

The rapid development of technology today demands an increase in the quality of the products made. This is not only focused on attractive design but also on how the weight of the product can be reduced, thus leading to economic benefits and so on. For example, in the automotive and transportation sectors, weight reduction in parts or components leads to considerable fuel savings and environmental protection by lowering CO_2 emissions [1]. Over several decades, the production of parts in the automotive industry has been dominated by high-strength steels and aluminum alloys, but recently the application of magnesium alloys has received a lot of attention from practitioners and engineers worldwide [2]. It has been generally known that magnesium in its pure form is the most plentiful metal but has low mechanical properties, making its use in structural applications unlikely [3]. Different magnesium alloys can aid weight reduction more significantly than existing steels and aluminum, owing to their low density [4]. In addition, they have outstanding specific strength (strength to density ratio), favorable machinability, castability [5,6], damping capacity, and dimensional stability [7], which are advantages. In contrast to their advantages, magnesium alloys have a relatively low modulus of elasticity [8], poor corrosion resistance due to magnesium's high reactivity [9], and low mechanical strength and wear resistance, thus making them limited in scope for use in structural and friction-resistant applications [10].

The good castability of magnesium alloys makes them much more suitable for manufacturing complex parts with high precision, thus reducing machining costs, which should be seriously considered. Accordingly, some varieties of magnesium alloys have been extensively employed by automotive manufacturers [11]. Although the efforts to improve the properties and strength of the AZ91 alloy have been vigorous, the outcomes are still far from satisfactory. Therefore, it is imperative to conduct intensive research to attain better mechanical properties for the AZ91 alloy, especially the elastic modulus, creep resistance [12,13], and fatigue strength [14,15]. Furthermore, automotive components or parts in particular are subjected to periodic cyclic loading, which induces their failure during service [16,17]. This means that the fatigue properties of AZ91 alloys also need more attention to meet the criteria that are necessary in the rapidly growing automotive industry. There have been several experiments conducted to analyze the fatigue behavior of the AZ91 alloy [18,19], either at low-cycle fatigue or high-cycle fatigue, but there is still a need to upgrade it accordingly to enable products made from magnesium alloys that are truly suitable for future use. Rettberg et al. [20] performed reasonable investigations of the low-cycle fatigue behavior of AM60 and AZ91 alloys. The results showed that AM60 had higher ductility and, hence, a higher increase in fatigue life than AZ91 at higher strain amplitudes. At the lowest strain of 0.2 percent, AZ91 was more susceptible to pore position and size initiation than AM60, ultimately impacting its fatigue life. The fatigue behavior of the as-cast AZ91 alloy was reported in [21,22], where the S–N curve results revealed that the region with a high number of cycles to failure was still relatively low, and the fatigue limit range was between 60 and 85 MPa. Meanwhile, the effect of the ECAP process on the fatigue behavior of the AZ91 alloy was investigated [23], and it was found that the effect of the ECAP was not so evident, where the amplitude stresses of the AZ91 alloy as-cast and the AZ91 alloy as-ECAP were 80 MPa and 85 MPa, respectively, at an endurance limit based on 10^7 cycles. That means that in the high-cycle fatigue range there is not much of an effect, even though the AZ91 alloy has been subjected to the ECAP process. In contrast, in the low-cycle fatigue region, the fatigue life considerably increases. By reviewing the existing results, it was determined that the fatigue behavior of the AZ91 alloy needs to be increased for future improvements since it is still relatively low.

An approach that has been undertaken as a solution for improving fatigue behavior is to transform the form of the magnesium materials from alloys to composites. Previous studies have reported that the incorporation of reinforcement materials has significantly improved the characteristics of reinforced Mg materials, involving mechanical strength [24], tribological behavior [25], and the modulus of elasticity [26,27]. Among the reinforcement materials that have been incorporated into the AZ91 magnesium alloy are carbon fibers [12], carbon nanotubes [28], alumina [29], tungsten disulfides [30], metal carbides SiC [31,32], and titanium diboride [33]. In terms of the fatigue behavior of magnesium composites, the incorporation of SiCp has been proven to improve their fatigue performance when compared to monolithic AZ91D [34]. Regarding AZ91 composites reinforced with short alumina fibers, it has been stated [35] that reinforcing the AZ91 alloy with a 0.25 volume fraction is effective in increasing its fatigue strength up to 85%. This increment can be understood as a better fatigue crack initiation resistance for the AZ91 composite as compared to the AZ91 alloy. A previous study addressing the LCF and HCF characteristics of a 23 vol% carbon fiber-reinforced AE42 Mg alloy identified that there was a two-fold increase in fatigue strength once its fatigue life reached 10^7 cycles at 250 °C, from initially 25 MPa to 52.7 MPa [17]. This result motivates the study of reinforced AZ91-C under fatigue loading.

In the present study, a short carbon fiber-reinforced AZ91 alloy was used since it has been found that incorporating short carbon fibers is an appropriate method for improving the hardness, strength, and wear characteristics of the matrix alloy AZ91 [36,37]. In addition, based on the results in the literature, there is still a research deficiency specifically on the behavior of the short carbon fiber-reinforced AZ91 alloy under cyclic loading, making it crucial and engaging for further investigation. Therefore, the goal of this research is to evaluate the behavior of the reinforced AZ91 alloy with highly fractionated short carbon fibers both under low-cyclic fatigue and high-cyclic fatigue at different temperatures.

2. Experimental Details

The main materials used were an AZ91 magnesium alloy and its composites. The reinforcing material used was a short carbon fiber with a high volume fraction. The specifications of the carbon fibers used in this study were approximately 100 μm in length with a diameter of approximately 7 μm and a volume fraction (v_f) of 0.23, which were quasi-isotropically dispersed within the AZ91 alloy matrix. The detailed chemical composition of the AZ91 alloy was Mg-9.05 Al, 0.88 Zn, 0.05 Si, 0.28 Mn, 0.001 Ni, and 0.004 Fe (wt%).

The matrix alloy and the carbon fiber-reinforced composite were produced using the squeeze casting process. A preform of short carbon fibers was prepared and preheated to 400 °C, and the matrix was superheated up to 730 °C to ensure well filling and fiber/matrix wetting upon squeezing the melt into the preform. The materials were cast into truck pistons and samples were machined from these piston blocks. A more detailed description of the production, characterization, and evaluation of these materials is reported in [1,2]. Figure 1 shows a graphical flow chart containing the main preparation and testing steps. The test samples were machined to be loaded in parallel to the strengthened plane. In the current study, fatigue testing was conducted at three different temperatures, namely 20, 150, and 250 °C, on the unreinforced and reinforced magnesium alloy AZ91.

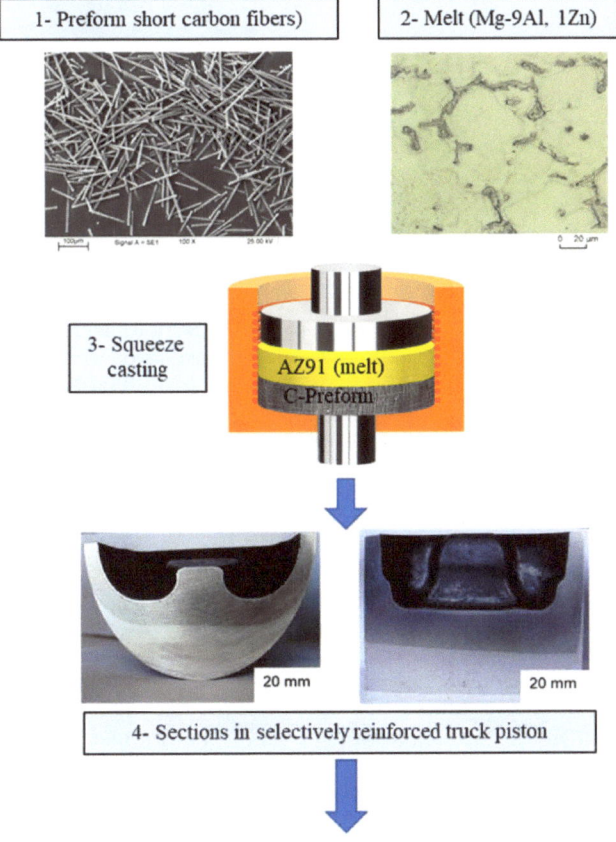

Figure 1. *Cont.*

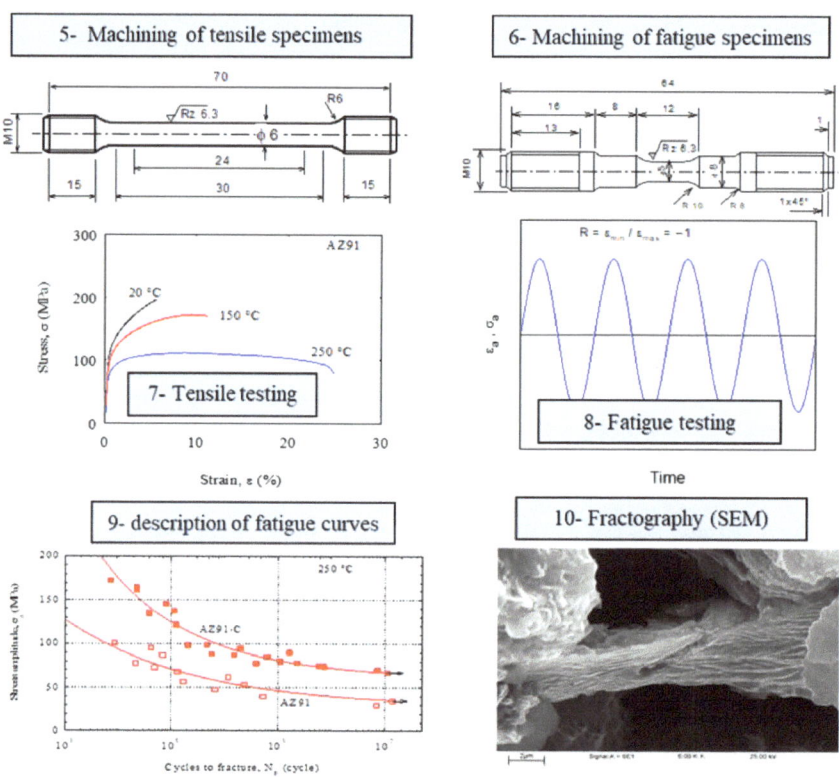

Figure 1. Flow chart with images showing the main sequence for conducting the investigation method.

The microstructures of the cast matrix alloy AZ91 and the composite AZ91-C were investigated using optical microscopy. Samples were molded for classical mechanical grinding and polishing. The polished specimens were prepared for optical microscopy by etching with 2% Nital and applying an etching time of 15 s. The light microscope Leica DM4000M (Leica Microsystems GmbH, Wetzlar, Germany) was used. The hardness of the studied materials was measured using a low-load Vickers hardness tester of Type HWDV-7S (TTS Unlimited, Osaka, Japan) with a load of 2 N for a dwell time of 15 s. An average Vickers hardness value of 10 indentations was calculated.

Samples of both tensile and alternating fatigue (R = −1) were examined in this study. Specifically, the tensile samples were prepared in a standard small-size sample according to ASTM E8 [24], while the fatigue samples were prepared following the ASTM E466 standard [25,26]. Two tensile test samples were tested for each temperature. The quasi-static tensile and cyclic tests were performed using a universal servo-hydraulic testing machine of MTS 810 (MTS Systems Corporation, Eden Prairie, MN, USA). The machine had a maximum capacity of 100 kN and a high rate of displacement of 100 mm/sec. This machine was fitted with a heating furnace that could be heated up to an upper temperature of 800 °C. Room temperature strain was measured with an extensometer that had an accuracy of 0.5 μm and a maximum displacement of 10 mm. For high-temperature testing, the extensometer used had the following criteria: it was equipped with an inductive rod and had a sensitivity of 1 μm and an upper displacement of 40 mm. The utilization of this extensometer was set at the strain-controlled LCF interval. The fatigue tests were performed at a frequency of 0.5 Hz with alternating strain control in the LCF interval of

N < 10^4 cycles, while the HCF interval was N > 10^4 cycles. Concurrently, stress-controlled examinations were carried out at a frequency and load cycles of 50 Hz and 2×10^7, respectively. The applied ratio of alternating strain and stress in all tests was R = -1 (minimum stress/maximum stress). The fatigue tests were performed until the sample was broken or stopped after exceeding a fatigue limit of 10^7 cycles. To construct the S–N curves, 16 to 17 test samples were used at the different stresses. Fracture investigation was carried out using a LEO Type 1450VP scanning electron microscope (SEM) with a voltage of 30 kV, fitted with an Oxford Type EDS detector to determine their composition. The sequence of the actions carried out in the study are systematically summarized as shown in Figure 1.

3. Results and Discussion

3.1. Microstructure and Hardness Measurement Results

AZ91 possesses some advantages in comparison to other types of magnesium alloys, such as high strength, outstanding castability, and being notably less expensive. In contrast, at temperatures greater than 140 °C, the creep resistance of AZ91 is poor [12]. This alloy has been selected for application in the manufacture of pistons made of composite materials after being reinforced with carbon fiber to improve strength and creep resistance. Figure 2 includes the casting microstructure structure of AZ91 and the reinforced AZ91-C. The cast microstructure of the matrix alloy (Figure 2a) shows a massive dendrite structure, which is observable primarily along the grain boundaries, especially at the grains' triple points. The quantitative grain size measurement indicates that the average grain size is 53 ± 11 μm. The grain boundaries, containing mainly a β–phase of $Mg_{17}(Al)_{12}$ and MgZn and the grain matrix, are composed of an α–Mg phase which contains the rest of the alloying Al additive mixed with magnesium [38,39]. A large overview of the composite material (AZ91-C) microstructure is presented in the as-polished section shown in Figure 2b. The fibers are randomly distributed in the matrix material and the fiber length is mostly laying on the reinforced plan. Figure 2c shows an optical image of etched samples, where no grain boundaries were revealed. This could be related to the presence of the high volume fraction of the fibers ($V_f = 0.23$) which act as nucleating agents, giving no chance to build the segregated structure at the grain boundaries. Some dispersed precipitates of intermetallic compounds are revealed in the matrix area and others can be seen at the fiber/matrix interface.

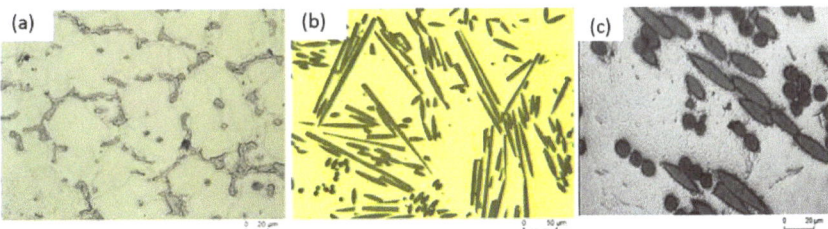

Figure 2. Microstructure of the investigated materials; (**a**) matrix (AZ91) [39] and (**b**) overview as polished composite (AZ91-C) [39] and (**c**) higher magnification for etched composite.

From the results obtained for the hardness test, the hardness of the unreinforced AZ91 was 69.2 ± 2 Hv, while the hardness of the reinforced AZ91-C was 111.53 ± 4 Hv. A remarkable increase in the hardness value of the unreinforced AZ91 of about 61.17% was found, owing to the influence of the short carbon fiber. Simultaneously, this proves that carbon fibers are efficient for becoming a potential type of reinforcement. In addition, these results are comparable to those obtained by other researchers [40,41]. It is worth mentioning that the applied hardness tester is a Vickers macro-hardness tester, and the diagonal of indentation ranged between 0.28 and 0.30 mm for the composites. The fiber diameter was about 7 μm. It means that the indentation diagonal of the composites is approximately equal to 40 times the fiber's diameter. In other words, the hardness measurements of the composite material are reliable.

3.2. Results of the Tensile Test

Figure 3 exhibits the stress–strain diagram of both unreinforced (AZ91) and reinforced (AZ91-C) at temperatures varying from room temperature to 300 °C. Both AZ91 and AZ91-C reveal a rapid increase in the flow stress during the initial stage of the deformation process, as shown in Figure 3. This is identified by the increase in the flow stress from 200 MPa for the unreinforced AZ91 to 265 MPa for the reinforced AZ91, which is clearly seen with a percentage of about 32.5% at 20 °C; this is the maximum stress achieved by both materials. This increase in stress experienced in the reinforced AZ91 can be directly understood as due to the addition of carbon fiber in the appropriate amount so that the carbon fiber can function properly [42]. In addition, the observed enhancement can be related to the onset of work hardening within the AZ91 alloy acting as the matrix, where the accumulation of continuous dislocations and kinks can promote the increase in strength of the reinforced AZ91 [43]. Another reason is the possibility of carbon fiber distribution occurring at the grain boundaries in the AZ91 alloy matrix, where grain growth is inhibited, allowing for the increased tensile stress of the reinforced AZ91 [44].

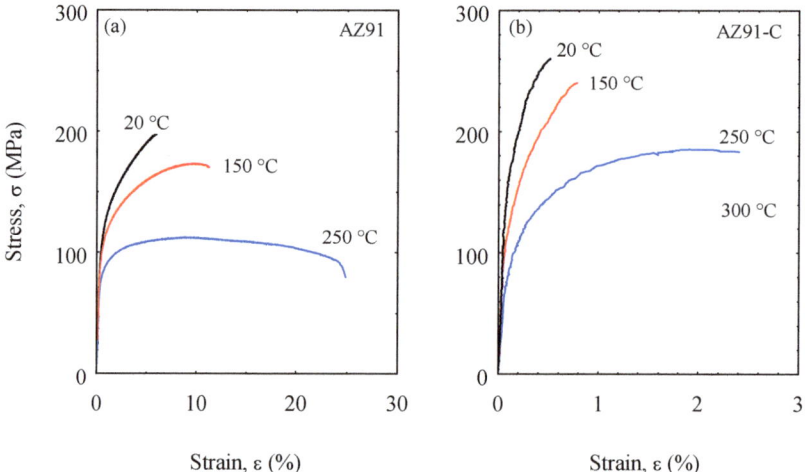

Figure 3. Comparison of stress–strain diagram of (**a**) unreinforced AZ91 and (**b**) reinforced AZ91-C at temperatures up to 300 °C.

Taking a further look, it is apparent that the tensile strength of both types of materials clearly decreases with increasing testing temperature (Figure 3). However, in general, the tensile strength of AZ91-C remains clearly higher than that of the unreinforced AZ91 at all temperatures. This means that an increase in temperature can cause a softening effect in both types of materials where dislocation rearrangement occurs, and concurrently, there is also a progressive dismantlement of the dislocations [45]. The uniqueness of the tensile test results seen in Figure 3 lies in the combination of opposing interactions in the tensile strength behavior of both materials, namely work hardening and dynamic softening. This indicates that temperature is an important factor that needs to be seriously considered in the strengthening mechanisms of materials. Even though the incorporation of short carbon fibers demonstrated the ability to enhance the tensile strength of reinforced AZ91 in general, there is a noticeable side effect in that the reinforced AZ91-C appeared to be more brittle and, thus, easily fractured during testing. This relates to its lower strain or ductility values if compared with the unreinforced AZ91 alloy. The short carbon fiber-reinforced AZ91 resulted in more limited plastic deformation at various temperatures.

Figure 4 compares the results of tensile yield stress ($\sigma_{0.2}$), ultimate tensile stress (σ_{UTS}), and strain (ε) between the matrix alloy AZ91 and the composite material AZ91-C at varying temperatures. The results presented are the average of the two tests, and the standard error has been calculated and presented in Figure 4. Table 1 includes the average values of the yield stress and the ultimate tensile strength for both the matrix alloy (AZ91) and the composite material (AZ91-C) as well as the improvement (in percent) that occurred at the different test temperatures. There is a noticeable improvement in the yield stress when reinforcing with carbon fibers. A downward trend is evident in the yield stress and the ultimate stress for both types of materials, but the reduction level in reinforced AZ91-C is somewhat greater than that of unreinforced AZ91 with increasing temperature. A high improvement in yield stress was attained at room temperature ($\approx$108%). With an increase in the test temperature, the increase in the yield stress changed to 73% and 64% at 150 °C and 250 °C, respectively. A lower improvement in the ultimate tensile strength was noticed, where the increase in strength ranged between 30% and 64%.

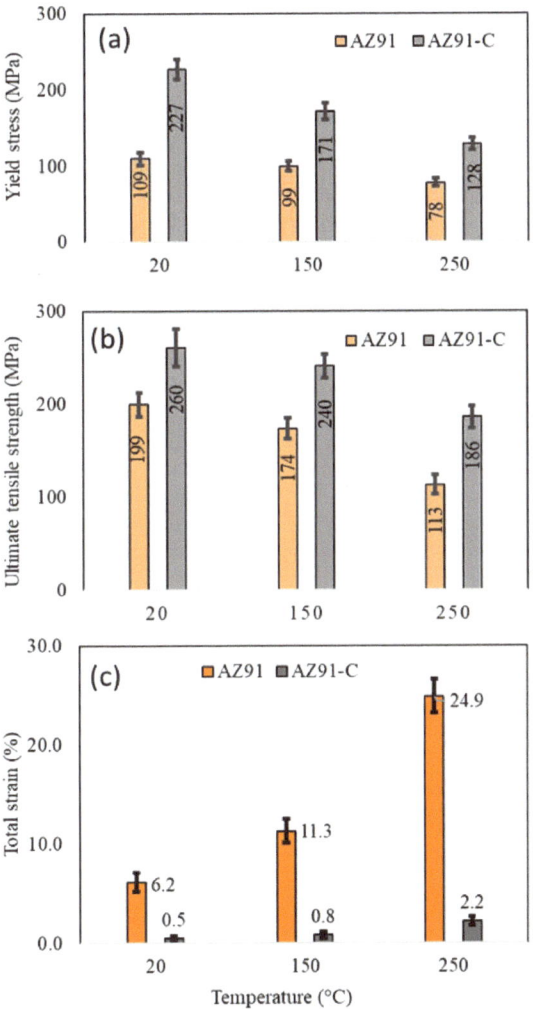

Figure 4. Tensile yield stress ($\sigma_{0.2}$) (**a**), ultimate tensile strength (σ_{UTS}) (**b**), and strain of the unreinforced AZ91 and reinforced AZ91-C (**c**) at temperatures 20, 150, and 250 °C.

Table 1. Improvement in tensile yield stress and ultimate tensile strength of AZ91 due to carbon short fiber reinforcement at different test temperatures.

	Yield Stress			Ultimate Tensile Strength		
Temperature	AZ91 (MPa)	AZ91-C (MPa)	Increase (%)	AZ91 (MPa)	AZ91-C (MPa)	Increase (%)
20 °C	109	227	107.89	199	260	30.61
150 °C	99	171	72.21	174	240	38.58
250 °C	78	128	65.16	113	186	64.92

The variation in yield strength improvement with temperature can be explained in light of the limited deformation range up to the yield stress of the composite. Moreover, the carbon fibers can be considered rigid bodies, where the strain of the carbon fibers up to fracture is very small (1.2–1.5%) [39]. At this deformation region, the contribution of the realignment of the fibers to carrying the applied load is very limited. There are two components that share the carrying of the applied stress: the matrix alloy (AZ91) which is greatly affected by the temperature and the rigid fibers which are unaffected by the temperature. Hence, the effect of temperature in the matrix material dominates the entire behavior of the composite materials. With the development of the deformation process, the fibers align with the applied load and contribute with the matrix in the deformation process, thus showing an improvement with temperature opposite to what has been obtained in the improvement of the yield stress.

The strengthening effect due to work hardening is more influential at low temperatures, while at higher temperatures it promotes softening due to dynamic recrystallization [46]. Regarding strain, the percentage strain of unreinforced AZ91 is greater than that of reinforced AZ91-C. This confirms that the unreinforced AZ91 has great ductility, while the reinforced AZ91-C is more brittle. However, one thing that needs to be highlighted is that the ductility of both materials increases with increasing temperature. Even with a general decrease in the materials' strength given an increase in the test temperature, the percent improvement in the composite's ultimate tensile strength increases with the increasing temperature. This indicates that the thermal stability of the reinforced materials (AZ91-C) is much higher than that of the matrix AZ91 alloy. The increase in the percent improvement of the ultimate tensile strength encourages the application of this composite material for high-temperature uses. Of equal importance, the key values obtained from the tensile test can be used as guidelines in the determination of the initial fatigue stress amplitude, which should be slightly higher than the yield stress.

3.3. Fatigue Testing Results

The fatigue (S–N) curves, as shown in Figure 5, are plotted based on the empirical arbitrary exponential formula [17] as described in Equation (1):

$$\sigma_a = a \frac{\exp(-\log N)}{b^2} + c. \qquad (1)$$

where σ_a is the stress amplitude, N is the number of cycles, a, b, and c are constants.

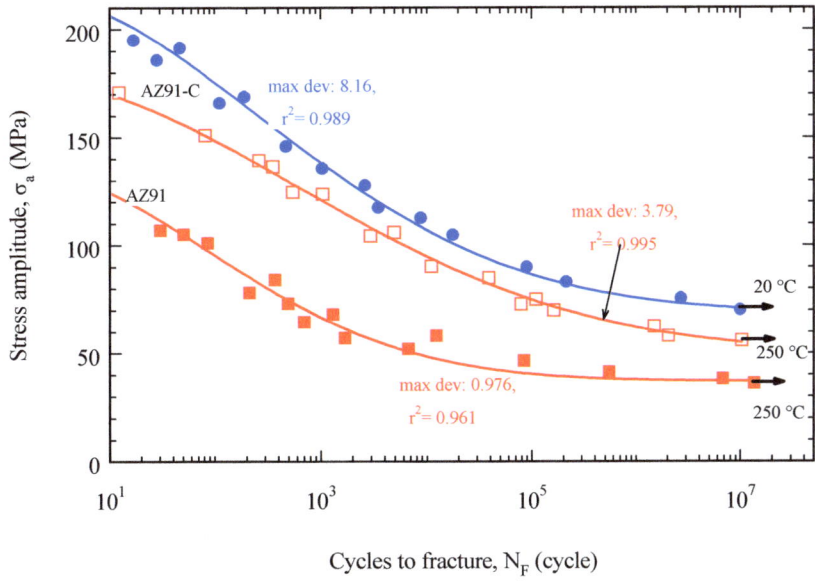

Figure 5. Fatigue S–N curves of AZ91-C and AZ91 at different temperatures.

Table 2 presents the constants obtained by fitting the fatigue curves at varying temperatures. The fatigue strength is calculated by the description of the curve as the fatigue stress at the attainable life (N = 10^7 cycles). For further details, the fatigue strength and the value of the different fitting parameters are given in Table 2.

Table 2. The fitting parameters of the S–N fatigue curves of unreinforced AZ91 and AZ91-C together with their fatigue strengths at N = 10^7 cycles.

Specimen	No. of Experiments	T (°C)	a	b	c	Max. Dev.	Standard Error, r^2	Fatigue Strength (MPa) at N = 10^7
AZ91	16	250	100	2.72	37	0.976	0.961	37
AZ91-C	16	20	150	3.42	68.7	8.16	0989	71.2
	17	250	128	3.90	50	3.79	0.995	55

The S–N fatigue curve fitting parameters at the test conditions are listed in Table 2. The fatigue strength can be estimated from the curve fitting as the fatigue stress at the highest attained fatigue life (N = 10^7 cycles). In addition, the fatigue resistance information is also provided in Table 2. The effect of reinforcement on fatigue strength can be judged at 250 °C. A percentage improvement in fatigue strength of about 92% was attained at 250 °C at the fatigue testing limit (at N = 10^7 cycles). As the aimed application of this material is at high temperatures, such as pistons for trucks, the reinforced material was tested at both room temperature and at 250 °C to evaluate the stability of this material at high temperatures. Even though the fatigue strength values in both the AZ91 reinforced and unreinforced alloys are low, the results obtained are still better than those experienced by the AE42 alloy and its composites [17]. There is an interesting phenomenon in the results obtained with the magnesium alloy composite (AZ91-C) in the present study, if compared with other aluminum alloys as reported in [47], where the fatigue strength of the AA6061 alloy was 100 MPa at room temperature and 10^7 cycles. Moreover, the fatigue strength of the AE42 alloy composite (AE42-C) was only 70.5 MPa under the same test conditions [17].

Figure 6 shows examples of the fatigue hysteresis loops at 250 °C under a strain amplitude of ε_a = 0.02 for the matrix alloy AZ91 (Figure 6a) and the composite material AZ91-C (Figure 6b). Figure 6a shows the three individual hysteresis loops of the unreinforced alloy AZ91 at 250 °C under a strain amplitude of ε_a = 0.02. There is some material softening after several cycles of 670. Softening can be noticed when decreasing the attained stress at the applied strain amplitude due to the high testing temperature. Applying the same strain amplitude on the reinforced AZ91-C materials (Figure 6b) shows a higher stress level that is nearly double of that which appeared for the unreinforced AZ91. And the fracture takes place very fast, after only three cycles.

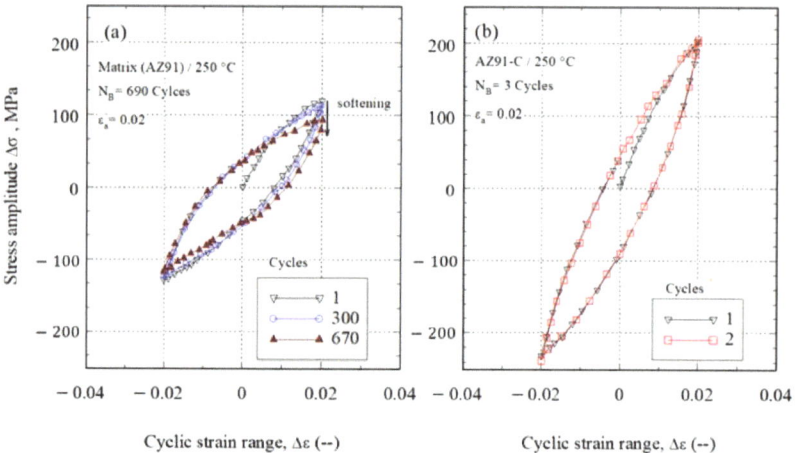

Figure 6. Hysteresis loops at 250 °C under strain amplitude of ε_a = 0.02 for (**a**) the matrix alloy AZ91 (**b**) and the composite material AZ91-C.

Cyclic straining of AZ91 at 250 °C (0.56 T_m) with a strain amplitude of ε_a = 0.02 resulted in cyclic softening affected by material recrystallization at this high temperature. Similar cyclic softening accompanied with a decrease in life has been detected on LCF testing of cast AZ91 at temperatures higher than 200 °C [48]. Such behavior is noted to decease the material's fatigue strength. Some of the reasons related to this softening are grain boundary sliding, formation of discontinuous precipitates, and softening of the β-Mg17Al12 phase at elevated temperatures [48].

3.4. Fractography (SEM)

Figure 7 presents the fatigue fracture features of an unreinforced AZ91 fatigue specimen at a temperature of 250 °C, a stress amplitude of 41 MPa, and a fatigue life of 5.4 × 10^5 cycles. By observing the fatigue fracture, it can be seen that the formation of fine serrations dominates almost all parts of the unreinforced AZ91 fatigue fracture surface in the HCF range and has a low stress amplitude of 41 MPa at 250 °C. The formation of fine serrations is caused by serration flow, which is closely related to dynamic strain aging (DSA). It has been explained by McCormick [49] that this dynamic strain aging occurs when the forest dislocation successfully restrains the movement of the dislocation, and finally, the solute atoms undergo diffusion towards the restrained dislocation. This phenomenon promotes an increase in the flow stress enhanced by the hindering factor of dislocation movement. Under some conditions, the restrained dislocation may release itself from the atmospheric environment of the solute when the flow stress successfully reaches its critical value. For this reason, a lower stress is required to evacuate the dislocation so that it returns to the following dislocation forest [49–51]. It was further found that the relationship between the serrated flow and the dynamic strain aging was induced by the factor of excess magnesium [52,53]. This serrated flow could also result from the repeated dynamic effects of pinning and unpinning [54,55]. It is also suggested that the serrations are formed due to the competition between the precipitate displacement caused by the dislocations and the dynamic strain aging [50,55].

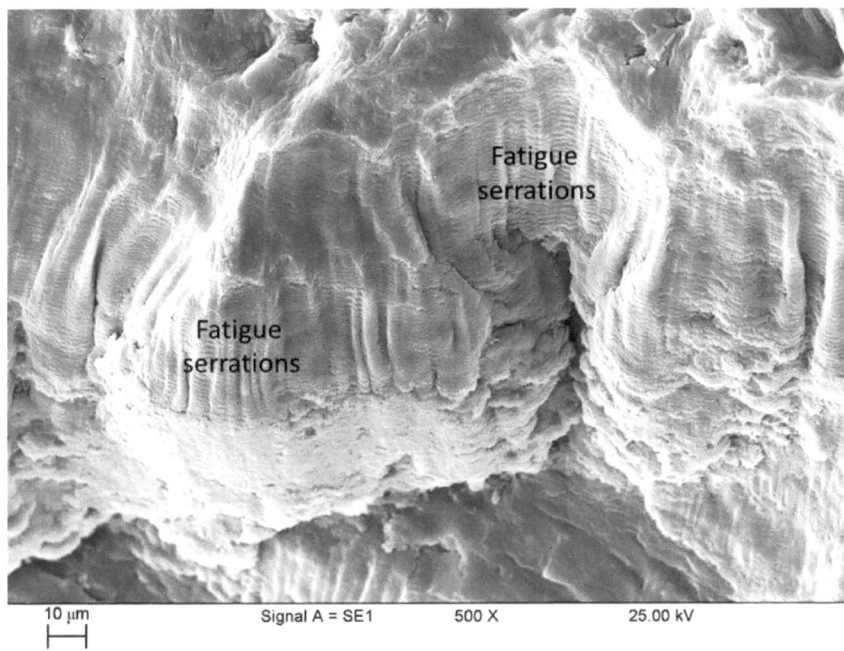

Figure 7. Fatigue fracture features of fractured fatigue specimen of the unreinforced AZ91 at 250 °C, stress amplitude of 41 MPa, and fatigue life of 5.4×10^5 cycles.

Figure 8 exhibits the perspicuous fatigue serrations and flattened areas by hammering the mating surfaces on an unreinforced AZ91 fatigue fracture specimen at 250 °C with a stress amplitude of 41 MPa and a fatigue life of 5.4×10^5 cycles. From the appearance of the fatigue fracture, two different contrast areas are evident. There is an area consisting of fatigue serrations that remain in their undamaged condition at the center, to the right of the fatigue fracture surface. On the other hand, there is also a small portion on the left side of the fatigue fracture surface that has evolved to be flattened due to the hammering of the opposite specimen's surface for the fatigue fracture surface in the unreinforced AZ91 [56]. If the flattened surface is further observed, cracks can be seen spanning the fatigue fracture's surface. The cracks cooperate with multiple grains on fatigue loading, producing a large reverse plastic zone. This eventually induces the synchronized creation of planar slip bands on multiple grains leading to a flattened crack [57].

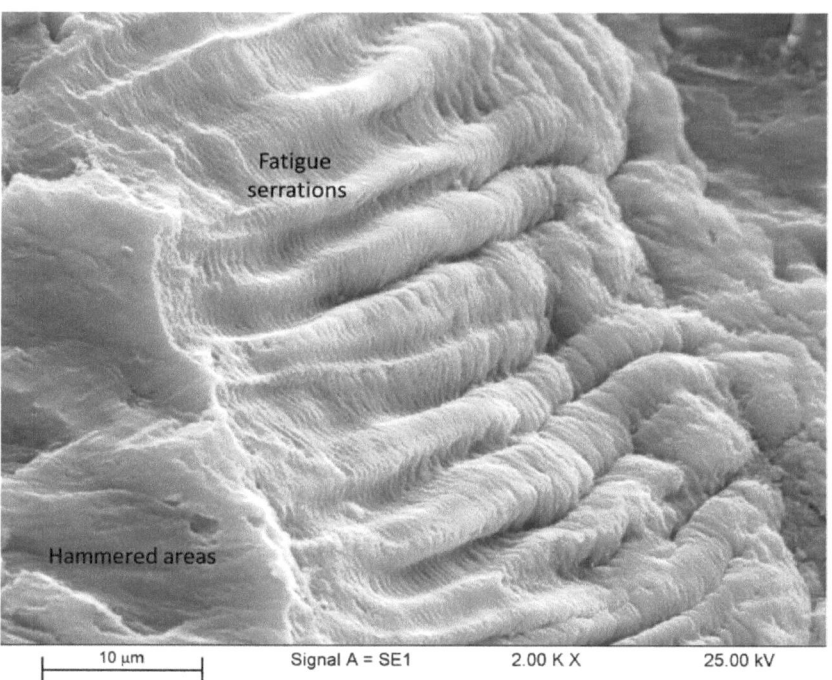

Figure 8. Perspicuous fatigue serrations and flattened areas by hammering the mating surfaces on an unreinforced AZ91 fatigue fracture specimen at 250 °C, stress amplitude of 41 MPa, and fatigue life of 5.4×10^5 cycles.

Figure 9 shows a magnified view of the unreinforced AZ91 fatigue specimen fracture surface at 250 °C, supplemented with EDS analysis of the ductile region with multiple dimples and cleavage surfaces containing cracks, indicating the brittle region. From the fatigue fracture surface, it is possible to distinguish ductile areas containing some dimples and brittle areas containing cleavages decorated with cracks. Crack initiation generally occurs in the softer region, referring to the ductile region, due to continuous cyclic loading. As a result, the fatigue fracture surface explicitly shows cracks propagating over the ductile fatigue serrations, which can be readily observed. Further examination of the ductile regions containing dimples (as indicated in point b) using EDS showed that the Mg concentration was very high, followed by the presence of Al in low concentrations. This means that the ductile region contains poor Zn and Zn compounds, which confirms that Zn-containing precipitates are not formed in the ductile region. In contrast, the EDS analysis of the brittle region (indicated by point c) clearly reveals that the highest peak is dominated by Mg, and there is an increase in Al concentration when compared to point B, but interestingly, Zn is identified in the brittle region. As is known, Zn and Mg have an equal valence electron number; this results in a very low concentration of electrons. Likewise, when viewed in terms of the atomic radius between Zn and Mg, where the atomic radius of Zn has a very close proximity to Mg, the presence of Zn can replace Mg atoms to encourage the occurrence of precipitated phases in the form of intermetallic compounds in brittle areas. The absence of Zn in the ductile region may refer to intermetallic $Mg_{17}Al_{12}$ formation [58], while at the brittle zone the phases formed are τ-$Mg_{32}(Al,Zn)_{49}$ and ϕ-$Mg_5Al_2Zn_2$ [59], where it is affected by the presence of Zn and also Al in low concentrations. Additionally, the cracks that appeared in the brittle area are the result of material decohesion occurring in the Zn-rich intermetallic precipitate phases.

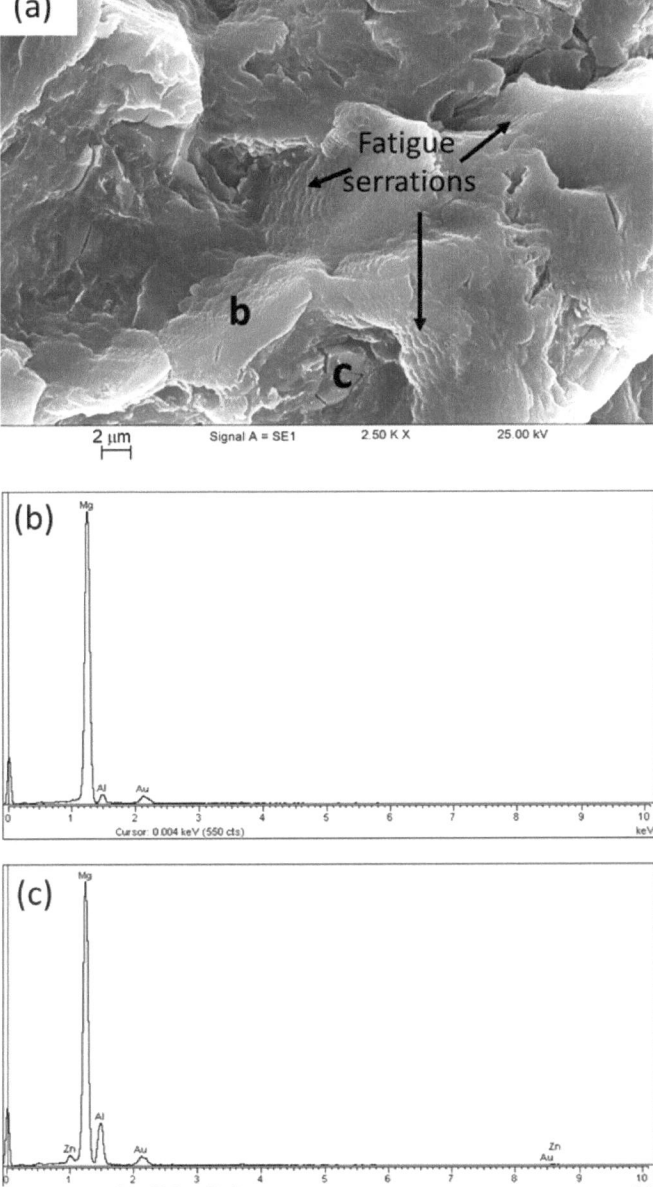

Figure 9. EDX spectrum of the unreinforced AZ91 fatigued specimen at 250 °C (**a**) Higher magnification, (**b**) EDX analysis spectrum of ductile regions with Zn-poor dimple, and (**c**) EDX spectrum on cleavage fracture with cracks at the Zn-containing region.

Figure 10 shows the features of the composite AZ91-C fatigue-fractured test sample at 20 °C. There is an observable phenomenon where the area consisting of fatigue serrations appears to be split due to cracks propagating in the perpendicular direction of the metal connection between the carbon fibers, so that the fatigue serrations appear to be separated, as specified in Figure 10. In addition, the presence of considerable dimple features under the fatigue serrations indicates the presence of large plastic deformation in the reinforced AZ91 sample [60]. Carbon fiber breaks are also seen located adjacent to the dimples. This can be understood due to the brittle nature of carbon fiber, which indicates its inability to respond to the deformation of the matrix material.

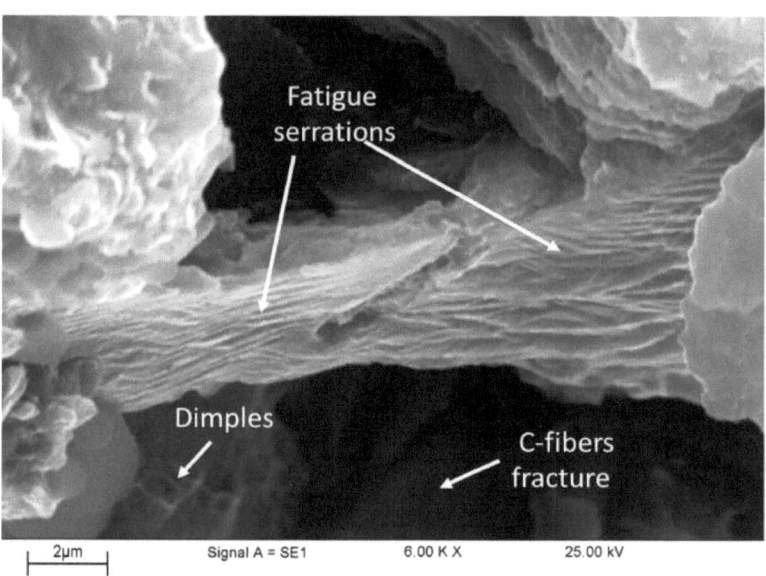

Figure 10. The features of the reinforced AZ91 fatigue fracture specimen at room temperature.

Figure 11 shows the features of the reinforced AZ91 fatigue-fractured specimen at 150 °C. The beginning state of carbon fiber is that it is encased in a metallic material that acts as a matrix bridge. However, being under continuous vibration stress causes crack initiation, which undergoes crack growth and propagation at the metal joints between the carbon fibers, resulting in fatigue fracture [17]. Then, a fractured metal joint is indicated by the formation of dimple features at the separation of the ductile metallic junction of the matrix alloy. There is a phenomenon of carbon detachment from the alloy matrix that causes it to stand independently without any support. The brittle nature of carbon fiber leads to fracture due to its inherent lack of response to the deformation of the matrix alloy. Li et al. stated that metal matrix composites reinforced with continuous fibers usually fracture at relatively low stress conditions without any damage found on the fiber surface after the manufacturing process [61]. In metal matrix composites, the mechanical properties are not only influenced by the constituents but, more importantly, the conditions between the fiber and the matrix, known as the interface [62]. If the interface has low strength, cracks will initiate in the brittle fibers which are more likely to shift and debond the fibers, leading to a single fiber pull, which has the effect of reducing stress concentration [63]. In this study, the fact that interface debonding prevailed under low stress conditions, as well as the ability of carbon fibers to withstand load, was not completely followed by lower flexural strength. In contrast to composites with moderate levels of interfacial strength, when micro-cracks start on the brittle carbon fibers, they will propagate and then terminate at the interface, i.e., on carbon fibers or a ductile matrix. Nevertheless, the main characteristic of composites with moderate interfacial bond strength is the occurrence of broken fiber bonds [64].

Figure 11. Fatigue fracture features of a fractured fatigue specimen of the reinforced AZ91 at 150 °C.

4. Conclusions

From the mechanical characterization of the unreinforced and reinforced AZ91 using hardness and tensile testing as well as the fatigue tests conducted and the fractographic investigations, the following conclusions have been drawn:

(1) The hardness value of the unreinforced AZ91 increased from 69.2 ± 2 Hv to 111.53 ± 4 Hv for the reinforced AZ91-C;

(2) A clear improvement in yield stress was achieved at room temperature (108%); this improvement changed to 73% and 64% at 150 °C and 250 °C, respectively. A relatively lower improvement in the ultimate tensile strength was noticed, where the increase in strength ranged between 30% and 64%;

(3) The fatigue strength at HCF range (over 10^7 cycles) at 250 °C for the AZ91 alloy was 37 MPa, while that for the reinforced AZ91-C reached 55 MPa. The reinforced AZ91-C displayed higher fatigue strength (71.2 MPa) at room temperature;

(4) The hysteresis loops for the strain-controlled fatigue test of AZ91 at 250 °C showed limited material softening;

(5) Diffused fatigue serrations with mixed ductile/brittle modes were observed on unreinforced AZ91's fracture surface;

(6) Fiber fracture and fiber decohesion were noticed on the composite's fracture surface both under tensile and fatigue loads. The presence of fatigue serrations and regions exhibiting restricted dimples was detected in metallic magnesium zones.

Author Contributions: Conceptualization, N.H.A., S.A. and M.M.Z.A.; methodology, S.A. and M.M.E.-S.S.; validation, S.A. and M.M.E.-S.S.; formal analysis, S.A. and M.M.Z.A.; investigation, S.A. and M.M.Z.A.; resources, N.H.A.; writing—original draft preparation, N.H.A. and M.M.E.-S.S.; writing—review and editing, M.M.E.-S.S., M.M.Z.A. and S.A.; supervision, N.H.A., S.A. and M.M.Z.A.; project administration, N.H.A. and S.A.; funding acquisition, S.A. and N.H.A. All authors have read and agreed to the published version of the manuscript.

Funding: Funded by the Deanship of Scientific Research at Imam Mohammad Ibn Saud Islamic University, Saudi Arabia, through grant No. (221414016).

Data Availability Statement: Data are available upon request through the corresponding author.

Acknowledgments: The authors extend their appreciation to the Deanship of Scientific Research, Imam Mohammad Ibn Saud Islamic University (IMSIU), Saudi Arabia, for funding this research study through grant No. (221414016).

Conflicts of Interest: The authors declare no conflict of interest.

References

1. Joost, W.J.; Krajewski, P.E. Towards Magnesium Alloys for High-Volume Automotive Applications. *Scr. Mater.* **2017**, *128*, 107–112. [CrossRef]
2. Demirci, E.; Yildiz, A.R. An Investigation of the Crash Performance of Magnesium, Aluminum and Advanced High Strength Steels and Different Cross-Sections for Vehicle Thin-Walled Energy Absorbers. *Mater. Test.* **2018**, *60*, 661–668. [CrossRef]
3. Kumar, D.S.; Sasanka, C.T.; Ravindra, K.; Suman, K.N.S. Magnesium and Its Alloys in Automotive Applications—A Review. *Am. J. Mater. Sci. Technol.* **2015**, *4*, 12–30. [CrossRef]
4. Kulekci, M.K. Magnesium and Its Alloys Applications in Automotive Industry. *Int. J. Adv. Manuf. Technol.* **2008**, *39*, 851–865. [CrossRef]
5. Tan, J.; Ramakrishna, S. Applications of Magnesium and Its Alloys: A Review. *Appl. Sci.* **2021**, *11*, 6861. [CrossRef]
6. Chalisgaonkar, R. Insight in Applications, Manufacturing and Corrosion Behaviour of Magnesium and Its Alloys—A Review. *Mater. Today Proc.* **2019**, *26*, 1060–1071. [CrossRef]
7. Ghaderi, S.H.; Mori, A.; Hokamoto, K. Analysis of Explosively Welded Aluminum-AZ31 Magnesium Alloy Joints. *Mater. Trans.* **2008**, *49*, 1142–1147. [CrossRef]
8. Cao, X.; Jahazi, M.; Immarigeon, J.P.; Wallace, W. A Review of Laser Welding Techniques for Magnesium Alloys. *J. Mater. Process. Technol.* **2006**, *171*, 188–204. [CrossRef]
9. Prasad, S.V.S.; Prasad, S.B.; Verma, K.; Mishra, R.K.; Kumar, V.; Singh, S. The Role and Significance of Magnesium in Modern Day Research-A Review. *J. Magnes. Alloy* **2022**, *10*, 1–61. [CrossRef]
10. Zhai, W.; Bai, L.; Zhou, R.; Fan, X.; Kang, G.; Liu, Y.; Zhou, K. Recent Progress on Wear-Resistant Materials: Designs, Properties, and Applications. *Adv. Sci.* **2021**, *8*, 2003739. [CrossRef]
11. Friedrich, H.; Schumann, S. Research for a "New Age of Magnesium" in the Automotive Industry. *J. Mater. Process. Technol.* **2001**, *117*, 276–281. [CrossRef]
12. Ataya, S.; El-Magd, E. Modeling the Creep Behavior of Mg Alloys with and without Short-Fiber Reinforcement. *Comput. Mater. Sci.* **2007**, *39*, 155–159. [CrossRef]
13. Srinivasan, A.; Ajithkumar, K.K.; Swaminathan, J.; Pillai, U.T.S.; Pai, B.C. Creep Behavior of AZ91 Magnesium Alloy. *Procedia Eng.* **2013**, *55*, 109–113. [CrossRef]
14. Lin, Y.C.; Chen, X.M.; Liu, Z.H.; Chen, J. Investigation of Uniaxial Low-Cycle Fatigue Failure Behavior of Hot-Rolled AZ91 Magnesium Alloy. *Int. J. Fatigue* **2013**, *48*, 122–132. [CrossRef]
15. Azadi, M.; Farrahi, G.H.; Winter, G.; Eichlseder, W. Fatigue Lifetime of AZ91 Magnesium Alloy Subjected to Cyclic Thermal and Mechanical Loadings. *Mater. Des.* **2014**, *53*, 639–644. [CrossRef]
16. Nascimento, L.; Yi, S.; Bohlen, J.; Fuskova, L.; Letzig, D.; Kainer, K.U. High Cycle Fatigue Behaviour of Magnesium Alloys. *Procedia Eng.* **2010**, *2*, 743–750. [CrossRef]
17. Alsaleh, N.A.; Ataya, S.; Latief, F.H.; Ahmed, M.M.Z.; Ataya, A.; Abdul-Latif, A. LCF and HCF of Short Carbon Fibers Reinforced AE42 Mg Alloy. *Materials* **2023**, *16*, 3686. [CrossRef]
18. Fintová, S.; Trško, L.; Chlup, Z.; Pastorek, F.; Kajánek, D.; Kunz, L. Fatigue Crack Initiation Change of Cast Az91 Magnesium Alloy from Low to Very High Cycle Fatigue Region. *Materials* **2021**, *14*, 6245. [CrossRef]
19. Chen, L.J.; Shen, J.; Wu, W.; Li, F.; Wang, Y.; Liu, Z. Low-Cycle Fatigue Behavior of Magnesium Alloy AZ91. *Mater. Sci. Forum* **2005**, *488–489*, 725–728. [CrossRef]
20. Rettberg, L.H.; Jordon, J.B.; Horstemeyer, M.F.; Jones, J.W. Low-Cycle Fatigue Behavior of Die-Cast Mg Alloys AZ91 and AM60. *Metall. Mater. Trans. A Phys. Metall. Mater. Sci.* **2012**, *43*, 2260–2274. [CrossRef]
21. Wolf, B.; Fleck, C.; Eifler, D. Characterization of the Fatigue Behaviour of the Magnesium Alloy AZ91D by Means of Mechanical Hysteresis and Temperature Measurements. *Int. J. Fatigue* **2004**, *26*, 1357–1363. [CrossRef]
22. Gu, X.N.; Zhou, W.R.; Zheng, Y.F.; Cheng, Y.; Wei, S.C.; Zhong, S.P.; Xi, T.F.; Chen, L.J. Corrosion Fatigue Behaviors of Two Biomedical Mg Alloys—AZ91D and WE43—In Simulated Body Fluid. *Acta Biomater.* **2010**, *6*, 4605–4613. [CrossRef] [PubMed]
23. Fintová, S.; Kunz, L. Fatigue Properties of Magnesium Alloy AZ91 Processed by Severe Plastic Deformation. *J. Mech. Behav. Biomed. Mater.* **2015**, *42*, 219–228. [CrossRef] [PubMed]
24. Ataya, S.; El-Magd, E. Quasi-Static Behavior of Mg-Alloys with and without Short-Fiber Reinforcement. *Theor. Appl. Fract. Mech.* **2007**, *47*, 102–112. [CrossRef]
25. Ataya, S.; Alsaleh, N.A.; El-Sayed Seleman, M.M. Strength and Wear Behavior of Mg Alloy AE42 Reinforced with Carbon Short Fibers. *Acta Metall. Sin. (English Lett.)* **2019**, *32*, 31–40. [CrossRef]
26. Aravindan, S.; Rao, P.V.; Ponappa, K. Evaluation of Physical and Mechanical Properties of AZ91D/SiC Composites by Two Step Stir Casting Process. *J. Magnes. Alloy* **2015**, *3*, 52–62. [CrossRef]
27. Lü, L.; Lai, M.O.; Gupta, M.; Chua, B.W.; Osman, A. Improvement of Microstructure and Mechanical Properties of AZ91/SiC Composite by Mechanical Alloying. *J. Mater. Sci.* **2000**, *35*, 5553–5561. [CrossRef]
28. Yuan, Q.H.; Zeng, X.S.; Liu, Y.; Luo, L.; Wu, J.B.; Wang, Y.C.; Zhou, G.H. Microstructure and Mechanical Properties of AZ91 Alloy Reinforced by Carbon Nanotubes Coated with MgO. *Carbon* **2016**, *96*, 843–855. [CrossRef]
29. Khandelwal, A.; Mani, K.; Srivastava, N.; Gupta, R.; Chaudhari, G.P. Mechanical Behavior of AZ_{31}/Al_2O_3 Magnesium Alloy Nanocomposites Prepared Using Ultrasound Assisted Stir Casting. *Compos. Part B Eng.* **2017**, *123*, 64–73. [CrossRef]

30. Huang, S.J.; Abbas, A. Effects of Tungsten Disulfide on Microstructure and Mechanical Properties of AZ91 Magnesium Alloy Manufactured by Stir Casting. *J. Alloys Compd.* **2020**, *817*, 153321. [CrossRef]
31. Wang, X.J.; Hu, X.S.; Wu, K.; Wang, L.Y.; Huang, Y.D. Evolutions of Microstructure and Mechanical Properties for SiCp/AZ91 Composites with Different Particle Contents during Extrusion. *Mater. Sci. Eng. A* **2015**, *636*, 138–147. [CrossRef]
32. Kumar, A.; Kumar, S.; Mukhopadhyay, N.K.; Yadav, A.; Sinha, D.K. Effect of TiC Reinforcement on Mechanical and Wear Properties of AZ91 Matrix Composites. *Int. J. Met.* **2022**, *16*, 2128–2143. [CrossRef]
33. Xiao, P.; Gao, Y.; Xu, F.; Yang, S.; Li, B.; Li, Y.; Huang, Z.; Zheng, Q. An Investigation on Grain Refinement Mechanism of TiB2 Particulate Reinforced AZ91 Composites and Its Effect on Mechanical Properties. *J. Alloys Compd.* **2019**, *780*, 237–244. [CrossRef]
34. Vaidya, A.R.; Lewandowski, J.J. Effects of SiCp Size and Volume Fraction on the High Cycle Fatigue Behavior of AZ91D Magnesium Alloy Composites. *Mater. Sci. Eng. A* **1996**, *220*, 85–92. [CrossRef]
35. Llorca, N.; Bloyce, A.; Yue, T.M. Fatigue Behaviour of Short Alumina Fibre Reinforced AZ91 Magnesium Alloy Metal Matrix Composite. *Mater. Sci. Eng. A* **1991**, *135*, 247–252. [CrossRef]
36. Ataya, S.; El-Sayed Seleman, M.M.; Latief, F.H.; Ahmed, M.M.Z.; Hajlaoui, K.; Soliman, A.M.; Alsaleh, N.A.; Habba, M.I.A. Wear Characteristics of Mg Alloy AZ91 Reinforced with Oriented Short Carbon Fibers. *Materials* **2022**, *15*, 4841. [CrossRef]
37. Ataya, S.; El-Sayed Seleman, M.M.; Latief, F.H.; Ahmed, M.M.Z.; Hajlaoui, K.; Elshaghoul, Y.G.Y.; Habba, M.I.A. Microstructure and Mechanical Properties of AZ91 Rein-Forced with High Volume Fraction of Oriented Short Carbon Fibers. *Materials* **2022**, *15*, 4818. [CrossRef]
38. Shastri, H.; Mondal, A.K.; Dutta, K.; Dieringa, H.; Kumar, S. Microstructural Correlation with Tensile and Creep Properties of AZ91 Alloy in Three Casting Techniques. *J. Manuf. Process.* **2020**, *57*, 566–573. [CrossRef]
39. Alrasheedi, N.H.; Ataya, S.; El-Sayed Seleman, M.M.; Ahmed, M.M.Z. Tensile Deformation and Fracture of Unreinforced AZ91 and Reinforced AZ91-C at Temperatures up to 300 °C. *Materials* **2023**, *16*, 4785. [CrossRef]
40. Chen, B.; Fu, J.; Zhou, J.; Ma, Y.; Qi, L. Influence of Microstructures on Mechanical Properties of the Csf/Mg Composite Fabricated by Liquid-Solid Extrusion Following Vacuum Pressure Infiltration. *J. Alloys Compd.* **2023**, *935*, 168083. [CrossRef]
41. Hou, L.G.; Wu, R.Z.; Wang, X.D.; Zhang, J.H.; Zhang, M.L.; Dong, A.P.; Sun, B.D. Microstructure, Mechanical Properties and Thermal Conductivity of the Short Carbon Fiber Reinforced Magnesium Matrix Composites. *J. Alloys Compd.* **2017**, *695*, 2820–2826. [CrossRef]
42. Xu, H.; Yang, Z.; Hu, M.; Ji, Z. Effect of Short Carbon Fiber Content on SCFs/AZ31 Composite Microstructure and Mechanical Properties. *Results Phys.* **2020**, *17*, 103074. [CrossRef]
43. Wang, M.; Jin, P.; Wang, J. Hot Deformation and Processing Maps of 7005 Aluminum Alloy. *High Temp. Mater. Process.* **2014**, *33*, 369–375. [CrossRef]
44. Mondal, A.K.; Blawert, C.; Kumar, S. Corrosion Behaviour of Creep-Resistant AE42 Magnesium Alloy-Based Hybrid Composites Developed for Powertrain Applications. *Mater. Corros.* **2015**, *66*, 1150–1158. [CrossRef]
45. WANG, J.; SHI, B.; YANG, Y. Hot Compression Behavior and Processing Map of Cast Mg-4Al-2Sn-Y-Nd Alloy. *Trans. Nonferrous Met. Soc. China* **2014**, *24*, 626–631. [CrossRef]
46. Fan, Y.; Deng, K.; Wang, C.; Nie, K.; Shi, Q. Work Hardening and Softening Behavior of Mg–Zn–Ca Alloy Influenced by Deformable Ti Particles. *Mater. Sci. Eng. A* **2022**, *833*, 142336. [CrossRef]
47. Murashkin, M.; Sabirov, I.; Prosvirnin, D.; Ovid'ko, I.; Terentiev, V.; Valiev, R.; Dobatkin, S. Fatigue Behavior of an Ultrafine-Grained Al-Mg-Si Alloy Processed by High-Pressure Torsion. *Metals* **2015**, *5*, 578–590. [CrossRef]
48. Mokhtarishirazabad, M.; Azadi, M.; Hossein Farrahi, G.; Winter, G.; Eichlseder, W. Improvement of High Temperature Fatigue Lifetime in AZ91 Magnesium Alloy by Heat Treatment. *Mater. Sci. Eng. A* **2013**, *588*, 357–365. [CrossRef]
49. McCormigk, P.G. A Model for the Portevin-Le Chatelier Effect in Substitutional Alloys. *Acta Metall.* **1972**, *20*, 351–354. [CrossRef]
50. Yazdani, S.; Vitry, V. Using Molecular Dynamic Simulation to Understand the Deformation Mechanism in Cu, Ni, and Equimolar Cu-Ni Polycrystalline Alloys. *Alloys* **2023**, *2*, 77–88. [CrossRef]
51. Cottrell, A.H.; Dexter, D.L. Dislocations and Plastic Flow in Crystals. *Am. J. Phys.* **1954**, *22*, 242–243. [CrossRef]
52. LIU, M.; JIANG, T.; WANG, J.; LIU, Q.; WU, Z.; YU, Y.; SKARET, P.C.; ROVEN, H.J. Aging Behavior and Mechanical Properties of 6013 Aluminum Alloy Processed by Severe Plastic Deformation. *Trans. Nonferrous Met. Soc. China* **2014**, *24*, 3858–3865. [CrossRef]
53. Cuniberti, A.; Tolley, A.; Riglos, M.V.C.; Giovachini, R. Influence of Natural Aging on the Precipitation Hardening of an AlMgSi Alloy. *Mater. Sci. Eng. A* **2010**, *527*, 5307–5311. [CrossRef]
54. Yazdani, S.; Mesbah, M.; Vitry, V. Molecular Dynamics Simulation of the Interaction between Dislocations and Iron–Vanadium Precipitates in Alpha Iron: Effect of Chemical Composition. *Crystals* **2023**, *13*, 1247. [CrossRef]
55. Wang, W.H.; Wu, D.; Chen, R.S.; Zhang, X.N. Effect of Solute Atom Concentration and Precipitates on Serrated Flow in Mg-3Nd-Zn Alloy. *J. Mater. Sci. Technol.* **2018**, *34*, 1236–1242. [CrossRef]
56. Daroonparvar, M.; Khan, M.U.F.; Saadeh, Y.; Kay, C.M.; Kasar, A.K.; Kumar, P.; Esteves, L.; Misra, M.; Menezes, P.; Kalvala, P.R.; et al. Modification of Surface Hardness, Wear Resistance and Corrosion Resistance of Cold Spray Al Coated AZ31B Mg Alloy Using Cold Spray Double Layered Ta/Ti Coating in 3.5 wt % NaCl Solution. *Corros. Sci.* **2020**, *176*, 109029. [CrossRef]
57. Kim, H.-K.; Lee, Y.-I.; Chung, C.-S. Fatigue Properties of a Fine-Grained Magnesium Alloy Produced by Equal Channel Angular Pressing. *Scr. Mater.* **2005**, *52*, 473–477. [CrossRef]
58. CANDAN, S.; CANDAN, E. Comparative Study on Corrosion Behaviors of Mg-Al-Zn Alloys. *Trans. Nonferrous Met. Soc. China* **2018**, *28*, 642–650. [CrossRef]

59. Cheng, K.; Sun, J.; Xu, H.; Wang, J.; Zhou, J.; Tang, S.; Wang, X.; Zhang, L.; Du, Y. On the Temperature-Dependent Diffusion Growth of ϕ-Mg5Al2Zn2 Ternary Intermetallic Compound in the Mg-Al-Zn System. *J. Mater. Sci.* **2021**, *56*, 3488–3497. [CrossRef]
60. Alam, M.E.; Hamouda, A.M.S.; Nguyen, Q.B.; Gupta, M. Improving Microstructural and Mechanical Response of New AZ41 and AZ51 Magnesium Alloys through Simultaneous Addition of Nano-Sized Al2O3 Particulates and Ca. *J. Alloys Compd.* **2013**, *574*, 565–572. [CrossRef]
61. Zhang, G.D.; Chen, R. Effect of the Interfacial Bonding Strength on the Mechanical Properties of Metal Matrix Composites. *Compos. Interfaces* **1993**, *1*, 337–355. [CrossRef]
62. Su, X.F.; Chen, H.R.; Kennedy, D.; Williams, F.W. Effects of Interphase Strength on the Damage Modes and Mechanical Behaviour of Metal–Matrix Composites. *Compos. Part A Appl. Sci. Manuf.* **1999**, *30*, 257–266. [CrossRef]
63. Vidal-Sétif, M.H.; Lancin, M.; Marhic, C.; Valle, R.; Raviart, J.-L.; Daux, J.-C.; Rabinovitch, M. On the Role of Brittle Interfacial Phases on the Mechanical Properties of Carbon Fibre Reinforced Al-Based Matrix Composites. *Mater. Sci. Eng. A* **1999**, *272*, 321–333. [CrossRef]
64. Wang, X.; Jiang, D.; Wu, G.; Li, B.; Li, P. Effect of Mg Content on the Mechanical Properties and Microstructure of Grf/Al Composite. *Mater. Sci. Eng. A* **2008**, *497*, 31–36. [CrossRef]

Disclaimer/Publisher's Note: The statements, opinions and data contained in all publications are solely those of the individual author(s) and contributor(s) and not of MDPI and/or the editor(s). MDPI and/or the editor(s) disclaim responsibility for any injury to people or property resulting from any ideas, methods, instructions or products referred to in the content.

Article

Assessment of the Interatomic Potentials of Beryllium for Mechanical Properties

Chengzhi Yang [1], Bin Wu [2,*], Wenmin Deng [1], Shuzhen Li [1], Jianfeng Jin [3] and Qing Peng [4,5,6,*]

1. Department of Physics, Beijing Normal University, Beijing 100875, China
2. College of Nuclear Science and Technology, Beijing Normal University, Beijing 100875, China
3. School of Materials Science and Engineering, Northeastern University, Shenyang 110819, China
4. State Key Laboratory of Nonlinear Mechanics, Institute of Mechanics, Chinese Academy of Sciences, Beijing 100190, China
5. School of Engineering Sciences, University of Chinese Academy of Sciences, Beijing 100049, China
6. Physics Department, King Fahd University of Petroleum & Minerals, Dhahran 31261, Saudi Arabia
* Correspondence: bwu6@bnu.edu.cn (B.W.); pengqing@imech.ac.cn (Q.P.)

Abstract: Beryllium finds widespread applications in nuclear energy, where it is required to service under extreme conditions, including high-dose and high-dose rate radiation with constant bombardments of energetic particles leading to various kinds of defects. Though it is generally known that defects give rise to mechanical degradation, the quantitative relationship between the microstructure and the corresponding mechanical properties remains elusive. Here we have investigated the mechanical properties of imperfect hexagonal close-packed (HCP) beryllium via means of molecular dynamics simulations. We have examined the beryllium crystals with void, a common defect under in-service conditions. We have assessed three types of potentials, including MEAM, Finnis–Sinclair, and Tersoff. The volumetric change with pressure based on MEAM and Tersoff and the volumetric change with temperature based on MEAM are consistent with the experiment. Through cross-comparison on the results from performing hydrostatic compression, heating, and uniaxial tension, the MEAM type potential is found to deliver the most reasonable predictions on the targeted properties. Our atomistic insights might be helpful in atomistic modeling and materials design of beryllium for nuclear energy.

Keywords: beryllium; molecular dynamics simulation; MEAM; Finnis–Sinclair and Tersoff potentials

1. Introduction

Thanks to its unique properties, including high specific strength, low density, high melting point, and particularly low neutron absorption and high neutron scattering cross sections [1], beryllium finds widespread applications in nuclear energy. For instance, beryllium is the top candidate for the first wall that directly faces the plasma in the ITER [2,3] and for neutron multiplier in the DEMO tokamak fusion reactors [4]; beryllium is also commonly employed as the moderator and reflector for fission reactor and spallation neutron sources [5,6]; finally, beryllium is broadly used as the convertor to yield neutrons via (p, n) reaction in the compact accelerator-based neutron sources that serve Boron Neutron Capture Therapy (BNCT) [7,8].

In these applications, beryllium is often required to function under extreme environments, including astonishingly high temperatures and pressure, and frequent energetic particle bombardments. For example, as a first-wall material for the tokamak fusion reactor, beryllium needs to withstand temperatures and pressures as high as 1500 K and 2 GPa [3], respectively. In compact accelerator-based neutron sources, beryllium constantly suffers from impingements of 1~10 MeV protons, leading to not only dramatic thermal gradient and stress, but also significant radiation damage. The mechanical integrity of beryllium under extreme environments is thus essential to support these critical applications.

There have been a handful of studies that conducted mechanical tests on beryllium since the late 1960s. It has been known that radiation damage generally leads to mechanical degradation [9,10]. The quantitative relationship between microstructure resulted from serving under extreme environments and mechanical performance is important in safety issues and materials design but remains to be uncovered. Beryllium is highly toxic, and its supplies are very limited. As a result, the numerical investigations are more feasible. Moreover, in parallel to experiments, numerical methods and simulations have been well established as the third pillar in science and engineering investigations [11,12]. Among the numerous numerical methods, the molecular dynamics simulations method has been developed to be a reliable and indispensable tool in atomistic scale in various investigations [13,14].

The interatomic potential is the key in molecular dynamics simulations. An accurate interatomic potential is a prerequisite for molecular dynamics simulations to produce reliable and material-specific results. To this end, there are first-principles quantum mechanical [15] and various kinds of interatomic potentials developed for beryllium, such as EAM [16,17], AMEAM [18], MEAM [19], Finnis–Sinclair [20], and Tersoff [21]. Nonetheless, which potential is more suitable to simulate beryllium under extreme environments encountered in the applications listed above is still open to question. Hence, in this paper, we select three representative types of potentials and cross-compare their predictions on the mechanical properties of beryllium crystals embedded with spherical void defects, which are commonly observed in beryllium due to energetic particle bombardments. Through cross-comparisons on the simulation predictions, we point out that the MEAM type potential is the most reliable one out of three chosen potentials.

2. Materials and Methods

We have performed molecular dynamics simulations using the LAMMPS software (LAMMPS 64-bit 24 March 2022) [22] developed by the Sandia National Laboratory, because it is free, open-source, and equipped with a variety of choices of interatomic potentials, which considerably facilitates their implementations and thus, the cross-comparison. We chose three types of potentials that were previously parameterized for beryllium as detailed below.

2.1. MEAM Type Potential

The total interatomic energy E for MEAM type potential is expressed as below, all equations presented below come from [19,23–25],

$$E = \sum_i \left[\frac{1}{2} \sum_{i \neq j} \phi(r_{ij}) + F[\overline{\rho_i}] \right] \quad (1)$$

where ϕ accounts for contribution from direct interaction between atoms i and j explicitly depending only on their distance r_{ij}, and F is the embedding function, whose input is the average electron density $\overline{\rho}$ at the position of atom i. The definition of ϕ is shown below,

$$\phi(r) = \frac{2}{Z} f_C \left(\frac{r_{cut} - r}{\delta} \right) \left\{ E^u(r) - F\left[\overline{\rho^0}(r)\right] \right\} \quad (2)$$

The first term Z equals 12 for the reference structure, a hexagonal close-packed structure (HCP) for beryllium. The second term $f_C(\frac{r_{cut}-r}{\delta})$ is the smooth cutoff function and takes the following expression,

$$f_C(x) = \begin{cases} 1 & x \geq 1 \\ [1-(1-x)^4]^2 & 0 < x < 1 \\ 0 & x \leq 0 \end{cases} \quad (3)$$

with
$$x = \frac{r_{cut} - r}{\delta} \quad (4)$$

where r_{cut} is the cutoff distance, and δ gives the cutoff region. Note that r_{cut} and δ are both tunable parameters. The third term $E^u(r)$ is the energy of the reference structure and takes the following expression,

$$E^u(r) = -E_c(1 + a^*)e^{-a^*} \quad (5)$$

with
$$a^* = \alpha(\frac{r}{r_e} - 1) \quad (6)$$

where E_c, r_e, and α are adjustable parameters.

Finally, the last term F from Equations (1) and (2) is the embedding function taking the following expression,

$$F(\bar{\rho}) = AE_c \frac{\bar{\rho}}{\rho_0} \ln \frac{\bar{\rho}}{\rho_0} f_C(\frac{r_{cut} - r}{\delta}) \quad (7)$$

where A is another adjustable parameter, and ρ_0 is equal to 12 for an HCP structure. On one hand, in Equation (2),

$$\bar{\rho} = \overline{\rho^0}(r) = Z\rho^{a(0)}(r) \quad (8)$$

where $\rho^{a(0)}$ is given in Equation (15). On the other hand, in Equation (1),

$$\bar{\rho} = \rho^{(0)} G(\Gamma) \quad (9)$$

with
$$G(\Gamma) = e^{\frac{\Gamma}{2}} \quad (10)$$

and
$$\Gamma = \sum_{h=1}^{3} t^{(h)} (\frac{\rho_i^{(h)}}{\rho_i^{(0)}})^2 \quad (11)$$

where $t^{(h)}$ are tunable parameters. The spherically symmetric partial electron density at the position of atom i is written below,

$$[\rho_i^{(0)}]^2 = [\sum_{i \neq j} S_{ij} \rho_j^{a(0)}(r_{ij})]^2 \quad (12)$$

Moreover, the counterparts characterizing angular contributions are given by similar formulas but weighted by the Cartesian projections of the distances between two involved atoms (denoted by superscripts u, v, and w) as follows,

$$[\rho_i^{(1)}]^2 = \sum_u \sum_{i \neq j} S_{ij} \rho_j^{a(1)}(r_{ij}) \frac{r_{ij}^u}{r_{ij}} \quad (13)$$

$$[\rho_i^{(2)}]^2 = \sum_{u,v} [\sum_{i \neq j} S_{ij} \rho_j^{a(2)}(r_{ij}) \frac{r_{ij}^u r_{ij}^v}{r_{ij}^2}]^2 - \frac{1}{3} \sum_{i \neq j} [S_{ij} \rho_j^{a(2)}(r_{ij})]^2 \quad (14)$$

$$[\rho_i^{(3)}]^2 = \sum_{u,v,w} [\sum_{i \neq j} S_{ij} \rho_j^{a(3)}(r_{ij}) \frac{r_{ij}^u r_{ij}^v r_{ij}^w}{r_{ij}^3}]^2 - \frac{3}{5} \sum_u [\sum_{i \neq j} S_{ij} \rho_j^{a(3)}(r_{ij}) \frac{r_{ij}^u}{r_{ij}}]^2 \quad (15)$$

Furthermore,
$$\rho^{a(h)}(r) = e^{-\beta^h(\frac{r}{r_e} - 1)} \quad (16)$$

where the decay lengths β^h are tunable parameters. The S_{ij} is a many-body screening function that quantifies the screening effect between two atoms i and j due to other atoms k in the system, and is defined as following,

$$S_{ij} = \prod_{k \neq i,j} f_c[\frac{C - C_{min}}{C_{max} - C_{min}}] \tag{17}$$

where C_{max} and C_{min} are the upper and lower bounds of C. In addition,

$$C = \frac{2(X_{ik} + X_{kj}) - (X_{ik} - X_{kj})^2 - 1}{1 - (X_{ik} - X_{kj})^2} \tag{18}$$

where $X_{ik} = (\frac{r_{ik}}{r_{ij}})^2$ and $X_{kj} = (\frac{r_{kj}}{r_{ij}})^2$.

The summary of the MEAM type potential parameterized for beryllium developed in ref. [26] is shown in Table S1 in Supplementary Information.

2.2. Tersoff Potential

The total interatomic energy for the Tersoff type potential is listed below; all equations presented below come from [21,27],

$$E = \frac{1}{2} \sum_i \sum_{j \neq i} f_C(r_{ij})[f_R(r_{ij}) + b_{ij} f_A(r_{ij})] \tag{19}$$

where r_{ij} denotes the distance r between atoms i and j. f_c is the sinusoidal cut-off function that takes the following expression,

$$f_C(r) = \begin{cases} 1 & r < R - D \\ \frac{1}{2} - \frac{1}{2}sin(\frac{\pi}{2}\frac{r-R}{D}) & R - D < r < R + D \\ 0 & r > R + D \end{cases} \tag{20}$$

where R and D are tunable parameters. In addition, $f_R(r)$ and $f_A(r)$ correspond to the repulsive and attractive components, respectively, and are shown below,

$$f_R(r) = Ae^{-\lambda_1 r} \tag{21}$$

$$f_A(r) = -Be^{-\lambda_2 r} \tag{22}$$

where A, B, λ_1, and λ_2 are adjustable parameters. Lastly, b_{ij} encodes an angular contribution that is written below,

$$b_{ij} = (1 + \beta^n \zeta_{ij}^n)^{-\frac{1}{2n}} \tag{23}$$

$$\zeta_{ij} = \sum_{k \neq i,j} f_C(r_{ik}) g[\theta(r_{ij}, r_{ik})] e^{\lambda_3^m (r_{ij} - r_{ik})^m} \tag{24}$$

$$g(\theta) = \gamma(1 + \frac{c^2}{d^2} - \frac{c^2}{[d^2 + (cos\theta - cos\theta_0)^2]}) \tag{25}$$

where $\theta(r_{ij}, r_{ik})$ represents the angle formed between atoms i, j, and k; n, β, m, λ_3, c, d, γ, and $cos\theta_0$ are the free parameters determined from fitting. The summary of the Tersoff type potential parameterized for beryllium developed in ref. [27] is shown in Table S2 in Supplementary Information.

2.3. Finnis–Sinclair Type Potential

Finnis–Sinclair type potential belongs to the embedded-atom method (EAM) potential. The total interatomic energy for the Finnis–Sinclair type potential can be written as follows, all equations presented below come from [20,28,29],

$$E = \frac{1}{2}\sum_{i,j} V(r_{ij}) - \sum_i f(\rho_i) \quad (26)$$

where $V(r_{ij})$ accounts for the direct interaction between two atoms i and j depending on their separation r_{ij}, and $f(\rho_i)$ accounts for the many-body effect as shown below,

$$f(\rho) = \sqrt{\rho(1+A\rho)} \quad (27)$$

where A is an adjustable parameter, and

$$\rho_i = \sum_j \Phi(r_{ij}) \quad (28)$$

The $V(r)$ and $\Phi(r)$ are cubic splines with the following expressions,

$$V(r) = \sum_{k=1}^{n} A_k (r_{ak} - r)^3 H(r_{ak} - r) \quad (29)$$

$$\Phi(r) = \sum_{k=1}^{m} B_k (r_{bk} - r)^3 H(r_{bk} - r) \quad (30)$$

where $H(x)$ is the Heavyside step function that equals 1 when x is greater than 0 and 0 otherwise. The n, m, A_k, r_{ak}, B, and r_{bk} are tunable parameters. The summary of the parameters for the Finnis–Sinclair type potential developed for beryllium [29] is listed in Table S3 in Supplementary Information.

2.4. Simulation Setup

The pristine beryllium characterizes a hexagonal close-packed (HCP) structure, whose unit cell is illustrated in Figure 1. There are two parameters in the unit cell determining the HCP structure, namely a and c. In the ordinary HCP crystalline, the ratio c/a equals $2\sqrt{6}/3$; however, experiments indicate that c/a for beryllium is slightly different from this value. The a and c/a for three types of potentials employed in this study are listed in Figure 2.

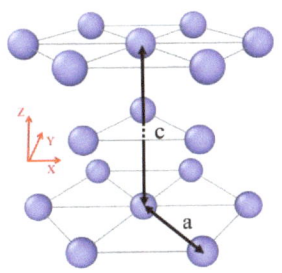

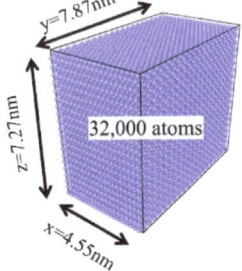

(a) The atomistic configurations
(b) The simulation box

Figure 1. (a) The atomistic configurations of atoms in a conventional unit cell of the hexagonal closed-packed crystalline structure of Be. (b) The simulation box with dimensions of the 32,000 Be atoms for this MD investigation.

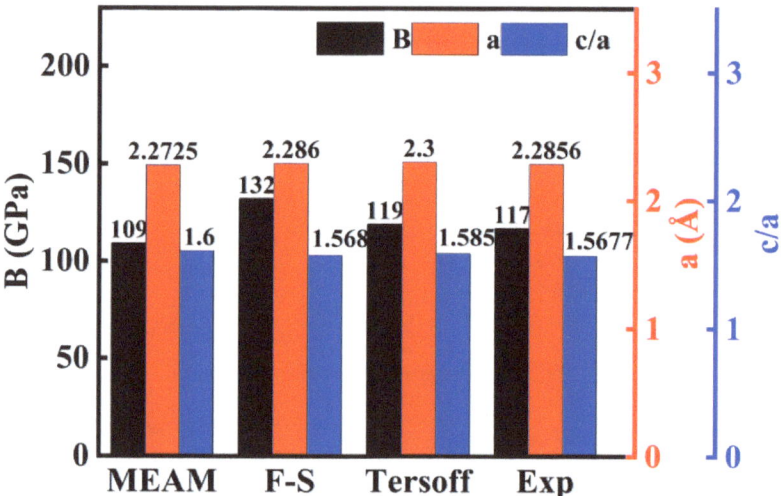

Figure 2. The lattice constants a (red right y-axis), c/a (blue right y-axis), and the bulk modulus, or B (black left y-axis) of pristine beryllium crystal derived from the MEAM, Finnis–Sinclair, Tersoff type potentials [26,27,29] compared to that from experiments [30,31].

The reference axes of our simulation domain are also illustrated in Figure 1. We first created a simulation box that measures 20 a in the x direction, 36.64 a in y the direction, and 20 c in the z direction, respectively, totaling 32,000 atoms; then, the system is relaxed at 300 K and 0 external pressure for 20 ps; after relaxation, the system was subjected to thermal expansion, compression, and elongation, respectively. To create the structures with defects, the atoms in the spherical region with a set radius centering in the middle of the crystalline were deleted before relaxation and mechanical loading. We have explicitly examined nine configurations. The numbers of deleted atoms, in an ascending order, are 12, 56, 159, 407, 775, 1339, 2114, 3148 and 4505, respectively. These systems are therefore named after the number of deleted atoms for convenience. The zero void size system (i.e., 0 atoms were deleted) refers to the perfect beryllium lattice used for reference for ease of the comparison. For mechanical loading, the system is elongated in the x direction, while the other two dimensions are kept at zero stresses to simulate the uniaxial tensile testing experiment.

3. Results

3.1. Equations of States

3.1.1. Hydrostatic Compression

Figure 3 illustrates the MD simulation results of volumetric change when the pristine beryllium is subject to hydrostatic compression using three types of potentials. Note that V represents the volume at the corresponding hydrostatic pressure, while V_0 is the volume at 0 pressure.

It can be seen that the volume is showing linear dependence on the hydrostatic pressure, while the slope of linearity varies among the three chosen types of potentials; the MEAM type potential shows the largest volumetric change, the Tersoff type is the second, and the Finnis–Sinclair type shows the least change. It is understandable that this slope reflects the bulk modulus, which is shown in Figure 2.

3.1.2. Thermal Expansion

Figure 4 illustrates the MD simulation results of volumetric change when the pristine beryllium is subject to heating using three types of potentials. Note that V represents the volume at the corresponding temperature, while V_0 is the volume at 0 K.

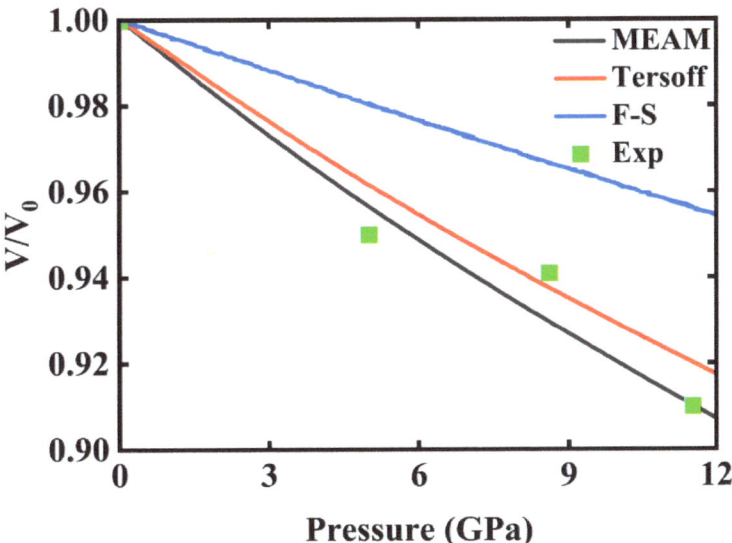

Figure 3. Normalized volumetric change as a function of pressure derived from MD simulations based on the three tested potentials, MEAM, Finnis–Sinclair (F–S), Tersoff, and compared to that from experiment [32].

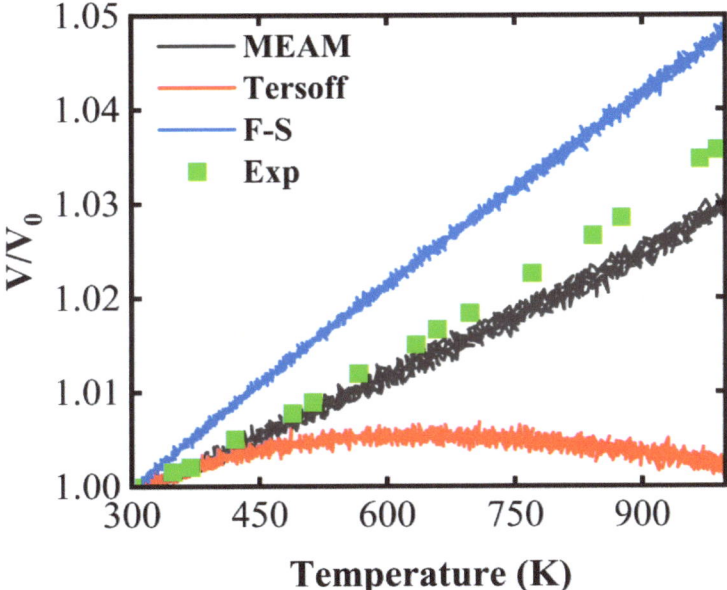

Figure 4. Normalized volumetric change as a function of temperature from MD simulations based on the three types of potentials, MEAM, Finnis–Sinclair (F–S), Tersoff, and that from Exp [33].

It can be seen that the three types of potentials show starkly distinct behaviors. First, in contrast to the case from hydrostatic compression, the Finnis–Sinclair type potential shows the largest volumetric change when the system is subject to heating. Moreover, the volume based on MEAM and Finnis–Sinclair type potential shows a linear dependence on temperature. Second, the volumetric curve based on the Tersoff type potential is very unusual. It starts to shrink at 500 K. The result based on MEAM potential is most similar to the result from Exp.

3.2. Uniaxial Tensile Response

In this section, we present the results from performing uniaxial tension on pristine beryllium crystals and those embedded with spherical void defects. Two dependent factors, namely the size of the void and temperature, along with the accompanying microstructural evolution are examined for three types of potentials.

3.2.1. MEAM Type Potential

Figure 5 illustrates the dependence on the size of the spherical void embedded in the beryllium crystals. The nine void-embedded systems (12, 56, 159, 407, 775, 1339, 2114, 3148, and 4505) and the primitive beryllium system (marked as "0") have been studied. For mechanical loading, the system is elongated in the x direction. It shows the stress–strain curves, where the color code is employed to represent the number of atoms removed from the crystals before relaxation and subsequent loading. These curves characterize similar features; the stress gradually builds up with strain until approaching the fracture point, after which there comes a sudden drop, suggesting a considerable stress release. The microstructural characteristics at the tagged points underlying the observed stress–strain relationship will be demonstrated later in this section. It can be readily seen that the larger size of the void necessarily leads to an earlier onset of fracture.

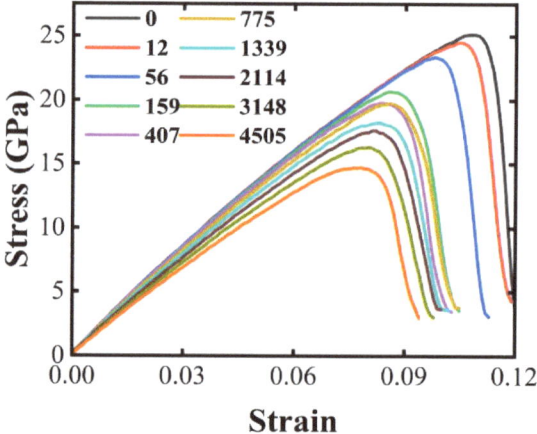

Figure 5. The strain–stress relationship of nine void-embedded beryllium systems (12, 56, 159, 407, 775, 1339, 2114, 3148, and 4505) under tensile tests compared that of pristine beryllium ("0" system). The voids are spheric. The zero void size system (0, grey line) refers to the perfect beryllium lattice used for reference for ease of the comparison. The system size is 32,000 lattice sites. The interatomic potential used is the MEAM potential. The temperature is 300 K.

Figure 6 illustrates the effect of the size of the spherical void on the properties of beryllium reflected by the strain–stress relationship. The tensile toughness, defined as the energy absorbed per unit volume before failure during uniaxial tensile testing, becomes smaller, and is shown in panel (a). The slope of the linear region on curves representing Young's modulus becomes smaller too, as shown in panel (b). Similarly, panel (c) and panel (b) demonstrate the stress and strain at the onset of fracture. The original data are listed in Table S4 in Supplementary Information.

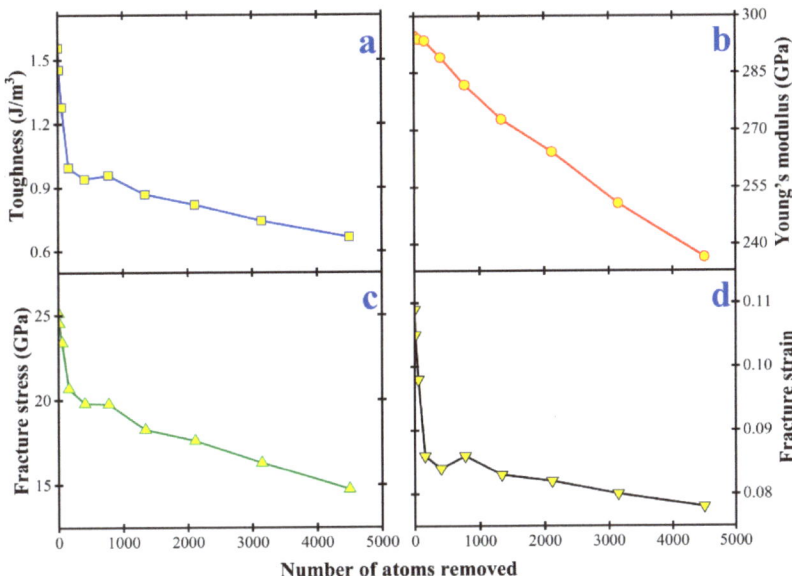

Figure 6. Effect of void size of spherical void on the mechanical performance of beryllium subject to uniaxial tension. The nine void-embedded beryllium systems (12, 56, 159, 407, 775, 1339, 2114, 3148, and 4505) are compared to the pristine one ("0"-sized void). (**a**) Tensile toughness; (**b**) Young's modulus; (**c**) Fracture stress; (**d**) Fracture strain. The system size is 32,000 lattice sites. The interatomic potential used is the MEAM potential. The temperature is 300 K.

To examine the dependence on temperature on the mechanical behaviors, we have investigated the uniaxial tensile tests with the same size (775) of the void defect fixed at various temperatures. We have considered four temperatures, 150 K, 300 K, 450 K, and 600 K to show the trend. The results are displayed in Figure 7. These four stress–strain curves suggest a general trend that the stress gradually builds up with strain until approaching the fracture point, after which there comes a sudden drop, suggesting a considerable stress release. The increase in temperature lowers the ultimate tensile strength, the fracture stress, and the fracture strain of the void-embedded hcp beryllium.

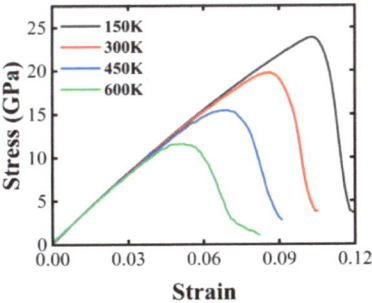

Figure 7. Effect of temperature on the strain–stress relationship of beryllium subject to uniaxial tension derived from MD simulations. The system size is 32,000 lattice sites. The interatomic potential used is the MEAM potential. The void size is 775 (this number of atoms were removed to form the void).

Figure 8 illustrates the effect of temperature on the properties of beryllium reflected by the strain–stress relationship. Similarly, the results indicate that a higher temperature leads to smaller toughness, smaller Young's modulus, and earlier onset of fracture (equivalently smaller fracture stress and strain). These quantities appear to be approximately linearly dependent on the temperature in the beryllium crystals. The original data are listed in Table S5 in Supplementary Information.

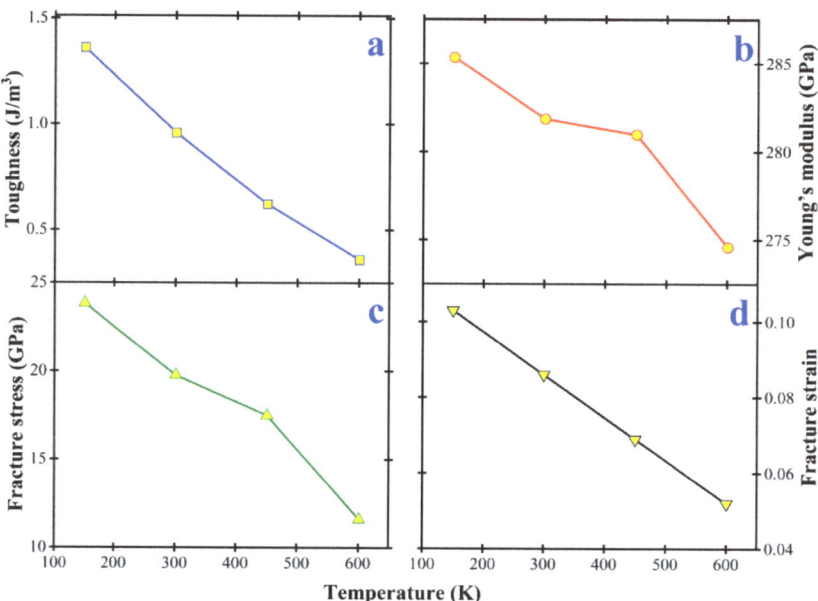

Figure 8. Effect of temperature on the mechanical performance of beryllium subject to uniaxial tension derived from MD simulations: (**a**) Toughness; (**b**) Young's modulus; (**c**) Fracture stress; (**d**) Fracture strain. The system size is 32,000 lattice sites. The interatomic potential used is the MEAM potential. The void size is 775 (this number of atoms were removed to form the void).

To investigate the fracture mechanism, we illustrate the microscopic characteristics in Figure 9. Panel (a) shows the strain–stress relationship and normalized energy change during the stretching process of strain. Panel (b) and panel (c) demonstrate the configurations of atoms on the plane slicing through the center of the void defect, where the color code is employed to represent the potential energy of each atom in panel (b) and the stress on each atom in the x direction in panel (c). Moreover, the circled numbers denote the stages throughout the loading process marked in panel (a). In addition, we know that cracks are more likely to occur where the stress is concentrated, and from Figure 9, the stress is concentrated in the middle, so the material is easy to break from the middle.

The first configuration corresponds to the onset of loading. The second configuration corresponds to the stage in the middle of loading before fracture, the color of panel (c) shows that the stress increases. The third configuration corresponds to the stage of the imminent fracture. Up to this stage, it is the point on the peak in panel (a) that some parts of panel (c) become deep red, showing that the stress is concentrated, and will crack. The fourth, fifth, and sixth configurations correspond to the process of fracture. The red color of panel (c) decreases, showing that the stress is released after the fracture.

3.2.2. Tersoff Type Potential

Figure 10 illustrates the dependence on the size of the spherical void embedded in the beryllium crystals based on the Tersoff potential. The color-code is consistent with that of Figure 5.

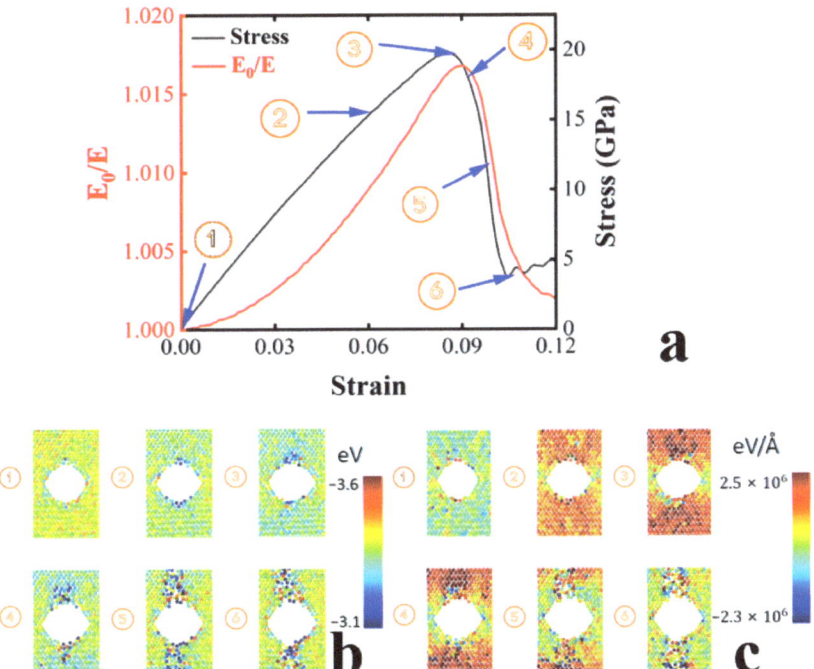

Figure 9. Microstructures underlying progress derived from MD simulations based on the MEAM type potential: (**a**) strain–stress relationship and strain–normalized energy relationship; (**b**) potential energy; (**c**) stress in the x direction. The system size is 32,000 lattice sites. The void size is 775 (this number of atoms were removed to form the void).

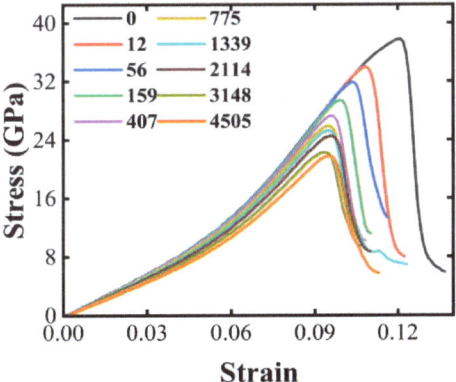

Figure 10. Effect of size of spherical void on the strain–stress relationship of beryllium subject to uniaxial tension derived from MD simulations based on the Tersoff type potential. The system size is 32,000 lattice sites. The temperature is 300 K.

Figure 11 illustrates the dependence on the size of the spherical void embedded in the beryllium crystals based on the Tersoff potential. The arrangement of panels is consistent with that of Figure 6. Similarly, we see toughness depicted in panel (a), Young's modulus shown in panel (b), and the fracture stress and strain illustrated in panel (c) and panel (d), all decrease with the enlargement of the spherical void defect. The original data are listed in Table S6 in Supplementary Information.

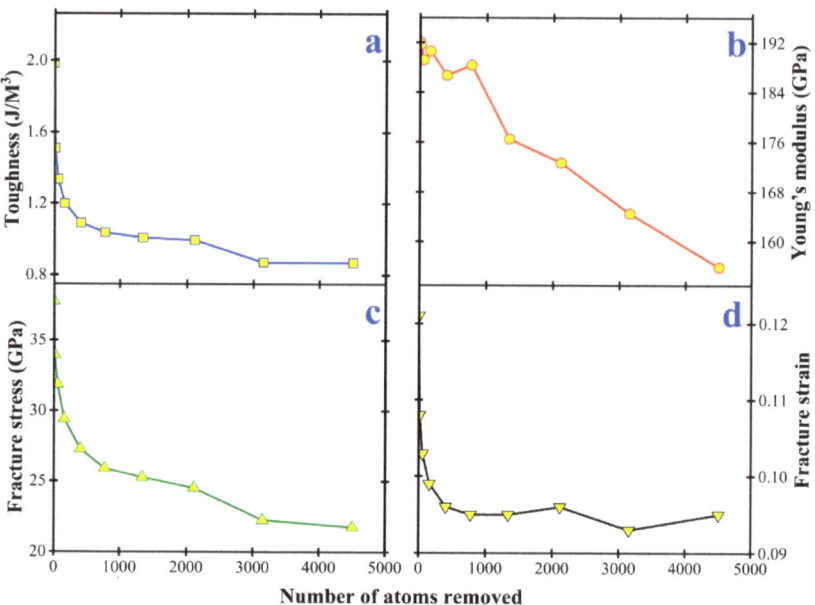

Figure 11. Effect of size of spherical void on the mechanical performance of beryllium subject to uniaxial tension derived from MD simulations based on the Tersoff type potential: (**a**) Toughness; (**b**) Young's modulus; (**c**) Fracture stress; (**d**) Fracture strain. The system size is 32,000 lattice sites. The temperature is 300 K.

Figure 12 illustrates the dependence on temperature with the size of the void defect fixed. The color-code is consistent with that of Figure 7.

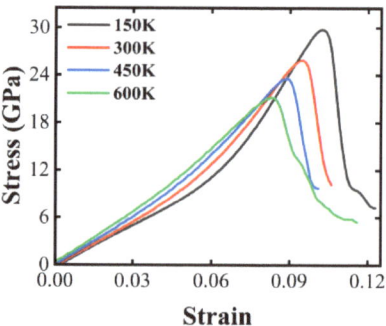

Figure 12. Effect of temperature on the strain–stress relationship of beryllium subject to uniaxial tension derived from MD simulations. The system size is 32,000 lattice sites. The interatomic potential used is the Tersoff potential. The void size is 775 (this number of atoms were removed to form the void).

Figure 13 illustrates the effect of temperature on the properties of beryllium reflected by the strain–stress relationship. The arrangement is consistent with that of Figure 8. The results on the fracture strain and stress, and toughness are similar. However, surprisingly, Young's modulus linearly increases with temperature. The original data is listed in Table S7 in Supplementary Information.

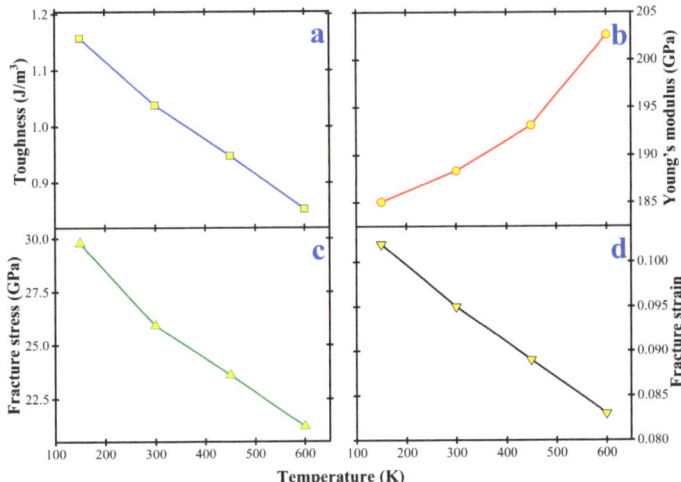

Figure 13. Effect of temperature on the mechanical performance of beryllium subject to uniaxial tension derived from MD simulations based on the Tersoff type potential: (**a**) Toughness; (**b**) Young's modulus; (**c**) Fracture stress; (**d**) Fracture strain. The system size is 32,000 lattice sites. The void size is 775 (this number of atoms were removed to form the void).

The microscopic characteristics throughout the fracture are illustrated in Figure 14. The arrangement and color-code are consistent with that of Figure 9. According to Figure 14, the stress distribution is relatively uniform, and small cracks may appear in some parts.

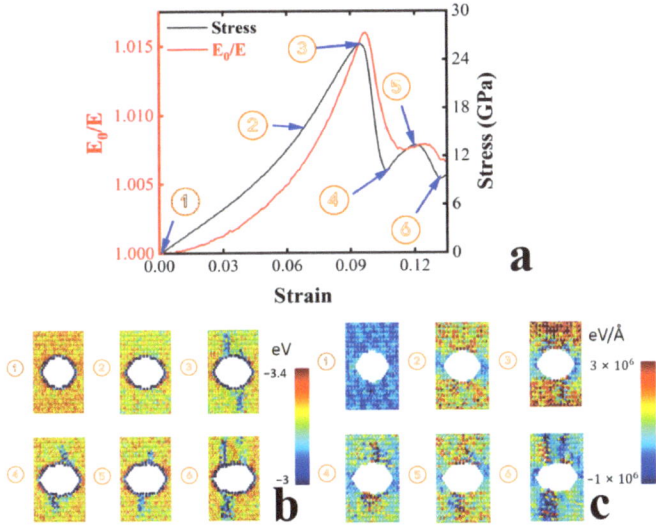

Figure 14. Microstructures underlying progress derived from MD simulations based on the Tersoff type potential: (**a**) strain–stress relationship and strain–normalized energy relationship; (**b**) potential energy; (**c**) stress in the x direction. The system size is 32,000 lattice sites. The void size is 775 (this number of atoms were removed to form the void).

In the first and second configurations, the system is stressed from its initial state. In the third configuration, corresponding to the stage at the imminence of fracture, the system began to develop some small cracks. In the fourth, fifth, and sixth configurations, the crack grows in both directions. Nonetheless, the system does not tear up.

3.2.3. Finnis–Sinclair Type Potential

Figure 15 illustrates the dependence on the size of the spherical void embedded in the beryllium crystals. The color-code is consistent with that of Figure 5. The observations are also qualitatively similar.

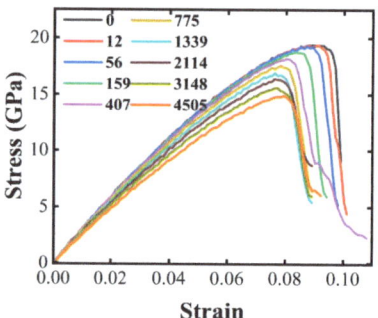

Figure 15. Effect of size of spherical void on the strain–stress relationship of beryllium subject to uniaxial tension derived from MD simulations based on the Finnis–Sinclair type potential. The system size is 32,000 lattice sites. The temperature is 300 K.

Figure 16 illustrates the effect of the size of the spherical void on the properties of beryllium reflected by the strain–stress relationship. The arrangement of panels is consistent with that of Figure 6. The results on toughness, Young's modulus and the fracture stress are similar. However, the fracture strain slightly increases when the atoms removed exceed 2000. The original data are listed in Table S8 in Supplementary Information.

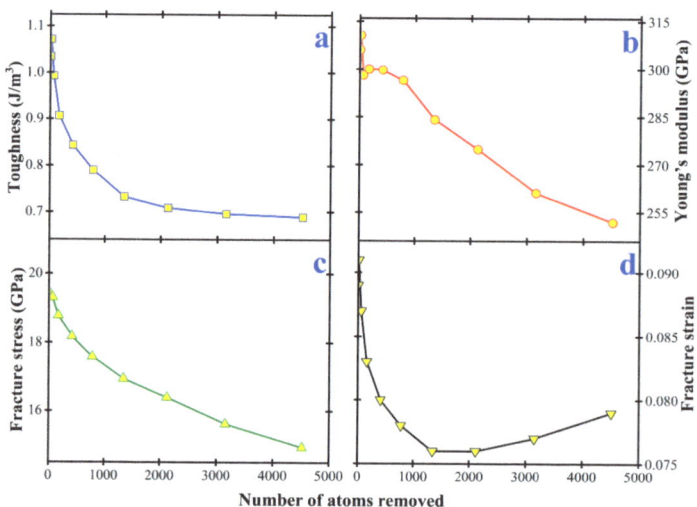

Figure 16. Effect of size of spherical void on the mechanical performance of beryllium subject to uniaxial tension derived from MD simulations based on the Finnis–Sinclair type potential: (**a**) Toughness; (**b**) Young's modulus; (**c**) Fracture stress; (**d**) Fracture strain. The system size is 32,000 lattice sites. The temperature is 300 K.

Figure 17 illustrates the dependence on temperature with the size of the void defect fixed. The color-code is consistent with that of Figure 7. The temperature has a slight effect on the strain-stress relationship.

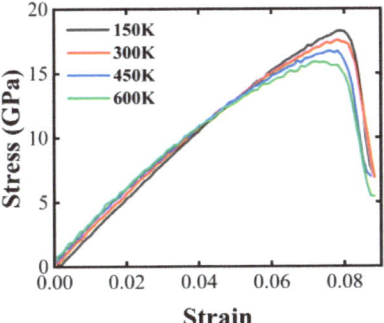

Figure 17. Effect of temperature on the strain–stress relationship of beryllium subject to uniaxial tension derived from MD simulations based on the Finnis–Sinclair type potential.

Figure 18 illustrates the effect of temperature on the properties of beryllium reflected by the strain–stress relationship. The arrangement is consistent with that of Figure 8. The results on toughness and fracture stress are similar. Moreover, Young's modulus and fracture strain show very weak dependence on temperature. The original data are listed in Table S9 in Supplementary Information.

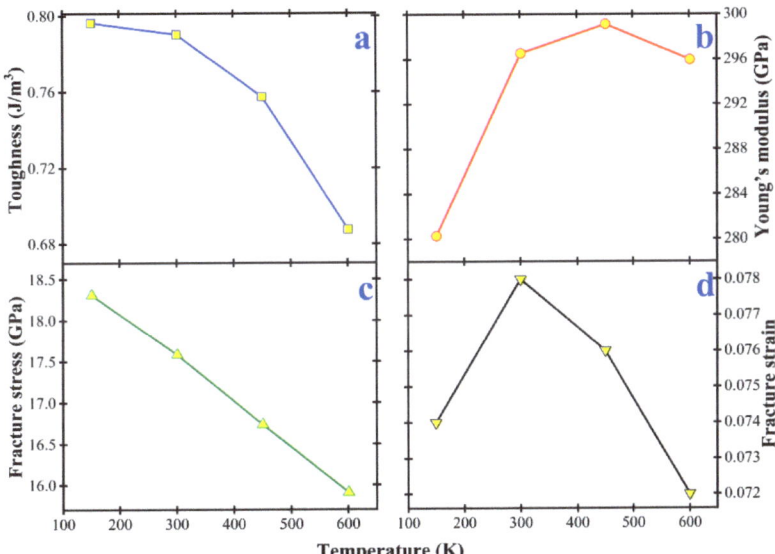

Figure 18. Effect of temperature on the mechanical performance of beryllium subject to uniaxial tension derived from MD simulations based on the Finnis–Sinclair type potential: (**a**) Toughness; (**b**) Young's modulus; (**c**) Fracture stress; (**d**) Fracture strain. The system size is 32,000 lattice sites. The void size is 775 (this number of atoms were removed to form the void).

The microscopic characteristics throughout the fracture are illustrated in Figure 19. The arrangement and color-code are consistent with that of Figure 9. According to Figure 19, the stress distribution is relatively uniform, and small cracks may appear in some parts.

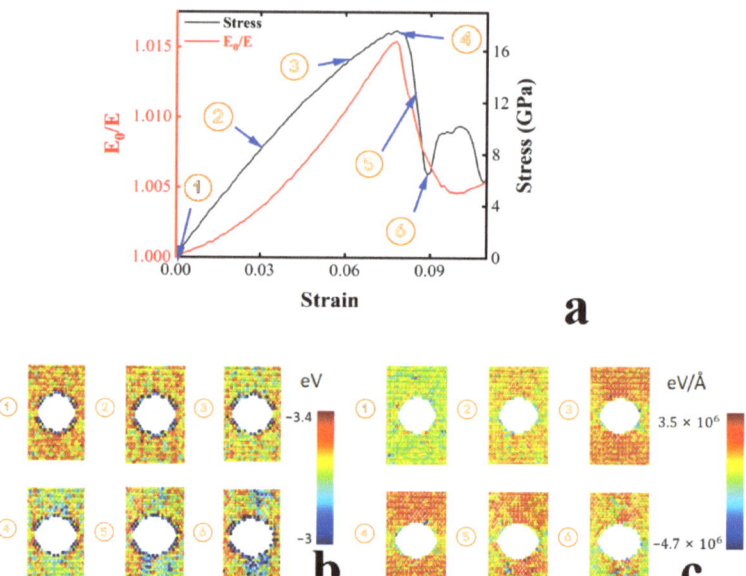

Figure 19. Microstructures underlying progress derived from MD simulations based on the Finnis–Sinclair type potential: (**a**) strain–stress relationship and strain–normalized energy relationship; (**b**) potential energy; (**c**) stress in the x direction. The system size is 32,000 lattice sites. The void size is 775 (this number of atoms were removed to form the void). The temperature is 300 K.

The first and second configurations are similar to those from prior potentials. Similarly, some cracks are created in the third configuration. However, these sheath cracks do not lead to the tearing up of the system.

4. Discussion

Figure 20 illustrates the strain–stress relationship of the three potentials, where the color code is employed to represent the type of potential. All three potentials are under the same temperature (300 K) and same size of the spherical void (775). Especially, the stress–strain relationship of MEAM and Finnis–Sinclair has some similarities.

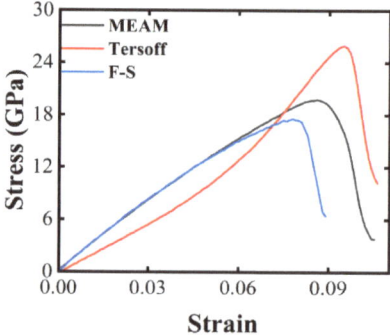

Figure 20. Strain–stress relationship of beryllium subject to uniaxial tension derived from MD simulations based on the three types of potentials: MEAM (black), Tersoff (red) and Finnis–Sinclair (blue). The system size is 32,000 lattice sites. The void size is 775 (this number of atoms were removed to form the void). The temperature is 300 K.

Figure 21 illustrates the effect of the size of the spherical void on the properties of beryllium, respectively, based on the three types of potential: MEAM, Tersoff, and Finnis–Sinclair type potential. In all three potentials, the toughness becomes smaller, and is shown in panel (a). The Young's modulus becomes smaller too, as shown in panel (b). Similarly, panel (c) and panel (b) demonstrate the fracture stress and strain at the onset of fracture.

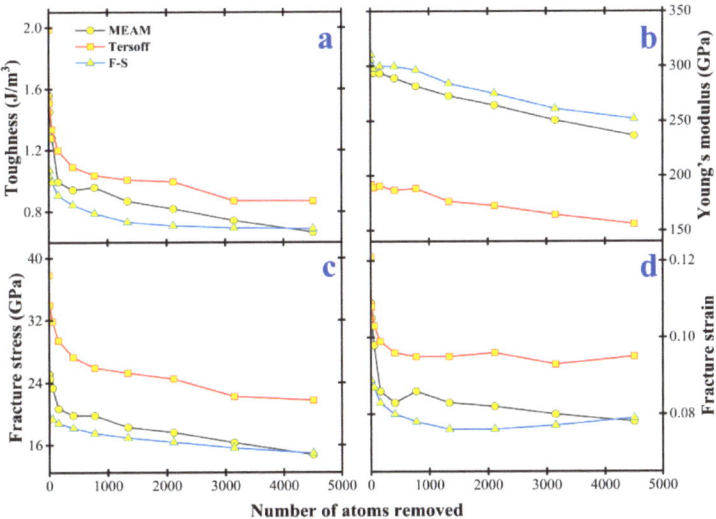

Figure 21. Comparison of three potentials on the mechanical performance of beryllium subject to uniaxial tension derived from MD simulations: (**a**) Toughness; (**b**) Young's modulus; (**c**) Fracture stress; (**d**) Fracture strain. The system size is 32,000 lattice sites. The void size is 775 (this number of atoms were removed to form the void). The temperature is 300 K.

However, there are subtle differences in the three types of potential. The toughness of the Tersoff potential is highest and the Finnis–Sinclair is lowest. Young's modulus of Finnis–Sinclair and MEAM is much higher than that of Tersoff. Fracture stress and strain are similar to toughness.

5. Conclusions

We have assessed interatomic potentials on the mechanical properties of HCP beryllium. We have explicitly examined the beryllium with defects of spherical void as well as the referring pristine perfect beryllium using three types of interatomic potentials, namely MEAM, Finnis–Sinclair, and Tersoff. Through systematic comparison, we gain atomistic insights into the relationship between the microstructure and mechanical performances.

The bulk modulus derived from hydrostatic compression suggests that the results derived from the MEAM and Tersoff type potentials are fairly close to that from the experiment. Among the thermal expansion curves, the result derived from the Tersoff type potential is unreasonable showing a negative thermal expansion coefficient beyond the temperature of 500 K.

From the uniaxial tension testing, the MEAM type predicts a clear fracture mechanism by tearing up along the direction perpendicular to that of stretching, whereas the Finnis–Sinclair and Tersoff types tend to accommodate the deformation more homogeneously without tearing up the system. Considering the fact that beryllium tends to be brittle after neutron irradiation, the predictions from the MEAM type potential are more reasonable.

In summary, through intercomparison of the three types of potentials previously developed for beryllium, this study concludes that the MEAM type potential yields the most reasonable predictions on the pristine beryllium and those with spherically shaped void defects. Our results might be useful in further atomistic investigation and material design on beryllium, a toxic yet important nuclear material.

Supplementary Materials: The following supporting information can be downloaded at: https://www.mdpi.com/article/10.3390/cryst13091330/s1, Supplementary Information includes nine tables, as (Tables S1–S3) the parameters of three potentials developed for beryllium; (Tables S4 and S5) effect of void size of spherical void and temperature on the mechanical performance of beryllium subject to uniaxial tension based on the MEAM potential; (Tables S6 and S7) effect of void size of spherical void and temperature on the mechanical performance of beryllium subject to uniaxial tension based on the Tersoff potential; (Tables S8 and S9) effect of void size of spherical void and temperature on the mechanical performance of beryllium subject to uniaxial tension based on the Finnis–Sinclair potential.

Author Contributions: Conceptualization, B.W., J.J. and Q.P.; methodology, J.J. and Q.P.; formal analysis, C.Y. and W.D.; investigation, C.Y., W.D. and S.L.; data curation, C.Y.; writing—original draft preparation, B.W.; writing—review and editing, B.W. and Q.P.; visualization, C.Y.; supervision, B.W. and Q.P.; project administration, B.W. All authors have read and agreed to the published version of the manuscript.

Funding: Q. P. would like to acknowledge the support provided by the National Natural Science Foundation of China (Grant No. 12272378) and the LiYing Program of the Institute of Mechanics, Chinese Academy of Sciences (Grant No. E1Z1011001).

Data Availability Statement: The data presented in this study are available upon reasonable requests.

Conflicts of Interest: The authors declare no conflict of interest. The funders had no role in the design of this study; in the collection, analyses, or interpretation of the data; in the writing of the manuscript, or in the decision to publish the results.

References

1. Zheng, L.; Liu, X.; Bao, S.; Zhang, J.; Zhong, J.; Ding, Y. Research Progress on Properties of Rare Metal Beryllium. *Chin. J. Rare Met.* **2023**, *24*, 292–302.
2. Barabash, V.; Eaton, R.; Hirai, T.; Kupriyanov, I.; Nikolaev, G.; Wang, Z.; Liu, X.; Roedig, M.; Linke, J. Summary of beryllium qualification activity for ITER first-wall applications. *Phys. Scr.* **2011**, *T145*, 014007. [CrossRef]
3. Tolias, P. Analytical expressions for thermophysical properties of solid and liquid beryllium relevant for fusion applications. *Nucl. Mater. Energy* **2022**, *31*, 101195. [CrossRef]
4. Vladimirov, P.V.; Chakin, V.P.; Dürrschnabel, M.; Gaisin, R.; Goraieb, A.; Gonzalez, F.A.H.; Klimenkov, M.; Rieth, M.; Rolli, R.; Zimber, N.; et al. Development and characterization of advanced neutron multiplier materials. *J. Nucl. Mater.* **2021**, *543*, 152593. [CrossRef]
5. DiJulio, D.D.; Lee, Y.J.; Muhrer, G.; Herwig, K.W.; Iverson, E.B. Impact of crystallite size on the performance of a beryllium reflector. *J. Neutron Res.* **2020**, *22*, 275–279. [CrossRef]
6. Muhammad, S.T.; Ahmad, S.-u.-I.; Chaudri, K.S.; Ahmad, A. Beryllium as reflector of MNSR. *Ann. Nucl. Energy* **2008**, *35*, 1708–1712. [CrossRef]
7. Hu, N.; Tanaka, H.; Akita, K.; Kakino, R.; Aihara, T.; Nihei, K.; Ono, K.; Baxter, D.; Gutberlet, T.; Kino, K.; et al. Accelerator based epithermal neutron source for clinical boron neutron capture therapy. *J. Neutron Res.* **2022**, *24*, 359–366. [CrossRef]
8. Magni, C.; Postuma, I.; Ferrarini, M.; Protti, N.; Fatemi, S.; Gong, C.; Anselmi-Tamburini, U.; Vercesi, V.; Battistoni, G.; Altieri, S.; et al. Design of a BNCT irradiation room based on proton accelerator and beryllium target. *Appl. Radiat. Isot.* **2020**, *165*, 109314. [CrossRef] [PubMed]
9. Pajuste, E.; Kizane, G.; Avotina, L.; Zariņš, A. Behaviour of neutron irradiated beryllium during temperature excursions up to and beyond its melting temperature. *J. Nucl. Mater.* **2015**, *465*, 293–300. [CrossRef]
10. Simos, N.; Elbakhshwan, M.; Zhong, Z.; Camino, F. Proton irradiation effects on beryllium—A macroscopic assessment. *J. Nucl. Mater.* **2016**, *479*, 489–503. [CrossRef]
11. Peng, Q.; Zhang, X.; Hung, L.; Carter, E.A.; Lu, G. Quantum simulation of materials at micron scales and beyond. *Phys. Rev. B* **2008**, *78*, 054118. [CrossRef]
12. Sun, Y.; Peng, Q.; Lu, G. Quantum mechanical modeling of hydrogen assisted cracking in aluminum. *Phys. Rev. B* **2013**, *88*, 104109. [CrossRef]

13. Peng, Q.; Ma, Z.; Cai, S.; Zhao, S.; Chen, X.; Cao, Q. Atomistic Insights on Surface Quality Control via Annealing Process in AlGaN Thin Film Growth. *Nanomaterials* **2023**, *13*, 1382. [CrossRef] [PubMed]
14. Peng, Q.; Meng, F.; Yang, Y.; Lu, C.; Deng, H.; Wang, L.; De, S.; Gao, F. Shockwave generates <100> dislocation loops in bcc iron. *Nat. Commun.* **2018**, *9*, 4880. [PubMed]
15. Ganchenkova, M.G.; Vladimirov, P.V.; Borodin, V.A. Vacancies, interstitials and gas atoms in beryllium. *J. Nucl. Mater.* **2009**, *386–388*, 79–81. [CrossRef]
16. Agrawal, A.; Mishra, R.; Ward, L.; Flores, K.M.; Windl, W. An embedded atom method potential of beryllium. *Model. Simul. Sci. Eng.* **2013**, *21*, 085001. [CrossRef]
17. Daw, M.S.; Baskes, M.I. Embedded-atom method: Derivation and application to impurities, surfaces, and other defects in metals. *Phys. Rev. B* **1984**, *29*, 6443–6453. [CrossRef]
18. Hu, W.Y.; Zhang, B.W.; Huang, B.Y.; Gao, F.; Bacon, D.J. Analytic modified embedded atom potentials for HCP metals. *J. Phys. Condens. Matter.* **2001**, *13*, 1193–1213. [CrossRef]
19. Baskes, M.I. Modified embedded-atom potentials for cubic materials and impurities. *Phys. Rev. B* **1992**, *46*, 2727–2742. [CrossRef] [PubMed]
20. Finnis, M.W.; Sinclair, J.E. A simple empirical N-body potential for transition metals. *Philos. Mag. A* **1984**, *50*, 45–55. [CrossRef]
21. Tersoff, J. New empirical approach for the structure and energy of covalent systems. *Phys. Rev. B* **1988**, *37*, 6991–7000. [CrossRef]
22. Plimpton, S. Fast Parallel Algorithms for Short-Range Molecular Dynamics. *J. Comput. Phys.* **1995**, *117*, 1–19. [CrossRef]
23. Baskes, M.I. Determination of modified embedded atom method parameters for nickel. *Mater. Chem. Phys.* **1997**, *50*, 152–158. [CrossRef]
24. Baskes, M.I.; Johnson, R.A. Modified embedded atom potentials for HCP metals. *Model. Simul. Mater. Sci. Eng.* **1994**, *2*, 147–163. [CrossRef]
25. Lee, B.-J.; Ko, W.-S.; Kim, H.-K.; Kim, E.-H. The modified embedded-atom method interatomic potentials and recent progress in atomistic simulations. *Calphad* **2010**, *34*, 510–522. [CrossRef]
26. Dremov, V.V.; Karavaev, A.V.; Kutepov, A.L.; Soulard, L.; Elert, M.; Furnish, M.D.; Chau, R.; Holmes, N.; Nguyen, J. Molecular Dynamics Simulation of Thermodynamic and Mechanical Properties of Be and Mg. In Proceedings of the Conference of the American-Physical-Society-Topical-Group on Shock Compression of Condensed Matter, Waikoloa, HI, USA, 24–29 June 2007.
27. Bjorkas, C.; Juslin, N.; Timko, H.; Vortler, K.; Nordlund, K.; Henriksson, K.; Erhart, P. Interatomic potentials for the Be-C-H system. *J. Phys. Condens. Matter* **2009**, *21*, 445002. [CrossRef]
28. Dai, X.D.; Kong, Y.; Li, J.H.; Liu, B.X. Extended Finnis–Sinclair potential for bcc and fcc metals and alloys. *J. Phys. Condens. Matter* **2006**, *18*, 4527–4542. [CrossRef]
29. Igarashi, M.; Khantha, M.; Vitek, V. N-body interatomic potentials for hexagonal close-packed metals. *Philos. Mag. B* **2006**, *63*, 603–627. [CrossRef]
30. Migliori, A.; Ledbetter, H.; Thoma, D.J.; Darling, T.W. Beryllium's monocrystal and polycrystal elastic constants. *J. Appl. Phys.* **2004**, *95*, 2436–2440. [CrossRef]
31. Petzow, G.; Aldinger, F.; Jönsson, S.; Welge, P.; van Kampen, V.; Mensing, T.; Brüning, T. *Ullmann's Encyclopedia of Industrial Chemistry*; Wiley-VCH: Hoboken, NJ, USA, 2005.
32. Ming, l.C.; Manghnani, M.H. Isotherma compression and phase-transition in beryllium to 28.3 GPa. *J. Phys. F Met. Phys.* **1984**, *14*, L1–L8. [CrossRef]
33. Gordon, P. A High Temperature Precision X-ray Camera: Some Measurements of the Thermal Coefficients of Expansion of Beryllium. *J. Appl. Phys.* **1949**, *20*, 908–917. [CrossRef]

Disclaimer/Publisher's Note: The statements, opinions and data contained in all publications are solely those of the individual author(s) and contributor(s) and not of MDPI and/or the editor(s). MDPI and/or the editor(s) disclaim responsibility for any injury to people or property resulting from any ideas, methods, instructions or products referred to in the content.

Article

Effect of Pore Defects on Very High Cycle Fatigue Behavior of TC21 Titanium Alloy Additively Manufactured by Electron Beam Melting

Qingdong Li [1], Shuai Liu [1], Binbin Liao [1], Baohua Nie [1,*], Binqing Shi [1], Haiying Qi [1], Dongchu Chen [2,*] and Fangjun Liu [3,*]

[1] School of Materials Science and Hydrogen Energy, Foshan University, Foshan 528000, China
[2] Guangdong Key Laboratory for Hydrogen Energy Technologies, Foshan 528000, China
[3] School of Mechatronic Engineering and Automation, Foshan University, Foshan 528000, China
* Correspondence: niebaohua@fosu.edu.cn (B.N.); chendc@fosu.edu.cn (D.C.); liufj@fosu.edu.cn (F.L.); Tel.: +86-075782700525 (B.N.); +86-075782700525 (D.C.); +86-075782700817 (F.L.)

Abstract: Titanium alloys additively manufactured by electron beam melting (EBM) inevitably obtained some pore defects, which significantly reduced the very high cycle fatigue performance. An ultrasonic fatigue test was carried out on an EBM TC21 titanium alloy with hot isostatic pressing (HIP) and non-HIP treatment, and the effect of pore defects on the very high cycle fatigue (VHCF) behavior were investigated for the EBM TC21 titanium alloy. The results showed that the S-N curve of non-HIP specimens clearly had a tendency to decrease in very high cycle regimes, and HIP treatment significantly improved fatigue properties. Fatigue limits increased from 250 MPa for non-HIP specimens to 430 MPa for HIP ones. Very high cycle fatigue crack mainly initiated from the internal pore for EBM specimens, and a fine granular area (FGA) was observed at the crack initiation site in a very high cycle regime for both non-HIP and HIP specimens. ΔK_{FGA} had a constant trend in the range from 2.7 MPa$\sqrt{m}$ to 3.5 MPa$\sqrt{m}$, corresponding to the threshold stress intensity factor range for stable crack propagation. The effect of pore defects on the very high cycle fatigue limit was investigated based on the Murakami model. Furthermore, a fatigue indicator parameter (FIP) model based on pore defects was established to predict fatigue life for non-HIP and HIP specimens, which agreed with the experimental data.

Keywords: titanium alloy; EBM; VHCF; HIP; pore defects

Citation: Li, Q.; Liu, S.; Liao, B.; Nie, B.; Shi, B.; Qi, H.; Chen, D.; Liu, F. Effect of Pore Defects on Very High Cycle Fatigue Behavior of TC21 Titanium Alloy Additively Manufactured by Electron Beam Melting. *Crystals* **2023**, *13*, 1327. https://doi.org/10.3390/cryst13091327

Academic Editors: Wangzhong Mu and Chao Chen

Received: 12 August 2023
Revised: 28 August 2023
Accepted: 28 August 2023
Published: 30 August 2023

Copyright: © 2023 by the authors. Licensee MDPI, Basel, Switzerland. This article is an open access article distributed under the terms and conditions of the Creative Commons Attribution (CC BY) license (https:// creativecommons.org/licenses/by/ 4.0/).

1. Introduction

Titanium alloys were widely used in aerospace due to their high specific strength and excellent corrosion resistance [1]. The components fabricated by additive manufacturing (AM) were more efficient, had shorter cycles, and free structural design [2–4]. Compared with selective laser melting (SLM), the EBM titanium alloy underwent a special thermal history due to the influence of high preheating temperature, resulting in an α + β dual-phase microstructure and low residual stress [5]. Fatigue fracture of these EBM titanium alloy components in aerospace still occurred when subjected to high frequency and low-stress cyclic loads during ultra-long life service (>10^7 cycles), i.e., VHCF, where fatigue cracks initiated from the specimen interior, and FGA was observed at the crack initiation site [6,7]. Therefore, the VHCF performance of titanium alloy components fabricated by EBM was a key factor for large-scale applications.

The widespread application of AM components was limited due to their inherent defects, such as pores, lack of fusion (LoF), and mirocracks. Fatigue properties of EBM titanium alloy components were still lower than that of the forgings ones, owing to the promotion of crack initiation by defects [8,9]. The Murakami model was widely used to evaluate the relationship between fatigue strength and defects in AM components [10].

The fatigue scatter and size effect of AM materials with defects were investigated based on the extreme value statistics theory and probability fatigue theory [11]. Furthermore, based on the weakest-link theory, a probability fatigue life model was established to analyze the location and size effects of defects [12]. The FIP model based on the stress intensity factor of defects had been applied to predict fatigue life for the titanium alloy [13].

In a very high cycle regime, fatigue crack initiation and fatigue life were much more sensitive to the defects in AM titanium alloys. The investigation by Günther [14] showed that very high cycle fatigue cracks initiated from pores and LoF for EBM titanium alloy, where FGA characteristics were observed around these defects. Fatigue performance was similar to that of the SLM specimens after stress-release treatment, indicating that the EBM specimens did not require stress-relieving annealing due to their low residual stress [8]. However, much research on very high cycle fatigue was focused on SLM titanium alloys [15–17]. The internal defects (pores, lack of fusion, etc.) and surface artificial defects dramatically reduced the very high cycle fatigue properties of the SLM titanium alloy, and fatigue strength models based on defect size were proposed based on the Murakami model [18].

HIP was widely applied to improve fatigue performance due to the reduction of defects in EBM titanium alloy components [19]. However, little attention was paid to the very high cycle fatigue behavior of the EBM titanium alloy treated by HIP. As for SLM titanium alloys, HIP significantly improved the very high cycle fatigue performance, which can be comparable to that of conventional forged ones, although HIP treatment coarsened the microstructure and reduced the tensile strength of the SLM Ti6Al4V titanium alloy [18,20]. A very high cycle fatigue crack initiated at the large α phase of SLM Ti6Al4V alloy was treated by HIP [14], suggesting that the reduction of the α phase size after HIP treatment can further improve very high cycle fatigue performance. Furthermore, it should be noted that the combination of HIP and surface machining can effectively enhance the fatigue performance of EBM titanium alloy components. HIP treatment did not generally improve the fatigue life of the AM titanium alloy without machining on the surface due to the effect of the surface roughness [9], while the specimens obtained excellent fatigue properties after HIP treatment and surface machining.

The very high cycle fatigue behavior of the EBM TC21 titanium alloy, which was developed for aerospace structural components, was investigated. Considering that LoF had a more negative impact on fatigue performance than pore defects [21], the EBM process with high energy input and low forming rate was adopted to restrain the dangerous LoF defects. Then, the effect of the internal pore in the EBM titanium alloy on very high cycle fatigue properties and the crack-initiation mechanism was discussed in comparison with that of HIP ones. Very high cycle fatigue strength and fatigue life of an EBM titanium alloy containing internal pore defects were quantitatively evaluated based on the Murakami model and FIP model, which was beneficial for the application of EBM titanium alloy components in aerospace engineering.

2. Experimental Procedures
2.1. Materials and Manufacturing

The TC21 titanium alloy was manufactured by EBM on an Arcam A2 machine (Arcam, Stockholm, Sweden) under a 5×10^{-3} Pa vacuum. The chemical composition (wt.%) of the powder is shown in Table 1. The diameter size of the particle was in the range of 50 to 125 μm. Preheating the chamber at 650 °C reduced the residual stress during EBM. To avoid the formation of dangerous LoF defects, an EBM process with high energy input and low forming rate was adopted to promote powder melting. EBM fabrication was operated at a voltage of 60 kV, beam current of 30 mA, spot size of 300 μm, and each layer thickness of 50 μm.

Table 1. Chemical composition of TC21 powder.

Al	Zr	Mo	Cr	Nb	Sn	O	H	Ti
6.26	2.13	2.38	0.96	1.84	1.87	0.092	0.0043	Bal.

The parts on the substrate were a length of 180 mm, height of 100 mm, and width of 15 mm. Fatigue specimens were cut from the building parts, and their longitudinal axis was parallel to the build direction. In order to eliminate the pore defects, HIP for EBM specimens were treated by a QIH-15 machine (ABB company, Vsters, Sweden) at 920 °C under 1000 bar for 2 h, and then the specimens were cooled using nitrogen gas.

Hardness measurements were carried out by a FM310 (Future-Tech, Tokyo, Japan) machine for both HIP and non-HIP specimens. Hardness values were obtained from five tests under a load of 0.5 kg for 15 s.

2.2. Ultrasonic Fatigue Test

Fatigue tests were carried out using a USF-2000 ultrasonic fatigue test machine (Shimadzu, Kyoto, Japan) with R = −1 and a frequency of 20 kHz [22]. The investigation [23] indicated that the frequency had little influence on ultrasonic fatigue tests under low-stress amplitude. Detailed information about the ultrasonic fatigue test machine was present in the reference [24]. To control the thermal effects of high-frequency fatigue tests, compressed air was used to cool the specimens.

Based on the elastic wave theory, the specimen was designed to work at resonance with the amplifier, and the geometries of the fatigue specimens are shown in Figure 1. According to Figure 1, the ultrasonic fatigue specimens were processed from the EBM parts. Thus, the rough surface of the EBM parts was removed by mechanical processing.

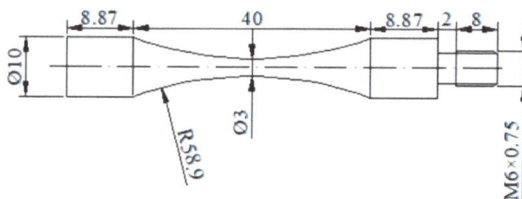

Figure 1. Shape and size of ultrasonic fatigue specimens.

Electro-polishing was treated on fatigue specimens to eliminate the machining influence on fatigue behavior. The electro-polishing process was under −20 °C and 20–25 V with 59% methanol, 35% n-butanol, and 6% perchloric acid. The fracture surfaces and microstructure were observed by CS3400 scanning electron microscopy (CamScan, Cambs, UK).

3. Results

3.1. VHCF Properties

Figure 2 shows the microstructure of the EBM TC21 alloy with non-HIP and HIP treatment, respectively. The microstructure for non-HIP treatment was basketweaved with a lamellar α phase, whereas the HIP treatment increased the size of the α phase. The mean hardness for non-HIP and HIP samples was 435 HV and 418 HV, respectively.

S-N curves of the EBM TC21 titanium alloy with the non-HIP and HIP treatment are shown in Figure 3. It showed that the curve of non-HIP specimens displayed a continuously descending characteristic, and fatigue life gradually increased with the decrease of fatigue stress. As the stress amplitude decreased, fatigue crack initiation characteristics changed from surface initiation to interior pore initiation without FGA and interior pore initiation with FGA. When the fatigue life exceeds 2×10^6 cycles, FGA characteristics can be observed around the pores at the fatigue crack initiation site. If the specimens did not break at

10^9 cycles, the stress amplitude can be defined as the very high cycle fatigue limit. Thus, EBM TC21 titanium alloy obtained a very high cycle fatigue limit of 250 MPa. Very high cycle fatigue S-N curve characteristics and fatigue limit of the EBM TC21 alloy were similar to that of EBM Ti6Al4V [14], but the fatigue limit was far lower than that of forged ones with a basketweave microstructure [25].

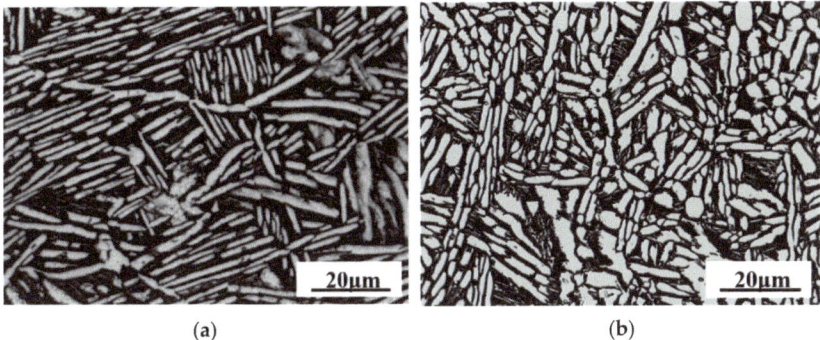

Figure 2. Microstructure of EBM titanium alloy: (**a**) non-HIP; (**b**) HIP.

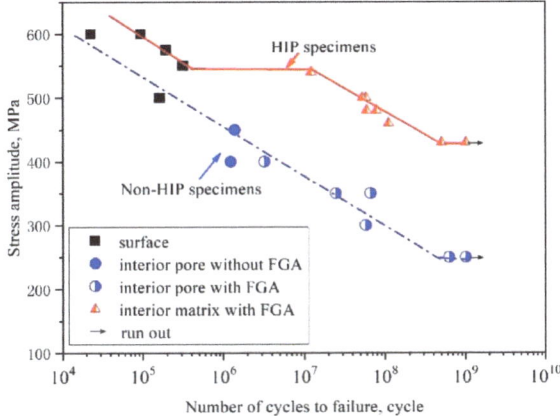

Figure 3. S-N curves for the non-HIP and HIP specimens.

Figure 3 also illustrated that HIP treatment dramatically improved fatigue properties in comparison with the non-HIP condition. The HIP specimens obtained a higher fatigue life at the same cyclic stress. It displayed a two-step curve for the HIP specimens, which was similar to that of the forged ones [25]. The VHCF limit was increased from 250 MPa for the HIP specimens to 430 MPa for the non-HIP ones. There was a transition platform from crack surface initiation under high stress to internal initiation under a 550 MPa stress amplitude. The fatigue life of surface-initiated cracks was lower than that of internally initiated cracks due to the environmental and stress concentration factors. However, the non-HIP specimens exhibited a continuous decease feature due to the presence of defects.

3.2. Fractograph

For EBM materials, insufficient energy input resulted in incomplete melting of the powder and LoF, whereas excessive energy input led to an unstable molten pool and vaporization, forming pore defects. As for the non-HIP specimens, the pore defects had a significant impact on the fatigue crack initiation characteristics for the EBM TC21 titanium alloy. Figure 4a shows that under 600 MPa stress amplitude, a crack initiated from the specimen surface without pore defects at the initiation site, as indicated by the red arrow.

When the stress amplitude decreased to 450 MPa, fatigue cracks initiated at a large pore inside the specimen (Figure 4b), where stage I propagation along the facet was observed near the pore. Under lower stress amplitude, fatigue cracks initiated at a small pore inside the specimen, and a fish-eye characteristic and FGA were observed near the pore (Figure 4c), which was a typical characteristic of VHCF of titanium alloys [6,7] and were consistent with that of the EBM Ti6Al4V alloy and SLM Ti6Al4V in the investigation by Günther [14]. It should be noted that even if the equivalent diameter of small pores in the EBM TC21 titanium alloy decreased to 12 μm, these small pores still promoted the VHCF crack initiation, and there were FGA characteristics around the pores (Figure 4d).

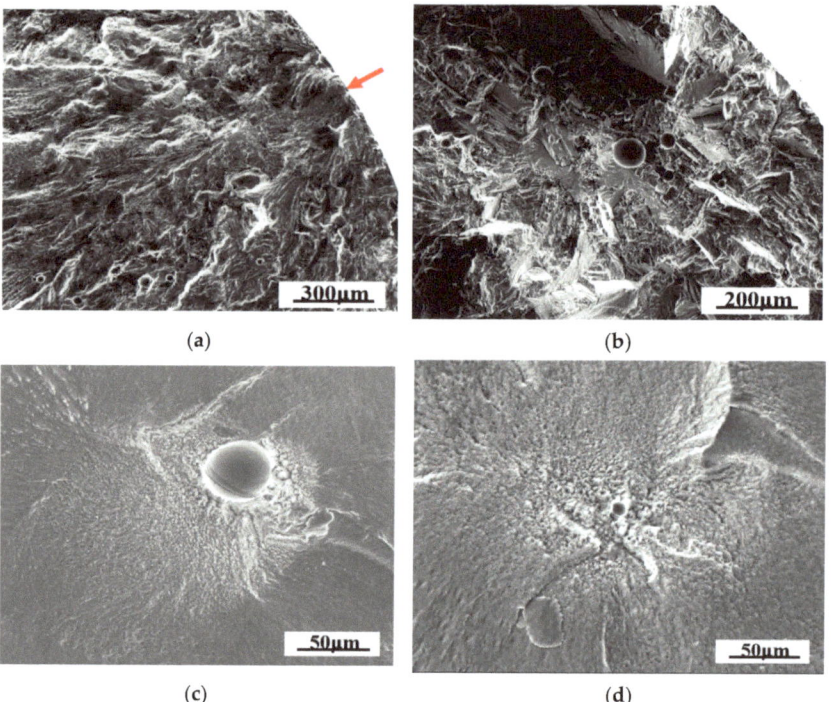

Figure 4. Fracture surface of non-HIP specimens: (**a**) σ_a = 600 MPa and Nf = 2.22 × 10^4 cycles; (**b**) σ_a = 450 MPa and Nf = 1.37 × 10^6 cycles; (**c**) σ_a = 300 MPa and N_f = 5.75 × 10^7 cycles; (**d**) σ_a = 250 MPa and N_f = 6.2 × 10^8 cycles. (Red arrow denoted the initiation site of surface crack).

As for the HIP specimens, the pores in EBM specimens were eliminated by HIP treatment. Under low stress, cracks initiated at the inner of the HIP specimen in the long-life regime, as indicated by the red ring (Figure 5a). An FGA characteristic without pore defects was present at the crack initiation site (Figure 5b), which was similar to the forged TC21 titanium alloy [25] and SLM Ti6Al4V alloy after HIP treatment [18,20]. In comparison, the investigation by Günther [14] suggested that VHCF cracks initiated at the large α phase of the SLM Ti6Al4V alloy treated by HIP with faceted fracture characteristics, which was attributed to the difference in microstructure.

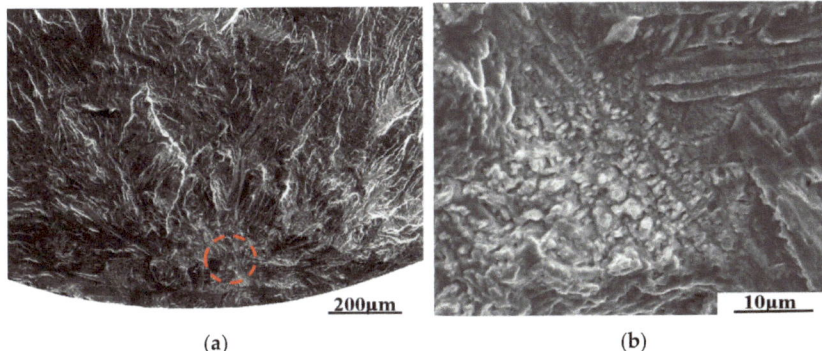

Figure 5. Fracture surface of HIP specimens at σ_a = 460 MPa and N_f = 1.09 × 10^8 cycles. (**a**) Crack initiation site, (**b**) FGA characteristic at crack initiation site. (Red ring denoted the initiation site of internal crack).

4. Discussion

4.1. Effect of Defect on Fracture Mechanism

Under high-stress amplitude, small pore defects were irrelevant to fatigue crack initiation, and the non-HIP specimens exhibited the surface crack initiation behavior (Figure 4a), and the slip band activity resulted in crack initiation. Fatigue crack initiated at the pore inside the specimens under low-stress amplitude. Stage I propagation along the facet was presented around the large pore under a 450 MPa stress amplitude (Figure 4b), while FGA was observed around the pore under a lower stress amplitude (Figure 4c,d). Figure 6 showed that the areas of the pore ranged from 100 μm^2 to 3840 μm^2, and there was no direct relationship between the area of the pore and fatigue life N_f. However, the area of FGA increased with fatigue life N_f. As FGA was formed by the discontinuous propagation of microcracks [6], it indicated that the consumption life in FGA was a major part of fatigue life.

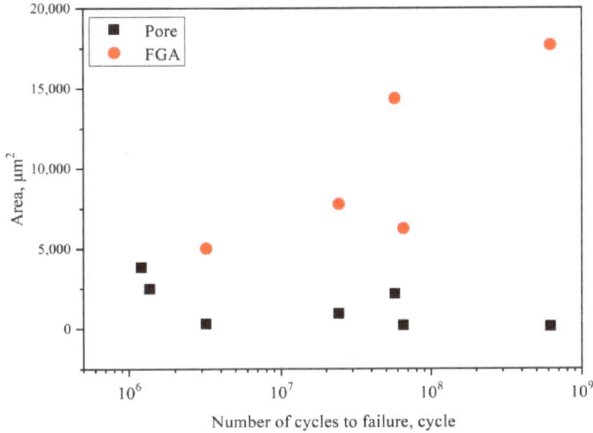

Figure 6. Relationship between the size of pore and fatigue life for EBM titanium alloy.

The stress intensity factor range ΔK_{pore} at the front of the pore inside the specimens can be expressed as [10]:

$$\Delta K_{pore} = 0.5\sigma_a\sqrt{\pi\sqrt{area_{pore}}} \tag{1}$$

where σ_a is the stress amplitude and $area_{pore}$ is the area of the pore at the crack initiation site. As for FGA, ΔK_{FGA} can be calculated by substituting $area_{FGA}$ into Equation (1).

Figure 7 shows that ΔK_{pore} was not directly related to fatigue life due to the random distribution of pore size but displayed a decreased trend. Similar to cast alloys, there exists the critical stress intensity factor K_{cr} with pores [26], and the value of K_{cr} for EBM TC21 titanium alloy was 0.78 MPa$\sqrt{\text{m}}$ in a very high cycle regime. The pore size was small enough, especially for HIP treatment, where K_{pore} was lower than K_{cr}; thus, the pore had no decisive effect on the cracks' initiation and fatigue cracks initiated from microstructure inhomogeneity for EBM specimens treated by HIP treatment (Figure 5).

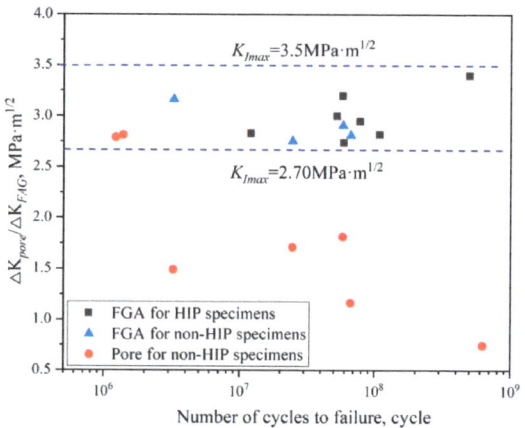

Figure 7. Relationship ΔK_{pore}, ΔK_{FGA} with fatigue life.

Furthermore, Figure 7 also shows that ΔK_{FGA} was independent of fatigue life from 10^6 cycles to 10^9 cycles and had a constant trend in the range from 2.7 MPa$\sqrt{\text{m}}$ to 3.5 MPa$\sqrt{\text{m}}$, which was considered as the stable crack growth threshold ΔK_{th} for titanium alloys [20,27]. The stable crack growth ΔK_{th} for forged the TC21 titanium alloy ranged from 3.8 MPa$\sqrt{\text{m}}$ to 4.5 MPa$\sqrt{\text{m}}$ [25]. The value of ΔK_{FGA} for the EBM TC21 titanium alloy was lower than that of the forged ones.

As for non-HIP specimens with 400 MPa/1.37×10^6 cycles and 450 MPa/1.26×10^6 cycles, the value of ΔK_{pore} reached that of ΔK_{FGA}, and fatigue cracks, which originated from these pores, could continuously propagate to failure. Therefore, the FGA was not observed around the pores for these two specimens. When ΔK_{pore} is lower than ΔK_{FGA}, fatigue cracks cannot propagate continuously, resulting in FGA. For the non-HIP specimen with 300 MPa/5.7×10^7 cycles, the equivalent diameter of FGA was approximately 105 µm, and the average propagation rate was 8.77×10^{-13} m/cycle, which was much lower than the Burgers vector. Thus, microcracks discontinuously propagated within FGA.

VHCF cracks initiated from the heterogeneous microstructure in the HIP specimens (Figure 5). Many models, such as numerous cyclic pressing (NCP) [28] and FGA [29], were established to reveal the FGA formation. A heterogeneous cyclic strain concentration occurred at the α/β interface due to the highly inhomogeneous dislocation arrangement [30]. A nanocrystalline layer was formed at the stress concentration area under very high cyclic loading, and the nanocrystalline boundaries were separated to form fine-grain characteristics [28,29].

4.2. Effect of Defect on Fatigue Strength

To quantitatively analyze the fatigue strength of the EBM titanium alloy with pore defects, the maximum pore size on the fracture surface was estimated by the statistical extreme value method. The pore sizes were assumed to follow the Gumbel distribution [31], and the pore size of EBM titanium alloys was plotted using the Gumbel extreme value analysis. In Figure 8, i/(N + 1) was the cumulative probability corresponding to the maximum pore size in each field of view and arranged in descending order. The maximum pore diameter of the EBM titanium alloy was evaluated as 96.2 μm.

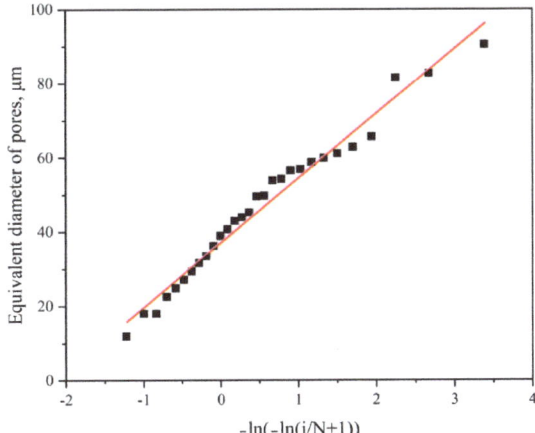

Figure 8. Statistics of pore diameter in additive manufacturing.

For EBM titanium alloys with internal defects, fatigue strength can be evaluated using the Murakami model [14]:

$$\sigma_w = \frac{1.56 \times (HV + 120)}{(\sqrt{area_{pore}})^{1/6}} \quad (2)$$

The parameter $area_{pore}$ was considered as the equivalent area of pores perpendicular to the direction of stress loading. Vickers hardness (HV) was the hardness of the material. The location parameter 1.56 was applicable to the internal defects.

Based on Equation (2), fatigue strengths of the EBM titanium alloy were 197 MPa, 230 MPa, and 394 MPa, respectively, corresponding to the maximum pore diameter of 96.2 μm, the average diameter of 60 μm, and the minimum diameter of 12 μm. The maximum size of the pore defect evaluated by the statistical extreme value method corresponded to the lowest limit of fatigue strength of the EBM titanium alloy. Compared with Figure 3, fatigue strength with the average pore size relatively approached the experimental data with an error of 8%. Fatigue strength corresponding to the minimum pore defect was much higher than the experimental data, while it was closer to the fatigue strength of the EBM titanium alloy after HIP treatment. It revealed that HIP reduced the pore size, thereby improving the fatigue strength of the EBM titanium alloy.

4.3. Effect of Defect on Fatigue Life

As mentioned above, the porosity defects significantly reduced the very high cycle fatigue life of the AM titanium alloy, and the defect factor should be considered for the fatigue life predicting model. The *FIP* model revealed the promoting effect of material defects and cyclic stress on fatigue damage and can be calculated as [26]:

$$FIP = \frac{\mu \sigma_a}{E}\left[1 + k\frac{\Delta K_{defect}}{\Delta K_{th}}\right] \quad (3)$$

where Schmid factor μ was equal to 0.408, and parameter k was equal to 1 [26]. ΔK_{defect} was estimated by Equation (1), and ΔK_{defect} was considered as the mean value of ΔK_{FGA}. As for HIP specimens, the pore defects were eliminated, and the K_{defect} was equal to zero.

The fatigue life was numerically fitted with *FIP* parameters based on fatigue data of non-HIP and HIP specimens, and the expression was

$$N_f = 7.86 \times 10^{-19}(FIP)^{-9.46} \tag{4}$$

The fitting curve between *FIP* parameters and fatigue life is shown in Figure 9. It showed that the prediction of fatigue life based on Equation (4) agreed well with the experimental data. The HIP specimens had a lower value of FIP and higher fatigue life than those of the non-HIP ones.

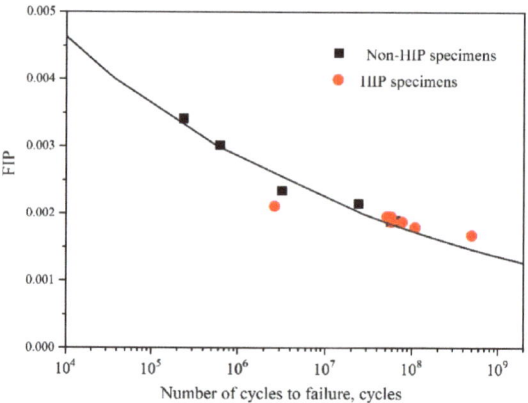

Figure 9. *FIP* versus the number of cycles to failure.

Based on the *FIP* model, the prediction of S-N curves for both non-HIP and HIP specimens is shown in Figure 10. The average diameter of the pores was calculated as 60 μm, and the prediction of fatigue life generally agreed with the experimental data. Obviously, EBM titanium alloys with a smaller defect obtained a longer fatigue life at the same stress level. For HIP treatment, pores in EBM titanium alloy specimens were considered to be healed with a diameter of zero. HIP treatment significantly improved fatigue life owing to the elimination of pore defects. Thus, the fatigue life of EBM titanium alloys with pore defects can be well predicted based on the FIP model.

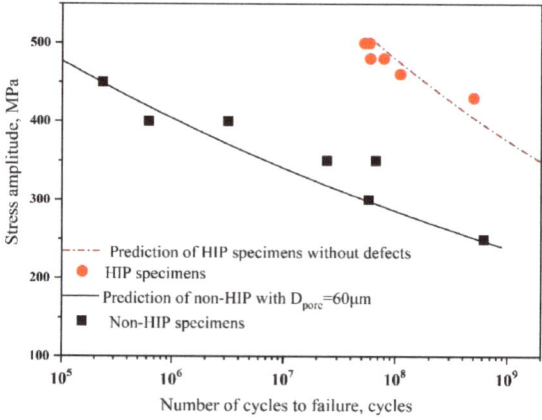

Figure 10. Prediction of fatigue life for the EBM TC21 titanium alloy based on the *FIP* model.

5. Conclusions

The VHCF of the EBM TC21 titanium alloy with HIP and non-HIP treatment was carried out by ultrasonic fatigue test, and the effect of pore defects on VHCF behavior was investigated based on the Murakami model and FIP model for the EBM TC21 titanium alloy. The conclusions are summarized as follows:

(1) The curve of the EBM TC21 titanium alloy displayed a continuously descending characteristic in a very high cycle regime, while HIP treatment significantly improved fatigue properties and illustrated two-step curve characteristics similar to the forged ones;

(2) The VHCF cracks were mainly initiated from the internal pore of the EBM TC21 titanium alloy, whereas cracks were initiated at the inner heterogeneous microstructure for HIP specimens in the very high cycle regime. FGA was observed at the crack initiation site in a very high cycle regime for both non-HIP and HIP specimens. The value of ΔK_{FGA} corresponded with the threshold of stable crack propagation;

(3) Based on the Murakami model, the lower limit of fatigue strength for the EBM TC21 titanium alloy was estimated by the statistical extreme value method. Fatigue strength with the average pore size relatively approached the experimental data with an error of 8%. Furthermore, a *FIP* model based on material defects was established to predict fatigue life for non-HIP and HIP specimens, which agreed well with the experimental data.

Author Contributions: B.N. and D.C. conceived and designed the research. F.L. fabricated titanium alloy specimens by EBM and characterized material defects. Q.L. and B.L. performed the fatigue test. S.L. and H.Q. analyzed experimental data. Q.L. and B.S. wrote the manuscript. All authors have read and agreed to the published version of the manuscript.

Funding: This research was funded by the R & D plan for key areas in Guangdong Province (2020B010186001), the Science and Technology Program of the Ministry of Science and Technology (G2022030060L), the Science and Technology Project in Guangdong (2020b15120093, 2020B121202002), the Overseas Famous Teacher Project of Guangdong (BGK46303), the R & D plan for key areas in Jiangxi Province (20201BBE51009, 20212BBE51012), and the Science and Technology Research Project of Foshan (2220001005305).

Data Availability Statement: The data presented in this study are available upon request from the corresponding author.

Conflicts of Interest: The authors declare no conflict of interest.

References

1. Zhao, Q.; Sun, Q.; Xin, S.; Chen, Y.; Wu, C.; Wang, H.; Xu, J.; Wan, M.; Zeng, W.; Zhao, Y. High-strength titanium alloys for aerospace engineering applications: A review on melting-forging process. *Mater. Sci. Eng. A* **2022**, *845*, 143260. [CrossRef]
2. Zhang, Y.; Zhang, H.; Xue, J.; Jia, Q.; Wu, Y.; Li, F.; Guo, W. Microstructure transformed by heat treatment to improve fatigue property of laser solid formed Ti6Al4V titanium alloy. *Mater. Sci. Eng. A* **2023**, *865*, 144363. [CrossRef]
3. Tran, T.Q.; Chinnappan, A.; Lee, J.K.Y.; Loc, N.H.; Tran, L.T.; Wang, G.; Kumar, V.V.; Jayathilaka, W.A.D.M.; Ji, D.; Doddamani, M.; et al. 3D printing of highly pure copper. *Metals* **2019**, *9*, 756. [CrossRef]
4. Wang, J.; Pan, Z.; Wang, Y.; Wang, L.; Su, L.; Cuiuri, D.; Zhao, Y.; Li, H. Evolution of crystallographic orientation, precipitation, phase transformation and mechanical properties realized by enhancing deposition current for dual-wire arc additive manufactured Ni-rich NiTi alloy. *Addit. Manuf.* **2020**, *34*, 101240. [CrossRef]
5. Galarraga, H.; Warren, R.; Lados, D.; Dehoff, R.; Kirka, M.; Nandwana, P. Effects of heat treatments on microstructure and properties of Ti-6Al-4V ELI alloy fabricated by electron beam melting (EBM). *Mater. Sci. Eng. A* **2017**, *685*, 417–428. [CrossRef]
6. Gao, T.; Xue, H.; Sun, Z.; Retraint, D.; He, Y. Micromechanisms of crack initiation of a Ti-8Al-1Mo-1V alloy in the very high cycle fatigue regime. *Int. J. Fatigue* **2021**, *150*, 106314. [CrossRef]
7. Liu, F.; Peng, H.; Liu, Y.; Wang, C.; Wang, Q.; Chen, Y. Crack initiation mechanism of titanium alloy in very high cycle fatigue regime at 400 °C considering stress ratio effect. *Int. J. Fatigue* **2022**, *163*, 107012. [CrossRef]
8. Hrabe, N.; Gnäupel-Herold, T.; Quinn, T. Fatigue properties of a titanium alloy (Ti-6Al-4V) fabricated via electron beam melting (EBM): Effects of internal defects and residual stress. *Int. J. Fatigue* **2017**, *94*, 202–210. [CrossRef]
9. Chern, A.; Nandwana, P.; Yuan, T.; Kirka, M.; Dehoff, R.; Liaw, P.; Duty, C. A review on the fatigue behavior of Ti-6Al-4V fabricated by electron beam melting additive manufacturing. *Int. J. Fatigue* **2019**, *119*, 173–184. [CrossRef]

10. Romano, S.; Brückner-Foit, A.; Brandão, A.; Gumpinger, J.; Ghidini, T.; Beretta, S. Fatigue properties of AlSi10Mg obtained by additive manufacturing: Defect-based modelling and prediction of fatigue strength. *Eng. Fract. Mech.* **2018**, *187*, 165–189. [CrossRef]
11. Niu, X.; Zhu, S.P.; He, J.C.; Liao, D.; Correia, J.A.; Berto, F.; Wang, Q. Defect tolerant fatigue assessment of AM materials: Size effect and probabilistic prospects. *Int. J. Fatigue* **2022**, *160*, 106884. [CrossRef]
12. He, J.C.; Zhu, S.P.; Luo, C.; Niu, X.; Wang, Q. Size effect in fatigue modelling of defective materials: Application of the calibrated weakest-link theory. *Int. J. Fatigue* **2022**, *165*, 107213. [CrossRef]
13. Li, H.; Tian, Z.; Zheng, J.; Huang, K.; Nie, B.; Xu, W.; Zhao, Z. A defect-based fatigue life estimation method for laser additive manufactured Ti-6Al-4V alloy at elevated temperature in very high cycle regime. *Int. J. Fatigue* **2023**, *167*, 107375. [CrossRef]
14. Günther, J.; Krewerth, D.; Lippmann, T.; Leuders, S.; Tröster, T.; Weidner, A.; Biermann, H.; Niendorf, T. Fatigue life of additively manufactured Ti–6Al–4V in the very high cycle fatigue regime. *Int. J. Fatigue* **2017**, *94*, 236–245. [CrossRef]
15. Jiang, Q.; Li, S.; Zhou, C.; Zhang, B.; Zhang, Y. Effects of laser shock peening on the ultra-high cycle fatigue performance of additively manufactured Ti6Al4V alloy. *Opt. Laser Technol.* **2021**, *144*, 107391. [CrossRef]
16. Du, L.; Qian, G.; Zheng, L.; Hong, Y. Influence of processing parameters of selective laser melting on high-cycle and very-high-cycle fatigue behaviour of Ti-6Al-4V. *Fatigue Fract. Eng. Mater. Struct.* **2021**, *44*, 240–256. [CrossRef]
17. Qian, G.; Li, Y.; Paolino, D.; Tridello, A.; Berto, F.; Hong, Y. Very-high-cycle fatigue behavior of Ti-6Al-4V manufactured by selective laser melting: Effect of build orientation. *Int. J. Fatigue* **2020**, *136*, 105628. [CrossRef]
18. Chi, W.; Wang, W.; Li, Y.; Xu, W.; Sun, C. Defect induced cracking and modeling of fatigue strength for an additively manufactured Ti-6Al-4V alloy in very high cycle fatigue regime. *Theor. Appl. Fract. Mec.* **2022**, *119*, 103380. [CrossRef]
19. Shui, X.; Yamanaka, K.; Mori, M.; Nagata, Y.; Kurita, K.; Chiba, A. Effects of post-processing on cyclic fatigue response of a titanium alloy additively manufactured by electron beam melting. *Mater. Sci. Eng. A* **2017**, *680*, 239–248. [CrossRef]
20. Sun, C.; Chi, W.; Wang, W.; Duan, Y. Characteristic and mechanism of crack initiation and early growth of an additively manufactured Ti-6Al-4V in very high cycle fatigue regime. *Int. J. Mech. Sci.* **2021**, *205*, 106591. [CrossRef]
21. Caivano, R.; Tridello, A.; Chiandussi, G.; Qian, G.; Paolino, D.; Berto, F. Very high cycle fatigue (VHCF) response of additively manufactured materials: A review. *Fatigue Fract. Eng. Mater. Struct.* **2021**, *44*, 2919–2943. [CrossRef]
22. Nie, B.; Liu, S.; Wu, Y.; Song, Y.; Qi, H.; Shi, B.; Zhao, Z.; Chen, D. Very High Cycle Fatigue Damage of TC21 titanium alloy under high/low two-step stress loading. *Crystals* **2023**, *13*, 139. [CrossRef]
23. Yang, K.; He, C.; Huang, Q. Very high cycle fatigue behaviors of a turbine engine blade alloy at various stress ratios. *Int. J. Fatigue* **2017**, *99*, 35–43. [CrossRef]
24. Bathias, C. Piezoelectric fatigue testing machines and devices. *Int. J. Fatigue* **2006**, *28*, 1438–1445. [CrossRef]
25. Nie, B.; Zhao, Z.; Chen, D.; Liu, S.; Lu, M.; Zhang, J.; Liang, F. Effect of basketweave microstructure on very high cycle fatigue behavior of TC21 titanium alloy. *Metals* **2018**, *8*, 401. [CrossRef]
26. Nie, B.; Zhao, Z.; Liu, S.; Chen, D.; Ouyang, Y.; Hu, Z.; Fan, T.; Sun, H. Very high cycle fatigue behavior of a directionally solidified Ni-base superalloy DZ4. *Materials* **2018**, *11*, 98. [CrossRef]
27. Zhao, A.; Xie, J.; Sun, C.; Lei, Z.; Hong, Y. Prediction of threshold value for FGA formation. *Mater. Sci. Eng. A* **2011**, *528*, 6872–6877. [CrossRef]
28. Hong, Y.; Liu, X.; Lei, Z.; Sun, C. The formation mechanism of characteristic region at crack initiation for very-high-cycle fatigue of high-strength steels. *Int. J. Fatigue* **2016**, *89*, 108–118. [CrossRef]
29. Zhang, H.; Yu, F.; Li, S.; He, E. Fine granular area formation by damage-induced shear strain localization in very-high-cycle fatigue. *Fatigue Fract. Eng. Mater. Struct.* **2021**, *44*, 2489–2502. [CrossRef]
30. Chai, G.; Zhou, N.; Ciurea, S.; Andersson, M.; Peng, R. Local plasticity exhaustion in a very high cycle fatigue regime. *Scr. Mater.* **2012**, *66*, 769–772. [CrossRef]
31. Du, L.; Pan, X.; Qian, G.; Zheng, L.; Hong, Y. Crack initiation mechanisms under two stress ratios up to very-high-cycle fatigue regime for a selective laser melted Ti-6Al-4V. *Int. J. Fatigue* **2021**, *149*, 106294. [CrossRef]

Disclaimer/Publisher's Note: The statements, opinions and data contained in all publications are solely those of the individual author(s) and contributor(s) and not of MDPI and/or the editor(s). MDPI and/or the editor(s) disclaim responsibility for any injury to people or property resulting from any ideas, methods, instructions or products referred to in the content.

Article

Nucleation of L1$_2$-Al$_3$M (M = Sc, Er, Y, Zr) Nanophases in Aluminum Alloys: A First-Principles ThermodynamicsStudy

Shuai Liu [1], Fangjun Liu [2], Zhanhao Yan [1], Baohua Nie [1,*], Touwen Fan [3], Dongchu Chen [1] and Yu Song [4,*]

[1] School of Materials Science and Hydrogen Energy, Foshan University, Foshan 528000, China; 2112206013@stu.fosu.edu.cn (S.L.); 20210580109@stu.fosu.edu.cn (Z.Y.); chendc@fosu.edu.cn (D.C.)
[2] School of Mechatronic Engineering and Automation, Foshan University, Foshan 528000, China; liufj@fosu.edu.cn
[3] Research Institute of Automobile Parts Technology, Hunan Institute of Technology, Hengyang 421002, China; 2021001018@hnit.edu.cn
[4] Shenzhen Rspower Technology Co., Ltd., Shenzhen 518000, China
* Correspondence: niebaohua@fosu.edu.cn (B.N.); songyu188200@foxmail.com (Y.S.); Tel.: +86-075782700525 (B.N.); +86-75583233458 (Y.S.)

Citation: Liu, S.; Liu, F.; Yan, Z.; Nie, B.; Fan, T.; Chen, D.; Song, Y. Nucleation of L1$_2$-Al$_3$M (M = Sc, Er, Y, Zr) Nanophases in Aluminum Alloys: A First-Principles ThermodynamicsStudy. *Crystals* **2023**, *13*, 1228. https://doi.org/10.3390/cryst13081228

Academic Editor: Wangzhong Mu

Received: 14 July 2023
Revised: 2 August 2023
Accepted: 7 August 2023
Published: 9 August 2023

Copyright: © 2023 by the authors. Licensee MDPI, Basel, Switzerland. This article is an open access article distributed under the terms and conditions of the Creative Commons Attribution (CC BY) license (https://creativecommons.org/licenses/by/4.0/).

Abstract: High-performance Sc-containing aluminum alloys are limited in their industrial application due to the high cost of Sc elements. Er, Zr, and Y elements are candidates for replacing Sc elements. Combined with the first-principles thermodynamic calculation and the classical nucleation theory, the nucleation of L1$_2$-Al$_3$M (M = Sc, Er, Y, Zr) nanophases in dilutealuminum alloys were investigated to reveal their structural stability. The calculated results showed that the critical radius and nucleation energy of the L1$_2$-Al$_3$M phases were as follows: Al$_3$Er > Al$_3$Y > Al$_3$Sc > Al$_3$Zr. The Al$_3$Zr phase was the easiest to nucleate in thermodynamics, while the nucleation of the Al$_3$Y and Al$_3$Er phases were relatively difficult in thermodynamics. Various structures of Al$_3$(Y, Zr) phases with the radius r < 1 nm can coexist in Al-Y-Zr alloys. At a precipitate's radius of 1–10 nanometers, the core–shelled Al$_3$Zr(Y) phase illustrated the highest nucleation energy, while the separated structure Al$_3$Zr/Al$_3$Y obtained the lowest one, and had thermodynamic advantages in the nucleation process. Moreover, the core–shelled Al$_3$Zr(Y) phase obtained a higher nucleation energy than Al$_3$Zr(Sc) and Al$_3$Zr(Er). Core–doubleshelled Al$_3$Zr/Er(Y) obtained a lower nucleation energy than that of Al$_3$Zr(Y) due to the negative ΔG_{chem} of Al$_3$Er and the negative Al$_3$Er/Al$_3$Y interfacial energy, and was preferentially precipitated in thermodynamics stability.

Keywords: Al$_3$Y; nucleation; first-principles; aluminum alloy

1. Introduction

Sc-containing aluminum alloys are ideal materials for key components in the aerospace, high-speed rail, and automobile industries due to their high strength, corrosion resistance, and formability [1–3]. L1$_2$-Al$_3$Sc nanoparticles precipitated in Sc-containing aluminum alloys can effectively inhibit the recrystallization process [4], thereby obtaining comprehensive properties such as high strength, toughness, and corrosion resistance. Moreover, Seidman et al. [5,6] developed a series of Al-Sc high-temperature aluminum alloys. However, due to the high diffusion rate of Sc atoms, L1$_2$-Al$_3$Sc nanoparticles are prone to coarsening, reducing their ability to inhibit recrystallization and high-temperature performance. On the other hand, Zr atoms can partially replace Sc atoms in the Al$_3$Sc nanophase, forming a core–shelled Al$_3$(Sc$_{1-x}$, Zr$_x$), namely an Al$_3$Sc core and Al$_3$Zr shell [7], where the Al$_3$Zr shell improves the coarsening resistance of the Al$_3$(Sc$_{1-x}$, Zr$_x$) nanophase due to the low diffusion rate of Zr atoms in the aluminum matrix [8].

However, the high cost of Sc elements greatly limits the engineering application of Sc-containing aluminum alloys. As members of the Sc element family, Er, Yb rare earth elements and the Y element have been considered ideal substitutes for Sc elements. Er and Yb elements were reported to form a core–shell structure of $Al_3(Er, Zr)$ [9,10] and $Al_3(Yb, Zr)$ [11,12] nanophases, which also effectively inhibited the recrystallization of aluminum alloys. Based on high-throughput first-principles calculations of the nucleation and growth for the $L1_2$ structure Al_3RE phases, Fan et al. [13,14] revealed the ΔG_V firstly decreased from Sc, Y to Ce, then increased linearly for RE elements, and the ΔG_V tended to increase linearly with the temperature. It was speculated that the Y element could replace the expensive Sc element. The investigation by Zhang et al. [15,16] showed that the precipitation phase was mainly the Al_3Y phase, which became the core of Al_3Zr and promoted the precipitation kinetics of solid solution Zr atoms, whereas a hybrid structure of $Al_3(Zr, Y)$ rather than the typical core–shelled structure was observed after long-term homogenization, where the Y and Zr elements were uniformly distributed in $Al_3(Y, Zr)$ nanoprecipitates through atom probe tomography (APT).

Some research has been conducted on the formation mechanism of the hybrid structure of $Al_3(Zr, Y)$. Zhang et al. [16] indicated that the hybrid structure of $Al_3(Zr, Y)$ was attributed to the strong interactions between the Y and Zr atoms, resulting in their co-precipitation. Based on first-principles calculations, Wang et al. [17] indicated that the doping of the Y element and the Zr element decreased the interface energies of the FCC-$Al(001)$/FCC-$Al_3Y(001)$ interface and formed a hybrid structure of $Al_3(Y, Zr)$ instead of an Al_3Y core + Al_3Zr shell structure. The author's previous research indicated that the interface energy of Al_3Zr/Al was lower than that of Al_3Zr/Al_3Y, and it was deduced that Al_3Zr tended to form a shell layer, while Al_3Y formed a core layer. However, the high coherent strain energy made the Al_3Y/Al_3Zr interface unstable, and it was difficult to form a stable cored Al_3Y/shelled Al_3Zr structure [18]. Although the author's previous investigation elucidated the reason for Al_3Y/Al_3Zr not having a core–shelled structure based on the coherent strain energy, there were several issues that needed to be answered, such as whether the hybrid structure of $Al_3(Zr, Y)$ was determined by atomic diffusion control or thermodynamic structure stability.

One view was that the formation of the core–shelled Al_3M phase was attributed to the differences in atomic diffusion rates. Al_3Sc and Al_3Er core structures were formed due to the fast diffusion rate of the Sc and Er atoms. The diffusion rate of the Zr element was slow, resulting in the formation of an Al_3Zr shell structure [19]. Furthermore, the core(Al_3Er)–double shell (Al_3Sc/Al_3Zr) structure of $L1_2$-$Al_3(Sc, Er, Zr)$ was precipitated in Al-Sc-Er-Zr alloys after homogenization at 400 °C [20]. Seidman et al. [20] suggested that the core–double-shelled $L1_2$-$Al_3(Sc, Er, Zr)$ can be attributed to their difference in diffusion rate, e.g., $D_{Er} > D_{Sc} > D_{Zr}$. Leibner et al. [21] found that there were two major groups of core–double-shelled $L1_2$-$Al_3(Sc, Er, Zr)$ observed after aging at 600 °C/4 h, one having the usual core (Al_3Er)–double shell (Al_3Sc/Al_3Zr) structure and the other having an unusual core (Al_3Sc)–double shell (Al_3Er/Al_3Zr) structure. They suggested that the segregation of the Sc atom to dislocations and the interaction between the solid solution atoms and the Sc atom promoted the formation of the unusual core (Al_3Sc)–double shell structure. It should be noted that the hybrid structure of $Al_3(Zr, Sc)$ [22] and $Al_3(Er, Zr)$ [23] was also observed in aluminum alloys. Therefore, the difference in diffusion rates between atoms does not explain the formation of core–shelled structures well.

The thermodynamic analysis of nanophases' nucleation based on first-principles calculations can provide insights into the phase transformation process of $L1_2$-Al_3M phases. Jiang et al. [24,25] used first-principles calculation methods to calculate the nucleation energies of the $Al_3(Er, Zr)$ and $Al_3(Sc, Zr)$ phases with different microstructures, revealing the thermodynamic stability of the $Al_3(Er, Zr)$ and $Al_3(Sc, Zr)$ phases during the homogenization precipitation. The nucleation properties calculated by Liu [26] showed that the core–shelled $Al_3(Er_{1-x}, Sc_x)$ obtained a highly stable structure due to its low nucleation energy, which was independent of the temperature and Sc/Er ratio. However, first-principles calculations of the nucleation and thermodynamic stability for the $Al_3(Y, Zr)$ phase were rarely reported. Furthermore, the author's investigation showed that Er atoms tended to segregate at the Al_3Y/Al_3Zr interface, and were inclined to form a core–double-shelled

Al$_3$(Y, Er, Zr) structure with an Al$_3$Y core, an Al$_3$Er inner shell, and an Al$_3$Zr outer shell [18]. The nucleation and thermodynamic stability of core–double-shelled Al$_3$(Y, Er, Zr) needed to be evaluated to develop Al-Y-Zr series alloys.

Combining with the calculation results of interface energies and the coherent strain energy in the previous research [18], the total nucleation energies of various structures of L1$_2$-Al$_3$M (M = Sc, Er, Zr, Y) phases were calculated based on first-principles thermodynamic calculation and classical nucleation theory. The critical nucleation energies and nucleation radii of Al$_3$M phases were calculated to compare the nucleation differences of Al$_3$M nanophases. The nucleation energies of various structures of ternary L1$_2$-Al$_3$(Y, Zr) phases were investigated to reveal the formation mechanism of Al$_3$(Y, Zr) with a hybrid structure. The nucleation calculation result of core–shelled Al$_3$(Y, Zr) was also compared with that of core–shelled Al$_3$(Sc, Zr) and core–shelled Al$_3$(Er, Zr). Furthermore, based on first-principles calculations, the nucleation energy of the core–double-shelled Al$_3$(Er, Y, Zr) phase was investigated to evaluate its thermodynamic stability. This paper aimed to reveal the internal formation mechanism of L1$_2$-Al$_3$M with a core–shelled structure from the perspective of first-principles thermodynamic calculations, and provided guidance for the development of new Al-Y-Zr series alloys

2. Computational Methods

Based on density functional theory (DFT) [27], first-principles calculations were carried out by VASP software [28]. The electron configuration was described by Al-3s^{2}3p^1, Sc-3s^{2}3p^{6}4s^{1}3d^2, Zr-4s^{2}4p^{6}5s^{1}4d^3, Er-6s^{2}5p^{6}5d^1, and Y-4s^{2}4p^{6}5s^{1}4d^2 valence states, respectively. The ion–electron interactions were described by the projection augmented wave (PAW) method within the frozen core approximation [29]. The exchange-correlation energy functional between electrons was described by the Perdew–Burke–Ernzerhof (PBE) [30,31] method of generalized gradient approximation (GGA). The kinetic energy cutoff of the plane wave basis and the size of the k-mesh for the Brillouin zone were tested for self-consistent convergence. The calculation of the bulk phase of L1$_2$-Al$_3$M (M = Sc, Er, Y, Zr) used conventional single cells. In each periodic direction of reciprocal space, the geometric structure was optimized by the Monkhorst–Pack k-point grids with linear k-mesh analytical values of less than 0.032π/Å. Using the linear tetrahedron method with the Blöchl correction, the total energy was calculated when the total energy converged to 10^{-4} eV/atom. The lattice constants (a) and bulk modulus (B) were predicted as fcc-Al (a = 4.042 Å and B = 78.2 GPa), L1$_2$-Al$_3$Sc (a = 4.103 Å and B = 86.4 GPa), L1$_2$-Al$_3$Zr (a = 4.108 Å and B = 102.3 GPa), and L1$_2$-Al$_3$Er (a = 4.232 Å and B = 78.5 GPa), respectively, which agreed well with Ref. [26].

Vibration entropy had a significant influence on the chemical formation energy ΔG_{chem} corresponding to the precipitation of the L1$_2$-Al$_3$M phase from the fcc-Al$_n$M solution matrix. The calculation of vibration entropy was based on the method of the density functional perturbation theory (DFPT) [32] under the simple harmonic approximation, and the phonon spectrum of Al$_3$M was calculated by using the 2 × 2 × 2 supercell model. The Al$_n$M was adopted by the 2 × 2 × 2 supercell Al matrix, and the M atom was doped and dissolved in the center. In this method, a small external disturbance was introduced, and the linear response of the system was calculated based on this disturbance. By calculating the response function, the perturbation expression of the vibration frequency can be derived, resulting in the vibration entropy difference of the Al$_3$M phase.

3. Results and Discussion

3.1. Nucleation of Binary L1$_2$-Al$_3$M Phases

According to the classical nucleation theory, the nucleation work consisted of two parts: the energy released by the precipitated phase from the Al matrix, and the energy from the new interface between the precipitated phase and the matrix. The precipitated phase was usually assumed to be a sphere with uniform density distribution. When the

L1$_2$-Al$_3$M nanophases are precipitated from the Al matrix, their precipitation radius R and nucleation work ΔG can be expressed as:

$$\Delta G = \frac{4\pi}{3}R^3 \cdot \Delta G_V + 4\pi R^2 \cdot \gamma \quad (1)$$

where γ is the interface energy per unit area after subtracting the coherent strain energy. The Al(001)/Al$_3$M(001)-contacting facet was the most energy-favored orientation [18,24,25], and the interface energy of Al(001)/Al$_3$M(001) was calculated to estimate the critical nucleation works and nucleation radius. ΔG_V is the volume-free energy per unit volume, which is defined as:

$$\Delta G_V = \Delta G_{chem} + G_s \quad (2)$$

where ΔG_{chem} is the chemical formation energy corresponding to the precipitation of the L1$_2$-Al$_3$M phase in the matrix; G_s is the coherent strain energy.

The chemical reaction equation of the Al$_3$M nanophase precipitation can be written as: Al$_n$M = Al$_3$M + (n − 3)Al; so its chemical energy is expressed as [33]:

$$\begin{aligned}\Delta G_{chem} &= G_{Al_3M} + (n-3)\mu_{Al} - G_{Al_nM} \\ &= (\Delta H_{Al_3M} - \Delta H_{Al_nM}) - T(\Delta S_{Al_3M} - \Delta S_{Al_nM})\end{aligned} \quad (3)$$

Here ΔH_{Al_3M} and ΔH_{Al_nM} are the formation enthalpies of L1$_2$-Al$_3$M and fcc-Al$_n$M, respectively, and the enthalpy can be approximately equal to the internal energy here because the volume–pressure term in the solid state can be ignored [33]; ΔS_{Al_3M} and ΔS_{Al_nM} are the formation entropy of L1$_2$-Al$_3$M and fcc-Al$_n$M, respectively. As the nucleation process of the Al$_3$M nanophases was sensitive to the temperature, the contribution of formation entropy should not be ignored. The contribution of formation entropy may become very important at high temperature. The entropy change in the alloy consisted of three parts: configuration entropy, hot electron entropy, and vibration entropy. In this calculation, the configuration entropy was generally negligible for a dilute alloy, which was clearly revealed for dilute Al-Sc-Zr alloys [24] and Al-Er-Zr alloys [25], and the hot electron entropy can be ignored for relatively low temperatures [34], so the vibration entropy was considered as contributing to entropy change.

According to the differentiation of Equation (1), the critical nucleation radius of $R*$ and the critical nucleation work $\Delta G_V(R*)$ can be obtained as:

$$R* = \frac{-2\gamma}{\Delta G_V} \quad (4)$$

$$\Delta G_V(R*) = \frac{16\pi}{3}\frac{\gamma^3}{\Delta G_V^2} \quad (5)$$

The corresponding differences in enthalpy and vibration entropy between the L1$_2$-Al$_3$M phases and the fcc-Al$_n$M solution matrix are shown in Table 1. The corresponding differences in enthalpy ($\Delta H_f^{Al_3M} - \Delta H_f^{Al_nM}$) were −0.718 eV/atom, −0.667 eV/atom, −0.823 eV/atom, and −0.902 eV/atom for the Al$_3$Sc phase, Al$_3$Zr phase, Al$_3$Er phase, and Al$_3$Y phase, respectively. The enthalpy difference of Al$_3$Sc was in good agreement with the calculated values of −0.72 eV/atom in the literature [33], and was higher than the calculated values of −0.776 eV/atom in the literature [24]. The corresponding enthalpy differences of the Al$_3$Zr and Al$_3$Er phases were −0.667 eV/atom and 0.867 ev/atom, respectively, which were also slightly higher than that of the investigation [24,25]. The enthalpy difference of the Al$_3$Y phase has not yet been documented, but the calculation result was −0.902 eV/atom.

In order to calculate the nucleation work of the Al$_3$M phase in the Al matrix, it was necessary to calculate the vibration entropy difference. The vibration entropy differences of the Al$_3$Sc phase, Al$_3$Zr phase, and Al$_3$Er phase were 3.35 k$_B$/atom, 4.01 k$_B$/atom, and

5.18 k$_B$/atom, respectively, which were higher than the calculated results in the literature (2.67 k$_B$/Sc [24], 2.72 k$_B$/Zr [24], and 3.53 k$_B$/Er [25]). The vibration entropy difference of the Al$_3$Y phase was 5.72 k$_B$/atom.

Table 1. The corresponding enthalpy difference and vibration entropy difference of L1$_2$-Al$_3$M phase precipitation.

	$\Delta H_f^{Al_3M} - \Delta H_f^{Al_nM}$ (eV/Atom)		$\Delta S_{vib}^{Al_3M} - \Delta S_{vib}^{Al_nM}$ (k$_B$/Atom)	
Al$_3$Sc-Al$_n$Sc	−0.718	−0.776 [24]	3.35	2.67 [24]
Al$_3$Zr-Al$_n$Zr	−0.667	−0.831 [24]	4.01	2.72 [24]
Al$_3$Er-Al$_n$Er	−0.823	−0.867 [25]	5.18	3.53 [25]
Al$_3$Y-Al$_n$Y	−0.902	-	5.72	-

The author's previous research calculated the coherent strain energy of L1$_2$-Al$_3$M/Al [18], where the coherent strain energies were 0.0035 ev/atom for Al$_3$Sc/Al, 0.0023 eV/atom for Al$_3$Zr/Al, 0.0088 eV/atom for Al$_3$Er/Al, and 0.0094 eV/atom for Al$_3$Y/Al. Based on Equations (1)–(3), the computation result at 673 K illustrated that the interface strains contributed to only ~8.5% of the volumetric formation energy for the Al$_3$Sc phase, the Al$_3$Zr phase, the Al$_3$Er phase, and the Al$_3$Y phase in Al. It indicated that the coherent strain energy of Al$_3$M/Al had little influence on the precipitation of Al$_3$M nanophases.

Combining with the Al/Al$_3$M interface energy [18], the critical nucleation radius and critical nucleation work of each phase at 673 K are shown in Table 2. For L1$_2$-Al$_3$M (M = Sc, Zr, Er, Y), the predicted critical nucleation radii were 5.95 Å, 3.89 Å, 9.57 Å, and 9.40 Å, for the Al$_3$Sc phase, the Al$_3$Zr phase, the Al$_3$Er phase, and the Al$_3$Y phase, respectively. The critical nucleation works were 2.01×10^{-19} J, 4.83×10^{-20} J, 6.94×10^{-19} J, and 6.84×10^{-19} J for the Al$_3$Sc phase, the Al$_3$Zr phase, the Al$_3$Er phase, and the Al$_3$Y phase, respectively. Among them, the calculated value of the critical nucleation radius of the Al$_3$Sc phase was slightly lower than the literature value [24], but the critical nucleation radii of the Al$_3$Zr phase and the Al$_3$Er phase were slightly higher than the value in Jiang's investigation [24,25]. On the other hand, the critical nucleation work of Al$_3$Sc was slightly less than the literature value [24], and the critical nucleation works of Al$_3$Zr and Al$_3$Er were slightly greater than the literature value [25]. The critical nucleation radius and critical nucleation work of the Al$_3$Y phase at 673 K has not been reported yet. The investigation of Fan et al. [14] showed that the critical nucleation radius of Al$_3$Y for the (100) plane was about 3 Å at 300K, which was lower than the calculation value in this research. The reason can be attributed to the different calculation methods of nucleation energy and the low temperature.

Table 2. Critical nucleation radius and critical nucleation work.

	Critical Nucleation Radius (Å)		Critical Nucleation Work (J)	
	Present	Ref.	Present	Ref.
Al$_3$Sc	5.95	6.6 [24]	2.01×10^{-19}	2.9×10^{-19} [24]
Al$_3$Zr	3.89	2.9 [24]	4.83×10^{-20}	2.9×10^{-20} [24]
Al$_3$Er	9.57	8.4 [25]	6.94×10^{-19}	5.4×10^{-19} [25]
Al$_3$Y	9.40	-	6.84×10^{-19}	-

Among the various L1$_2$-Al$_3$M phases, the Al$_3$Zr phase obtained the smallest critical nucleation radius and lowest critical nucleation work, whereas the Al$_3$Er and Al$_3$Y phases obtained similar nucleation characteristics, and displayed the largest critical nucleation radius and highest critical nucleation work. The critical nucleation radius and nucleation work of Al$_3$Sc were lower than those of the Al$_3$Er and Al$_3$Y phases, which agreed well with Fan's calculation [14]. It indicated that Al$_3$Zr had thermodynamic advantages in the nucleation process, while the Al$_3$Er and Al$_3$Y phases were relatively difficult to nucleate but had advantages in precipitation kinetics.

3.2. Nucleation and Stability of Multicomponent L1$_2$-Al$_3$M Phases

As described in Section 3.1, the thermodynamic priority order of precipitation was: Al$_3$Zr > Al$_3$Sc > Al$_3$Er > Al$_3$Y. The lowest interface energy of Al$_3$Zr/Al suggested that the Al$_3$Zr phase always tended to wrap outside the precipitation phase during the precipitation process. Due to the low interfacial energy of L1$_2$-Al$_3$Zr/Al$_3$Sc and L1$_2$-Al$_3$Zr/Al$_3$Er, once a core–shell structure was formed, the core–shelled Al$_3$Sc (Zr) and Al$_3$Er (Zr) were stable structures. However, the previous research showed that Al$_3$(Y, Zr) transformed from a core–shelled structure into a hybrid structure during homogenization at high temperatures [15,16]. In this section, the nucleation of multicomponent L1$_2$-Al$_3$(N, Zr) (N = Y, Sc, Er) phases were investigated based on first-principles thermodynamic calculations. The nucleation of possible ternary L1$_2$-Al$_3$(Y, Zr) phases included the core–shelled structures (the Al$_3$Y-core + Al$_3$Zr-shell structure, denoted as L1$_2$-Al$_3$Zr(Y)), the hybrid structure (denoted as L1$_2$-Al$_3$(Zr$_x$, Y$_{1-x}$), and the separate nucleation of binary L1$_2$-Al$_3$Zr and L1$_2$-Al$_3$Y (denoted as L1$_2$-Al$_3$Zr/Al$_3$Y). Moreover, the nucleation calculation result of core–shelled Al$_3$Zr(Y) was also compared with that of core–shelled Al$_3$Zr(Sc,) and core–shelled Al$_3$Zr(Er).

Based on the classical nucleation theory, the structure stability of L1$_2$ nanoparticles with the different structures can be evaluated through the total nucleation energy $\Delta G_{Al_3(N,Zr)}$ (N = Y, Sc, Er), and the expressions are given as [24]:

$$\Delta G_{Al_3Zr(N)} = \frac{4\pi}{3}[(R^3 - r^3)\cdot \Delta G_V^{Al_3Zr} + r^3 \cdot \Delta G_V^{Al_3N}] + 4\pi(r^2 \cdot \gamma_{Al_3Zr/Al_3N} + R^2 \cdot \gamma_{Al_3Zr/Al}) \quad (6)$$

$$\Delta G_{Al_3N+Al_3Zr} = \frac{4\pi}{3}(r^3 \times \Delta G_V^{Al_3N} + r^3 \times \Delta G_V^{Al_3Zr}) + 4\pi(r^2 \times \gamma_{Al/Al_3N} + r^2 \times \gamma_{Al/Al_3Zr}) \quad (7)$$

$$\Delta G_{Al_3(N_x, Zr_{1-x})} = \frac{4\pi}{3}R^3 \cdot \Delta G_V^{Al_3(N_x, Zr_{1-x})} + 4\pi R^2 \cdot \gamma_{Al_3(Nx, Zr1-x)} \quad (8)$$

Here R is the radius of ternary L1$_2$-Al$_3$(N, Zr), and r is the radius of the binary Al$_3$N. Assuming that all the solute atoms had completely precipitated from the Al matrix, the R and r values of ternary L1$_2$-Al$_3$(N, Zr) with a core–shelled structure depended on the relative precipitation amount of solute atoms N and Zr. $\Delta G_V^{Al_3N}$ and $\Delta G_V^{Al_3Zr}$ are the volumetric formation energy of the L1$_2$-Al$_3$N phase and the Al$_3$Zr phase in aluminum alloys. γ_{Al_3Zr/Al_3N} and $\gamma_{Al_3Zr/Al}$ are the interface energies of the Al$_3$Zr/Al$_3$N and Al$_3$Zr/Al interface in aluminum alloys. The Al$_3$Zr(001)/Al$_3$N(001)-contacting facets were considered to be the most energy-favored orientation, and the interfaces' energies were calculated in the authors' previous investigation [18]. It should be noted that the interfaces' energies were generally oerestimated at the actual precipitation temperature due to the density functional principles of the ground state. $\Delta G_V^{Al_3(N_x, Zr_{1-x})}$ and $\gamma_{Al_3(N_x, Zr_{1-x})}$ are the volumetric formation energy and the interface energy of the hybrid structure of $Al_3(N_x, Zr_{1-x})$. However, it was difficult to directly calculate the value of $\Delta G_V^{Al_3(N_x, Zr_{1-x})}$, which was estimated by the composition-weighted summation of $\Delta G_V^{Al_3N}$ and $\Delta G_V^{Al_3Zr}$ [24]. Similarly, $\gamma_{Al_3(N_x, Zr_{1-x})}$ was evaluated by the composition-weighted summation of $\gamma_{Al_3N/Al}$ and $\gamma_{Al_3Zr/Al}$.

Furthermore, the authors' previous investigation indicated that Er atoms tended to segregate at the Al$_3$Y/Al$_3$Zr interface, and were inclined to form a core–double-shelled Al$_3$Y/Al$_3$Er/Al$_3$Zr structure [18], denoted as Al$_3$Zr/Er(Y), and its nucleation energy and thermodynamic stability can be evaluated as:

$$\Delta G_{Al_3Zr/Er(Y)} = \frac{4\pi}{3}[(R_2^3 - R_1^3) \times \Delta G_V^{Al_3Zr} + (R_1^3 - r^3) \times \Delta G_V^{Al_3Er} + r^3 \times \Delta G_V^{Al_3Y}]$$
$$+ 4\pi(r^2 \times \gamma_{Al_3Y/Al_3Er} + R_1^2 \times \gamma_{Al_3Zr/Al_3Er} + R_2^2 \times \gamma_{Al_3Zr/Al}) \quad (9)$$

Here R_1 and R_2 are the radii of the first and second shells of the core–double-shelled Al$_3$Zr/Er (Y), respectively; r is the radius of the Al$_3$Y core layer.

Under the conditions of homogenization temperature (T = 673 K) and the equal solute atomic ratio (the atomic ratio of Y to Zr was 1), the total nucleation energies (ΔG) of the various possible structures for the L1$_2$-Al$_3$(Y, Zr) phase were calculated as a function of the precipitate radius (R), and the results are plotted in Figure 1. It showed that the nucleation energy of various structures of Al$_3$(Y, Zr) increased with the radius of the precipitated phase. At a radius of 0–1 nanometers, there was no significant difference in the free energy of each phase; thus, several structures of Al$_3$(Y, Zr) phases can coexist in the early stage of homogenization. To some extent, the 0–1 nanometer precipitation stage corresponded to the early aging stage of atomic clusters, and did not form a stable microstructure.

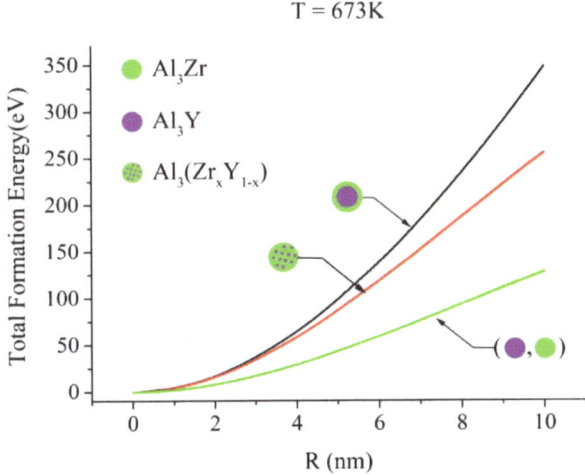

Figure 1. Relationship between nuclear energy and the radii of various structures of Al$_3$(Y, Zr) at the homogenization temperature of 673 K and the equal stoichiometric ratio.

At a radius of 1–10 nanometers, the difference in the total nucleation energy among different structures became increasingly significant. The core–shelled Al$_3$Zr(Y) phase illustrated the highest nucleation work among various precipitate structures, indicating that the core–shelled Al$_3$Zr(Y) phase precipitated without advantage in thermodynamics. However, the separated nucleation of binary L1$_2$-Al$_3$Zr/Al$_3$Y obtained the lowest nucleation energy, suggesting that L1$_2$-Al$_3$Zr/Al$_3$Y had thermodynamic advantages in the nucleation process. Due to the low segregation energy of Zr elements at the Al$_3$Y interface, it was beneficial to drive the segregation of Zr elements at the Al$_3$Y interface [18]. Gao et al. [16] studied the early precipitation phase structure of Al-0.08Y-0.30Zr alloy at 350 °C for 10 min, and the results showed that the precipitation phase was mainly the Al$_3$Y phase, which became the core of Al$_3$Zr and promoted the precipitation kinetics of solid solution Zr atoms. However, it was difficult to form a stable core–shelled Al$_3$Y/Al$_3$Zr owing to the large coherency strain energy and high mismatch between Al$_3$Y and Al$_3$Zr [18]. Thus, the separated structure of L1$_2$-Al$_3$Zr/Al$_3$Y was considered to be the thermodynamically stable structure. The investigation by Gao et al. [16] showed that after isothermal aging at 400 °C for 200 h, the Y and Zr atoms in the Al-Y-Zr alloy were almost uniformly distributed within the precipitate phase, indicating a separated structure of L1$_2$-Al$_3$Zr/Al$_3$Y, and did not exhibit a clear core–shelled structure, which confirmed the first-principles calculation results in this paper.

Figure 2 shows the total nucleation energies (ΔG) for three kinds of core–shelled structures, Al$_3$Zr(Sc), Al$_3$Zr(Er), and Al$_3$Zr(Y), under the condition of homogenization at

673 K and the complete precipitation of Sc, Y, Er, and Zr in equal proportion, respectively. The nucleation energies of core–shelled Al$_3$Zr(Y) and Al$_3$Zr(Sc) increased with the radius of the precipitated phase, whereas the nucleation energies of Al$_3$Zr(Er) were negative, and decreased with the radius of the precipitated phase. The calculations of Al$_3$Zr(Sc) and Al$_3$Zr(Er) were similar to the investigation by Jiang et al. [24,25]. The order of the nucleation energies was: Al$_3$Zr(Y) > Al$_3$Zr(Sc) > Al$_3$Zr(Er). The core–shelled Al$_3$Zr(Y) phase obtained the highest nucleation energy, indicating that it was inclined to form a separated structure, L1$_2$-Al$_3$Zr/Al$_3$Y, which was very consistent with the experimental observation [16]. The core–shelled Al$_3$Zr(Sc) and Al$_3$Zr(Er) were thermodynamically stable structures owing to their low nucleation energies, which were confirmed by the experimental observation in Al-Sc-Zr alloys [8] and Al-Er-Zr alloys [9]. In comparison with Al$_3$Zr(Y) and Al$_3$Zr(Sc), although the Al$_3$Er/Al$_3$Zr interface had a higher coherent strain energy than that the of Al$_3$Sc/Al$_3$Zr interface [18], core–shelled Al$_3$Zr(Er) obtained a low nucleation energy due to its low chemical energy ΔG_{chem} and the Al$_3$Er/Al$_3$Zr interface energy. Thus, the nucleation energy of Al$_3$M nanophases depended on their chemical energy ΔG_{chem} and the interface energy.

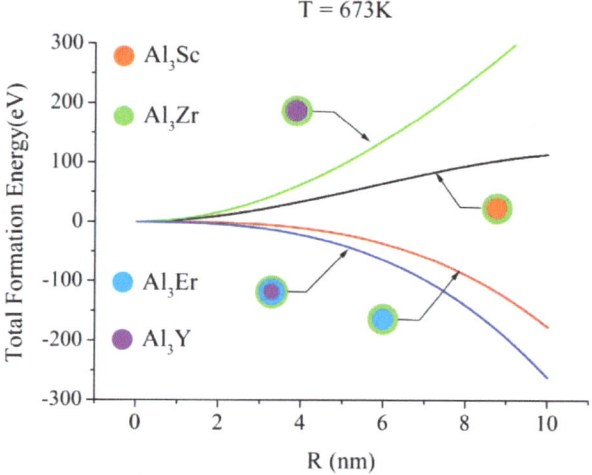

Figure 2. Relationship between nuclear energy and the radii of various structures of L1$_2$-Al$_3$(N, Zr) (N = Er, Y, Sc) at the homogenization temperature of 673 K and the equal stoichiometric ratio.

The nucleation energies (ΔG) of core–double-shelled Al$_3$Zr/Er(Y) were carried out under the condition of homogenization at 673 K and the complete precipitation of Y, Er, and Zr in equal proportion. The nucleation energy of Al$_3$Zr/Er(Y) was negative and significantly decreased with the precipitation radius, as shown in Figure 2. The nucleation energy of Al$_3$Zr/Er(Y) was far lower than that of core–shelled Al$_3$Zr(Y), and obtained high thermodynamic stability, preferentially precipitating in thermodynamics. Core–double-shelled Al$_3$Zr/Er(Y) was inclined to form in Al-Y-Er--Zr alloys, as the Er atom tended to segregate at the Al$_3$Y/Al$_3$Zr interface [18]. The segregation of the Er atom dramatically decreased the nucleation energy due to the decrease in ΔG_{chem} and strain energy G_S, as illustrated in Al$_3$Zr(Er), although the high interfacial energy of Al$_3$Y/Al$_3$Er replaced the relatively low interface energy of Al$_3$Y/Al$_3$Zr. Interestingly, the nucleation energy of Al$_3$Zr/Er(Y) was even lower than that of Al$_3$Zr(Er) due to the addition of the Y atom, which can be attributed to the negative interface energy of Al$_3$Er/Al$_3$Y and low coherent strain energy G_s. Similarly, in the Al-Sc-Zr aluminum alloy, the addition of the Er atom formed a core–double-shelled Al$_3$Zr/Sc (Er) instead of forming separated Al$_3$(Sc, Zr) and Al$_3$(Er, Zr) [20], which was attributed to the decreased nucleation energy of Al$_3$(Sc,

Zr) nanoparticles by its low chemical energy ΔG_{chem}. Therefore, the design of the core–double-shelled Al$_3$Zr/Er(Y) nanophase can provide guidance for the development of new Al-Er-Y-Zr alloys.

4. Conclusions

Based on the first-principles thermodynamic calculation, the nucleation energies of the L1$_2$-Al$_3$M (M = Sc, Zr, Er, Y) nanophases in aluminum alloys were studied combined with classical nucleation theory. The conclusions were as follows:

(1) The critical radius and nucleation work of the L1$_2$-Al$_3$M precipitate phase were as follows: Al$_3$Er > Al$_3$Y > Al$_3$Sc > Al$_3$Zr. The Al$_3$Zr phase was the easiest to nucleate in thermodynamics, while the nucleation of the Al$_3$Y and Al$_3$Er phases were relatively difficult in thermodynamics.

(2) Various structures of Al$_3$(Y, Zr) phases with the radius r < 1 nm can coexist in Al-Y-Zr alloys. At a precipitate's radius of 1–10 nanometers, the core–shelled Al$_3$Zr(Y) phase illustrated the highest nucleation energy, while the separated structure, Al$_3$Zr/Al$_3$Y, obtained the lowest one, and had thermodynamic advantages in the nucleation process.

(3) Core–double-shelled Al$_3$Zr/Er(Y) obtained a lower nucleation energy than that of Al$_3$Zr(Y) due to the negative ΔG_{chem} of Al$_3$Er and the negative Al$_3$Er/Al$_3$Y interface energy, and preferentially precipitated in thermodynamics stability.

Author Contributions: B.N. and Y.S. conceived and designed the research; S.L., F.L. and Z.Y. performed the first-principles calculation; T.F. and D.C. analyzed the experimental data; S.L. wrote the manuscript. All authors have read and agreed to the published version of the manuscript.

Funding: This research was funded by the R & D plan for key areas in Guangdong Province (2020B01 0186001), the Science and Technology Program of the Ministry of Science and Technology (G2022030060L), the Science and technology project in Guangdong (2020b15120093, 2020B121202002), the Science and technology research project of Foshan (1920001000412, 2220001005305), and the R and D plan for key areas in Jiangxi Province (20201BBE51009, 20212BBE51012).

Data Availability Statement: The data presented in this study are available on request from the corresponding author.

Conflicts of Interest: The authors declare no conflict of interest.

References

1. Liu, L.; Xu, G.; Deng, Y.; Yu, Q.; Li, G.; Zhang, L.; Liu, B.; Fu, L.; Pan, Q. Existing form of Sc in metal-inert gas welded Al-0.60 Mg-0.75 Si alloy and its role in welding strength. *Mater. Charact.* **2023**, *197*, 112649. [CrossRef]
2. Ye, J.; Pan, Q.; Liu, B.; Hu, Q.; Qu, L.; Wang, W.; Wang, X. Effects of co-addition of minor Sc and Zr on aging precipitates and mechanical properties of Al-Zn-Mg-Cu alloys. *J. Mater. Res. Technol.* **2023**, *22*, 2944–2954. [CrossRef]
3. Deng, P.; Mo, W.; Ouyang, Z.; Tang, C.; Luo, B.; Bai, Z. Mechanical properties and corrosion behaviors of (Sc, Zr) modified Al-Cu-Mg alloy. *Mater. Charact.* **2022**, *196*, 112619. [CrossRef]
4. Zha, M.; Tian, T.; Jia, H.L.; Zhang, H.M.; Wang, H.Y. Sc/Zr ratio-dependent mechanisms of strength evolution and microstructural thermal stability of multi-scale hetero-structured Al–Mg–Sc–Zr alloys. *J. Mater. Sci. Technol.* **2023**, *140*, 67–78. [CrossRef]
5. De Luca, A.; Seidman, D.N.; Dunand, D.C. Effects of Mo and Mn microadditions on strengthening and over-homogenization resistance of nanoprecipitation-strengthened Al-Zr-Sc-Er-Si alloys. *Acta Mater.* **2019**, *165*, 1–14. [CrossRef]
6. Booth-Morrison, C.; Mao, Z.; Diaz, M.; Dunand, D.C.; Wolverton, C.; Seidman, D.N. Role of silicon in accelerating the nucleation of Al$_3$(Sc, Zr) precipitates in dilute Al-Sc-Zr alloys. *Acta Mater.* **2012**, *60*, 4740–4752. [CrossRef]
7. Senkov, O.N.; Shagiev, M.R.; Senkova, S.V.; Miracle, D.B. Precipitation of Al$_3$(Sc, Zr) particles in an Al-Zn-Mg-Cu-Sc-Zr alloy during conventional solution heat treatment and its effect on tensile properties. *Acta Mater.* **2008**, *56*, 3723–3738. [CrossRef]
8. Forbord, B.; Lefebvre, W.; Danoix, F.; Hallem, H.; Marthinsen, K. Three dimensional atom probe investigation on the formation of Al$_3$(Sc, Zr)-dispersoids in aluminium alloys. *Scripta Mater.* **2004**, *51*, 333–337. [CrossRef]
9. Wu, H.; Wen, S.P.; Huang, H.; Li, B.L.; Wu, X.L.; Gao, K.Y.; Wang, W.; Nie, Z.R. Effects of homogenization on precipitation of Al$_3$(Er, Zr) particles and recrystallization behavior in a new type Al-Zn-Mg-Er-Zr alloy. *Mater. Sci. Eng. A* **2017**, *689*, 313–322. [CrossRef]
10. Leibner, M.; Vlach, M.; Kodetová, V.; Kudrnová, H.; Veselý, J.; Zikmund, S.; Čížek, J.; Melikhova, O.; Lukáč, F. Effect of deformation on evolution of Al$_3$(Er, Zr) precipitates in Al-Er-Zr-based alloy. *Mater. Charact.* **2022**, *186*, 111781. [CrossRef]

11. Peng, G.; Chen, K.; Fang, H.; Chen, S. A study of nanoscale Al$_3$(Zr, Yb) dispersoids structure and thermal stability in Al–Zr–Yb alloy. *Mater. Sci. Eng. A* **2012**, *535*, 311–315. [CrossRef]
12. Pang, H.C.; Shang, P.J.; Huang, L.P.; Chen, K.H.; Liu, G.; Xiong, X. Precipitates and precipitation behavior in Al–Zr–Yb–Cr alloys. *Mater. Lett.* **2012**, *75*, 192–195.
13. Hu, T.; Ruan, Z.; Fan, T.; Wang, K.; He, K.; Wu, Y. First-principles investigation of the diffusion of TM and the nucleation and growth of L1$_2$ Al$_3$TM particles in Al alloys. *Crystals* **2023**, *13*, 1032. [CrossRef]
14. Fan, T.; Ruan, Z.; Zhong, F.; Xie, C.; Li, X.; Chen, D.; Tang, P.; Wu, Y. Nucleation and growth of L1$_2$-Al$_3$RE particles in aluminum alloys: A first-principles study. *J. Rare Earths* **2023**, *41*, 1116–1126. [CrossRef]
15. Zhang, Y.; Gu, J.; Tian, Y.; Gao, H.; Wang, J.; Sun, B. Microstructure evolution and mechanical property of Al–Zr and Al–Zr–Y alloys. *Mater. Sci. Eng. A* **2014**, *616*, 132–140. [CrossRef]
16. Gao, H.; Feng, W.; Wang, Y.; Gu, J.; Zhang, Y.; Wang, J.; Sun, B. structure and compositional evolution of Al$_3$(Zr, Y) precipitates in Al-Zr-Y alloy. *Mater. Charact.* **2016**, *121*, 195–198. [CrossRef]
17. Wang, Y.; Miao, Y.; Peng, P.; Gao, H.; Wang, J.; Sun, B. Ab initio investigation on preferred orientation at the Al/Al$_3$(Zr, Y) interface in Al–Zr–Y alloy. *J. Appl. Phys.* **2022**, *131*, 225111. [CrossRef]
18. Song, Y.; Zhan, S.; Nie, B.; Liu, S.; Qi, H.; Liu, F.; Fan, T.; Chen, D. First-principle investigation of the interface properties of the core-shelled L1$_2$-Al$_3$M (M = Sc, Zr, Er, Y) phase. *Crystals* **2023**, *13*, 420. [CrossRef]
19. Dorin, T.; Babaniaris, S.; Jiang, L.; Cassel, A.; Eggeman, A.; Robson, J. Precipitation sequence in Al-Sc-Zr alloys revisited. *Materialia* **2022**, *26*, 101608. [CrossRef]
20. Booth-Morrison, C.; Dunand, D.C.; Seidman, D.N. Coarsening resistance at 400 °C of precipitation-strengthened Al–Zr–Sc–Er alloys. *Acta Mater.* **2011**, *59*, 7029–7042. [CrossRef]
21. Leibner, M.; Vlach, M.; Kodetová, V.; Veselý, J.; Čížek, J.; Kudrnová, H.; Lukáč, F. On the Sc-rich core of Al$_3$(Sc, Er, Zr) precipitates. *Mater. Lett.* **2022**, *325*, 132759. [CrossRef]
22. Qian, W.; Zhao, Y.; Kai, X.; Gao, X.; Huang, L.; Miao, C. Characteristics of microstructureand mechanical evolution in 6111Al alloy containing Al$_3$(Er, Zr) nanoprecipitates. *Mater. Charact.* **2021**, *178*, 111310. [CrossRef]
23. Zhang, J.; Hu, T.; Yi, D.; Wang, H.; Wang, B. Double-shell structure of Al$_3$(Zr, Sc) precipitate induced by thermomechanical treatment of Al–Zr–Sc alloy cable. *J. Rare Earths* **2019**, *37*, 668–672. [CrossRef]
24. Zhang, C.; Jiang, Y.; Cao, F.; Hu, T.; Wang, Y.; Yin, D. Formation of coherent, core-shelled nano-particles in dilute Al-Sc-Zr alloys from the first-principles. *J. Mater. Sci. Technol.* **2019**, *35*, 930–938. [CrossRef]
25. Zhang, C.; Yin, D.; Jiang, Y.; Wang, Y. Precipitation of L1$_2$-phase nano-particles in dilute Al-Er-Zr alloys from the first-principles. *Comp. Mater. Sci.* **2019**, *162*, 171–177. [CrossRef]
26. Liu, X.; Wang, Q.; Zhao, C.; Li, H.; Wang, M.; Chen, D.; Wang, H. Formation of ordered precipitates in Al-Sc-Er-(Si/Zr) alloy from first-principles study. *J. Rare Earths* **2023**, *9*, 609–620. [CrossRef]
27. Nityananda, R.; Hohenberg, P.; Kohn, W. Inhomogeneous electron gas. *Resonance* **2017**, *22*, 809–811. [CrossRef]
28. Kresse, G.; Furthmüller, J. Efficiency of ab-initio total energy calculations for metals and semiconductors using a plane-wave basisset. *Comp. Mater. Sci.* **1996**, *6*, 15–50. [CrossRef]
29. Kresse, G.; Joubert, D. From ultrasoft pseudopotentials to the projector augmented-wave method. *Phys. Rev. B* **1999**, *59*, 1758–1775. [CrossRef]
30. Perdew, J.P.; Burke, K.; Ernzerhof, M. Generalized Gradient Approximation Made Simple. *Phys. Rev. Lett.* **1998**, *77*, 3865–3868. [CrossRef]
31. Budimir, M.; Damjanovic, D.; Setter, N. Piezoelectric Response and Free Energy Instability in the Perovskite Crystals BaTiO$_3$, PbTiO$_3$ and Pb(Zr, Ti)O$_3$. *Phys. Rev. B* **2006**, *73*, 4106. [CrossRef]
32. Baroni, S.; De Gironcoli, S.; Dal Corso, A.; Giannozzi, P. Phonons and related crystal properties from density-functional perturbation theory. *Rev. Mod. Phys.* **2001**, *73*, 515. [CrossRef]
33. Mao, Z.; Chen, W.; Seidman, D.N.; Wolverton, C. First-principles study of the nucleation and stability of ordered precipitates in ternary Al–Sc–Li alloys. *Acta Mater.* **2011**, *59*, 3012–3023. [CrossRef]
34. Swan-Wood, T.L.; Delaire, O.; Fultz, B. Vibrational entropy of spinodal decomposition in FeCr. *Phys. Rev. B* **2005**, *72*, 024305. [CrossRef]

Disclaimer/Publisher's Note: The statements, opinions and data contained in all publications are solely those of the individual author(s) and contributor(s) and not of MDPI and/or the editor(s). MDPI and/or the editor(s) disclaim responsibility for any injury to people or property resulting from any ideas, methods, instructions or products referred to in the content.

Article

Effect of Cooling Rate on Crystallization Behavior during Solidification of Hyper Duplex Stainless Steel S33207: An In Situ Confocal Microscopy Study

Yong Wang [1,2] and Wangzhong Mu [2,*]

1. Key Laboratory for Ferrous Metallurgy and Resources Utilization of Ministry of Education & Hubei Provincial Key Laboratory for New Processes of Ironmaking and Steelmaking, Wuhan University of Science and Technology, Wuhan 430081, China; wangyong6@wust.edu.cn
2. Department of Materials Science and Engineering, KTH Royal Institute of Technology, SE-100 44 Stockholm, Sweden
* Correspondence: wmu@kth.se

Abstract: Hyper duplex stainless steel (HDSS) is a new alloy group of duplex stainless steels with the excellent corrosion resistance and mechanical properties among the existing modern stainless steels. Due to the incorporation of the high content of alloying elements, e.g., Cr, Ni, Mo, etc., the crystallization behavior of δ-ferrite from liquid is of vital importance to be controlled. In this work, the effect of the cooling rate (i.e., 4 °C/min and 150 °C/min) on the nucleation and growth behavior of δ-ferrite in S33207 during the solidification was investigated using a high-temperature confocal scanning laser microscope (HT-CLSM) in combination with electron microscopies and thermodynamic calculations. The obtained results showed that the solidification mode of S33207 steel was a ferrite–austenite type (FA mode). L→δ-ferrite transformation occurred at a certain degree of undercooling, and merging occurred during the growth of the δ-ferrite phase dendrites. Similar microstructure characteristics were observed after solidification under two different cooling rates. The variation in the area fraction of δ-ferrite with different temperatures and time intervals during the solidification of S33207 steels was calculated at different cooling rates. The post-microstructure as well as its composition evolution were also briefly investigated. This work shed light on the real-time insights for the crystallization behavior of hyper duplex stainless steels during their solidification process.

Keywords: solidification; cooling rate; hyper duplex stainless steel; high-temperature confocal scanning laser microscope

1. Introduction

Duplex stainless steel (DSS) has shown increasing demands for its application across important industrial areas, such as deep-sea pipelines and submarines [1,2] and petrochemical industries [1,3], due to their excellent corrosion resistance and mechanical properties [4,5]. The desired properties of DSS are mainly corrected with the precise control of the fractions of ferrite and austenite. Despite the very successful applications and experience of DSSs, the development of highly alloyed DSSs has been very active since there are still areas where the corrosion resistance of the current DSSs has been insufficient for a long service life or at higher temperatures. Based on this challenge, the flagship hyper DSS, UNS S33207 (marked as S33207 from herein), which has the highest corrosion resistance and strength among the existing modern DSSs has since been developed [6–9].

DSSs are designed to solidify with ferrite as the parent phase, with subsequent austenite formation occurring in the solid state, implying that the solidification process plays an important role in the dual-phase microstructure control [10,11]. That is to say, the changes in the fractions of ferrite and austenite in these DSSs are due to the different solidification conditions. It has been recognized that the cooling rate has an important effect on the microstructure evolution of the DSSs [12,13]. Zhu et al. [14] revealed that the austenite fraction in DSS 2205 after the sub-rapid solidified process was substantially decreased compared to the slow cooling process, and the hardness and wear-resisting properties of steel were improved. High-temperature confocal laser scanning microscopy (HT-CLSM) has provided a convenient possibility of making an in situ observations of the solidification and post-microstructure evolution of various alloy grades, including low carbon steels [15–17], austenite stainless steels [18–20], and duplex stainless steels [21–23]. Sun et al. [21] assessed the characteristics of the $\delta \rightarrow \gamma$ phase transformations at different cooling rates on the surface of DSS S31308 by HT-CLSM. They found that the γ-cells preferred to precipitate along the δ/δ grain boundaries with a flaky pattern, and that their fronts were jagged in shape under the slow cooling rate but were in a needle-like feature at the rapid cooling rate. Moreover, they reported that higher N contents promoted the nucleation and growth of the γ-phase during the $\delta \rightarrow \gamma$ transformation by increasing both the starting and finishing temperatures of the phase transformation [23]. Shin et al. [24] assessed the effects of the cooling rate after heat treatment on the pitting corrosion of DSS S32750, and they found that the ferrite volume fraction increased accordingly with the increase in the cooling rate. In a former study by the current authors, the solidification behavior of the different grades of DSSs have been investigated in situ [25] using a combinational approach of HT-CLSM and differential scanning calorimetry (DSC) [26]. Till now, the $\delta \rightarrow \gamma$ transformation has been reported in previous works, but the evolution mechanism of δ-ferrite during the solidification process is still not clear regarding the DSSs. Moreover, there have been no in situ observations made for the solidification process of hyper duplex stainless steel S33207 under different cooling conditions.

Previous studies have demonstrated that the solidification process of DSSs has a great influence on their performance. Changing the cooling rate is one of the important methods to control the fractions of the ferrite and austenite phases. S33207 hyper DSS has quite high Cr contents (32%) compared to the other grades of super DSSs, which results in a high corrosion resistance, yield strength, and superior fatigue features. Therefore, understanding its solidification behavior is of great interest for the subsequent annealing homogenization treatment process. In this work, the characteristics of the δ formation and growth during the solidification process were observed on the sample surface of DSS S33207 at different cooling rates using the HT-CLSM. The chemical element distribution of the solidified samples was also characterized. This work is of great significance to both understand and control the solidification process of Hyper DSS S33207.

2. Materials and Methods

Differential thermal analysis (DTA), equipped in a Netzsch STA 449 F1 Jupiter®, was used to detect the melting and solidification temperatures during the heating and cooling stages to ultimately guide the design of the in situ observation experimental conditions. The DTA facility was calibrated using pure metals (Ag, Ni, and Fe) prior to its use, and approximately a 100 mg of specimen was used for the measurement. HT-CLSM (Lasertec, 1LM21H) with a He–Ne laser beam, equipped with a power of 1.5 mW and a wavelength of 632.8 nm, was used to analyze the crystallization during the solidification of DSS 3207 under different cooling conditions. The chemical composition of the proposed DSS sample is listed in Table 1. For all HT-CLSM experiments, the diameter of the sample was 4 mm and the thickness was 1.5 mm, respectively. All the slices were grounded by 800, 1200, and 2000 SiC papers, following which they were then polished using 3 μm and 1 μm diamond pastes, respectively. The processed sample was filled with an Al_2O_3 crucible (Φ 5.5 mm O.D., Φ 4.5 mm I.D., and 5.0 mm height) and was then put into the ellipsoidal chamber of the HT-CLSM instrument. The chamber was cleaned thoroughly through a vacuum cycle (of less than 4.5×10^{-5} torr) followed by purging with a high purity Ar (purity > 99.9999%) passing through a 300 °C Mg column. The temperature was measured using a Type B (PtRh30%–PtRh6%) thermocouple at the bottom of the holder of the crucible.

The steel samples were first heated to 1500 °C at a rate of 20 °C/min and were then kept for 2 min for melt stabilization, followed by cooling at two different cooling rates of 4 °C/min (0.07 °C/s) and 150 °C/min (2.5 °C/s), respectively. The solidified samples after HT-CLSM were directly analyzed using a scanning electron microscope (SEM, S3700N-Hitachi, Japan) equipped with an energy dispersive spectrometer (EDS, Bruker, Germany). The nitrogen content of the samples before and after HT-CLSM was measured through LECO analysis. Standard stainless steels with 0.1% and 0.2% N content were used, respectively.

Table 1. Chemical compositions of the S33207 steel samples (mass percent, %).

Steel	C	Si	Mn	S	Cr	Ni	Mo	Cu	N	Ti	Nb	Al	V	O	Fe
S33207	0.015	0.250	0.700	0.001	31.200	7.000	3.460	0.200	0.470	0.010	0.010	0.010	0.070	0.004	Bal.

3. Results and Discussion

3.1. In situ Observations of the Solidification of S33207 at Different Cooling Rates

The solidification phase diagram and phase fraction during the solidification of S33207 steel was calculated using the Scheil solidification and equilibrium modules in Thermo-Calc 2022a, the results are shown in Figure 1. The Scheil solidification module can calculate the actual solidification path of DSS S33207 under a non-equilibrium state and provide a basis for the solidification sequence in situ observation using the HT-CLSM. According to Figure 1, the solidification sequence of S33207 steel calculated by both modules belonged to the ferrite–austenite mode (FA): L(liquid)→L + δ (ferrite)→L + δ + γ (austenite). This indicated that the δ-ferrite phase started to form first followed by the δ→γ transformation at the terminal stage of the solidification process. The morphologies of the crystallized phases can vary depending on the cooling rate, which has been discussed in detail in the following part.

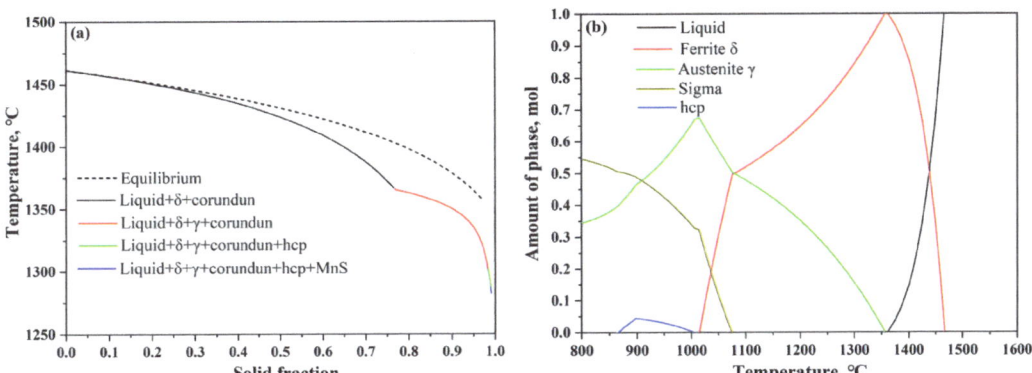

Figure 1. (a) Scheil solidification and (b) equilibrium calculations of the phase formations under different temperatures in the S33207 steel.

DTA results of the DSS33207 with the different solidification rates of 4 °C/min and 150 °C/min are presented in Figure 2. For both measurements, the same heating of 20 °C/min was used to heat the specimen till 1550 °C, and two cooling rates with the same ones used for the HT-CLSM was performed here. The phase transition temperatures of the heating and cooling stages are summarized in Table 2. In Figure 2, it is shown that the liquidus and solidus temperatures during heating are almost overlapping since the heating rate was the same. The on-set and peak temperatures of the solidification varied a lot due to the influence of the solidification cooling rate. Slower cooling led to the much higher solidification temperatures obtained (i.e., 1469 and 1466 °C for the cooling rate of 4 °C/min; 1452 and 1402 °C for the cooling rate of 150 °C/min) in the DSS3207.

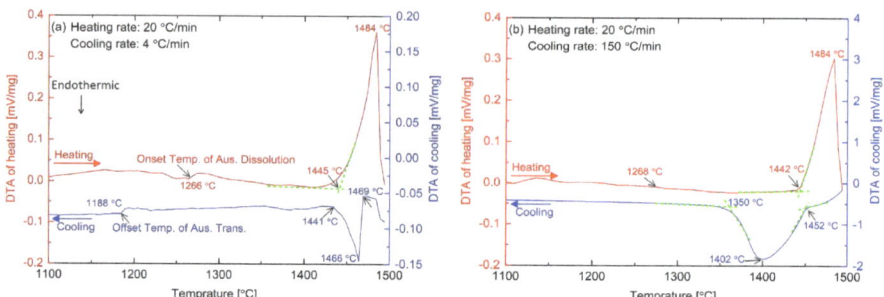

Figure 2. DTA results of the DSS33207 with the solidification rate of (**a**) 4 °C/min and (**b**) 150 °C/min, respectively.

Table 2. Phase transition temperatures during the heating and cooling processes.

Specimen	Heating		Cooling	
	$T_{sol.}$ (°C)	$T_{liq.}$ (°C)	$T_{on.Soli.}$ (°C)	$T_{Peak.Soli.}$ (°C)
Slow cooling	1445	1484	1469	1466
Fast cooling	1442	1484	1452	1402

Figure 3 presents several representative micrographs of phase formation during the solidification process of the DSS33207 through in situ observations at a cooling rate of 4 °C/min. When the molten steel was undercooled to 1482.7 °C, the L→δ transformation began to occur, and the formation of the cellular δ phase was observed on the surface of the sample. It should be pointed out that a thin δ layer formed on the outside of the liquid steel at the beginning of the solidification process (Figure 3a). It is generally believed that the crucible provided a core for the heterogeneous nucleation of the new phase formation during the solidification observation by the HT-CLSM. Due to the slow cooling rate of liquid steel, pro-eutectoid δ-ferrite solidified in the form of cellular crystals. With the decrease in temperature, the number of δ-ferrite nuclei increased and the grains gradually grew up and coarsened. When the δ-ferrite cell grew to some extent, several cells gradually approached and merged into a large irregular δ cell, and there was no obvious boundary between these cells after merging (Figure 3b). With further cooling, more δ cells merged together and the remaining liquid between these δ cells became less and less (Figure 3c). The growth of δ-ferrite was completed when the area of δ cells was not obviously increased based on the HT-CLSM observations. Some amount of liquid was still present after the ferrite growth was complete (Figure 3d), which has been reported in previous works [27,28].

The solidification process of the δ-ferrite phase with a cooling rate of 150 °C/min is shown in Figure 4. It should be mentioned that the focus of the initial stage of δ-ferrite nucleation from the liquid steel was not clear enough during the HT-CLSM observations made at this high cooling rate. Therefore, the formation temperature of δ-ferrite on the liquid surface was not determined correctly. More δ-ferrite cells can be formed simultaneously at the beginning of the solidification process compared to that of the low cooling rate. This indicated that a fast cooling rate favored primary δ-ferrite nucleation. The increase of the cooling rate during the solidification process of steel increased the undercooling of the liquid steel's composition, which promoted the increase in the nucleation rate of the δ-ferrite phase. It is believed to be beneficial to the refinement of δ cell crystals. Additionally, the growth temperature of the δ-ferrite phase (Figure 3b) formed at a higher temperature with a slow cooling rate. This tendency was the same as the DTA results (Figure 2) as well as the previous work [29], where a higher cooling rate resulted in a higher undercooling degree of melt and a lower crystallization temperature of δ-ferrite. The δ-ferrite grew quickly and a similar merging phenomenon was observed with a temperature decrease. In addition, a clear interphase boundary that usually separated these phases was observed (Figure 4b).

Under a high cooling rate, an obvious δ-ferrite growth layer was observed outside of the original δ cells due to the unstable growth characteristics of the δ cells. The liquid area was pushed by the growth of δ-ferrite during the late stage of the solidification process and then a volume shrinkage occurred between the δ-ferrite phase boundaries (Figure 4c). The growth rate of δ-ferrite decreased due to less liquid having been left at this stage. With a further decrease in the temperature, the area of δ-ferrite slowly increased and was kept at a stable value (Figure 4d). This remaining liquid area was the place where the localized transformation occurred from δ-ferrite to γ-austenite at a lower temperature. Generally, the concentration of the segregated elements increased greatly in the remaining liquid, where serious segregation took place [27]. This segregation can result in the decrease in the solidifying temperatures for steel. With time having passed, the δ-ferrite to γ-austenite transformation began at the δ-ferrite boundaries (Figure 4e,f). This finding is similar to that of Li et al. [19], who reported that δ→γ transformation occurred accompanied with a significant volume shrinkage.

However, the start temperature of the δ to γ transformation was not able to be obtained clearly due to the limitation of resolution of uneven surface after solidification. It was reported that the δ–γ phase transformation started earlier and occurred at a higher temperature in the S32101 DSS based on the concentric solidification technique with the increase in the cooling rate [22]. However, Sun et al. [21] found that starting temperature of the δ–γ transformation at the slow cooling rate was higher than that at the rapid cooling rate in an S31308 DSS. Additionally, the slow cooling rate more strongly favored the nucleation and growth of the γ-phase than the rapid cooling rate due to the fact that the higher diffusion rates of elements and longer diffusion times were obtained at a lower cooling rate.

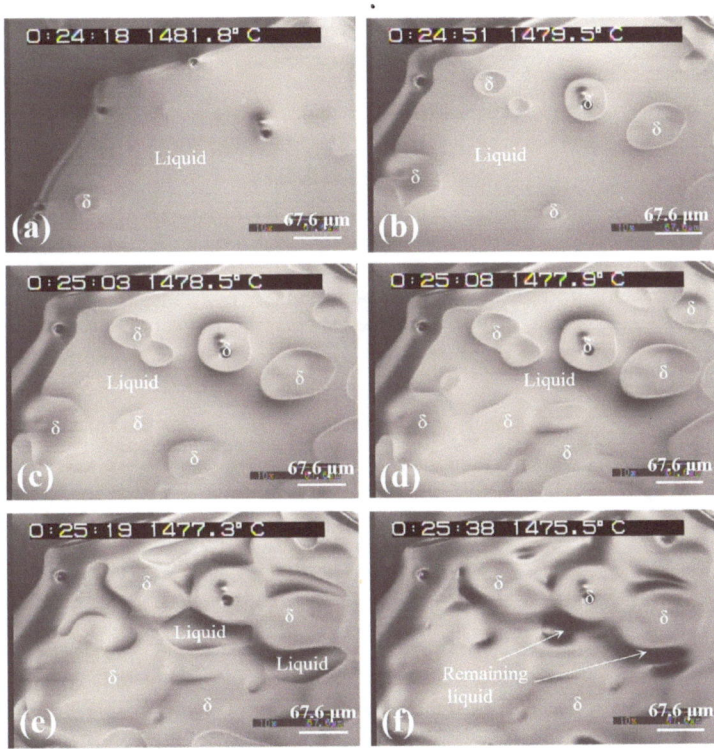

Figure 3. Growth process and characteristics of the δ-ferrite phase at a cooling rate of 4 °C/min at different temperatures of (**a**) 1481.8 °C, (**b**) 1479.5 °C, (**c**) 1478.5 °C, (**d**) 1477.9 °C, (**e**) 1477.3 °C, and (**f**) 1475.5 °.

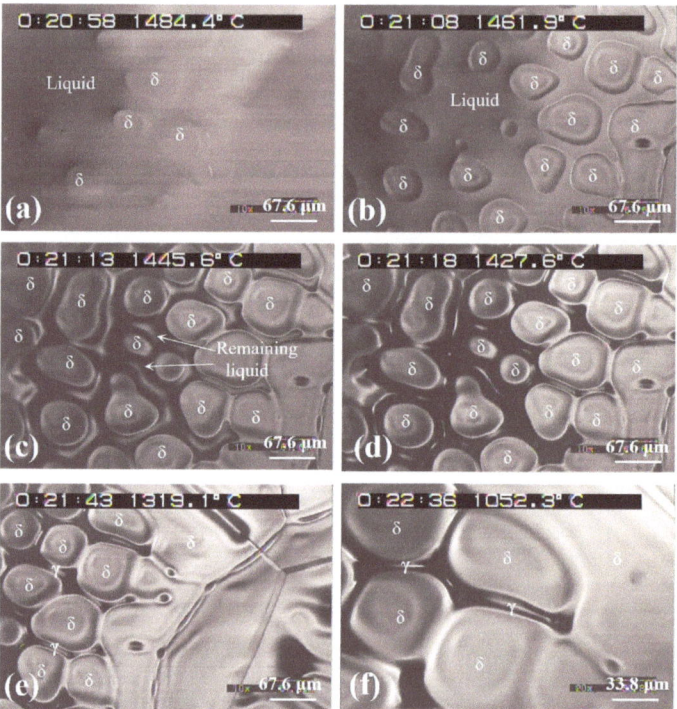

Figure 4. Growth process and characteristics of the δ phase at a cooling rate of 150 °C/min at different temperatures of (**a**) 1484.4 °C, (**b**) 1461.9 °C, (**c**) 1445.6 °C, (**d**) 1427.6 °C, (**e**) 1319.1 °C, and (**f**) 1052.3 °C.

The area fraction of ferrite in multiple sets of video screenshots during the solidification process at different cooling rates and temperatures were evaluated by Image J software, and the Avrami equation was used to fit the relationship between the area fraction of ferrite and the time and temperature; the results are shown in Figure 5. The Avrami Equations (1) and (2) [30,31] describe the crystallization of undercooled liquids into a solid state:

$$f_\delta = 1 - \exp(-k \cdot t^n) \tag{1}$$

$$t_{max} = [(n-1)/(n \cdot k)]^{1/n} \tag{2}$$

where is the area fraction of δ-ferrite, n is the Avrami coefficient, k is the overall growth rate constant, and t is the solidification time after the nucleation of δ-ferrite, t_{max}, the time required for maximum crystallization rate of ferrite.

It can be seen from Figure 5a that the area fraction of δ-ferrite increases with time, and it has a higher growth rate at a high cooling rate. In addition, the growth rate of δ-ferrite showed smaller values at the initial and late stages of solidification and a higher value at the stable stage of the solidification process. As mentioned before, the formation point of δ-ferrite in liquid was not obtained correctly, and thus the area fraction of δ-ferrite curve was obtained using the Avrami fitting function when the fitted linear correlation coefficient (R^2) was greater than 95%. Furthermore, the area fraction of δ-ferrite was less than one due to the adverse effects of its undulating cellular morphology on the depth of the observation field in the HT-CLSM image, which was more prominent under the high cooling rate. The fitted equations between the area fraction of δ-ferrite (f_δ) and time in the cases of the cooling rates of 4 °C/min and 150 °C/min are expressed by Equations (3) and (4), respectively. The rate constant k represents the velocity at which liquid transforms to solid. According to the fitted equations, the larger growth rate constant can be obtained under the fast cooling rate, which indicates the larger growth rate of δ-ferrite. This can be explained by the different number density of the nucleation sites and total growth times for the δ-ferrite. Specifically, the time required for the maximum crystallization rate of ferrite under the slow cooling rate obtained from Equation (2) was calculated to be 113 s, which is almost five times longer than that under the fast cooling rate (23 s).

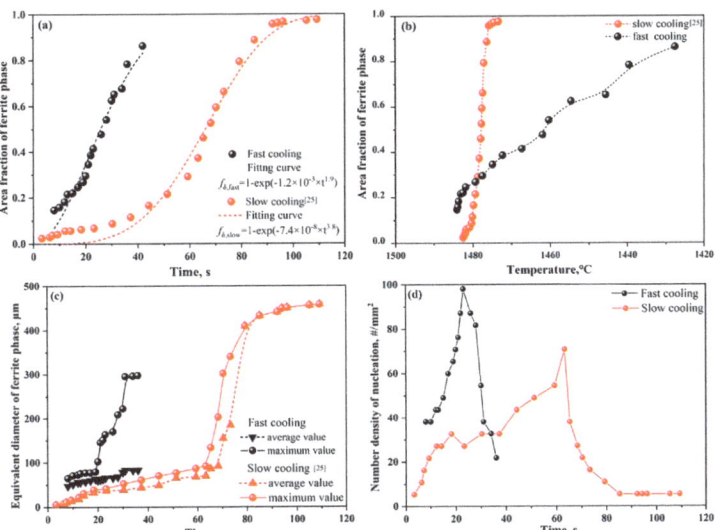

Figure 5. Relationships between the area fraction of δ-ferrite with the time (**a**) and temperature (**b**), relationships between the average and maximum diameters of δ-ferrite with time (**c**), and relationships between the number density of nucleation sites with time (**d**). The experimental data of the slow cooling condition in (**a**) to (**c**) were taken and adapted from Ref. [25].

$$f_{\delta,\text{fast}} = 1 - \exp(-1.2 \times 10^{-3} \times t^{1.9}) \tag{3}$$

$$f_{\delta,\text{slow}} = 1 - \exp(-7.4 \times 10^{-8} \times t^{3.4}) \tag{4}$$

The growth of δ-ferrite finished in a narrow temperature range (of less than 10 °C) at the cooling rate of 4 °C/min, while a much wider temperature was obtained at the cooling rate of 150 °C/min (Figure 5b). Notably, δ-ferrite began to form under a higher temperature with the growth ending at a lower temperature in the case of fast cooling compared to that of slow cooling. It is known that undercooling phenomenon usually occurs in the solidification process of steel, meaning that the ferrite formation temperature should be lower than the liquidus temperature of steel. From the in situ HT-CLSM observation results of the solidification process, the δ-ferrite formation temperature was slightly higher but has the same tendency as the results measured by DTA, which has been presented in Figure 2. This can be attributed to the details of the different instrumentation settings (sample size, thermocouple position, etc.) between the DTA and HT-CLSM measurements.

In terms of the maximum and average diameter changes of δ-ferrite (Figure 5c), the maximum diameter of δ-ferrite slowly increased at the beginning and began to increase rapidly when the δ-ferrite grew to a certain extent. The maximum and average diameter changes of δ-ferrite showed a similar tendency in the case of slow cooling, while the average diameter of δ-ferrite gradually increased during the whole solidification period under a fast cooling rate. As a result, the average diameter of δ-ferrite at the end of the solidification stage was approximately 80 μm at the cooling rate of 150 °C/min compared to 460 μm at the cooling rate of 4 °C/min, respectively. It can be indicated that the faster cooling rate can result in a smaller size of δ-ferrite. Moreover, the larger number density of the nucleation sites of δ-ferrite at the beginning of solidification can be found in the case of a high cooling rate, as shown in Figure 5d. The peak of the number density of the nucleation sites curves indicates that no new nucleation sites can be formed after that point, and the existing nucleation phases start to grow and merge to form bigger cells. In combination with the evolution of the number density of nucleation sites and the diameter of δ-ferrite changing during the solidification of steel, we can conclude that a slow cooling rate favored the growth of δ-ferrite, whereas a high cooling rate favored the nucleation of δ-ferrite.

In conclusion, the grain size will be finer as the cooling rate increases, the morphology comprises smaller grains, and the solidification behavior will commerce faster. This information can guide the actual continuous casting process of DSSS33207s.

3.2. Microstructure and Composition Evolution on the Surface of the Specimen after Solidification

In this section, the microstructure, as well as the chemical composition of the S33207 steel after solidification with different cooling rates, is briefly presented in this section.

3.2.1. Features on the Surface of the Specimen with a Slow Cooling Rate

Figure 6 presents the typical morphology of the S33207 steel after solidification with a cooling rate of 4 °C/min. It is shown that the solidified surface shows a 'bumpy' morphology, which is presented in Figure 6a; however, there is no clear segregation of the chemical elements, with all the elemental maps presented in Figure 6b–i showing an almost homogeneous distribution. In order to examine the detailed microstructure and chemical composition evolution, a line scan analysis of the main elements in the steel with the same cooling condition has been presented in Figure 7. To see these trends more clearly, the atomic % of each main element was used in this line scan. It is indicated that there is a main variation in the presence of Fe, Cr, and O at each specific location on the solidified surface, and the variation trend of Fe and Cr always corresponded with each other at the same location. This fact can lead to a basic understanding in that the 'bumpy' surface morphology formed is either due to the chemical element variation or precipitate and inclusion (i.e., oxide and nitride) formation in the bulk under the surface. Detailed considerations can be investigated further in future work.

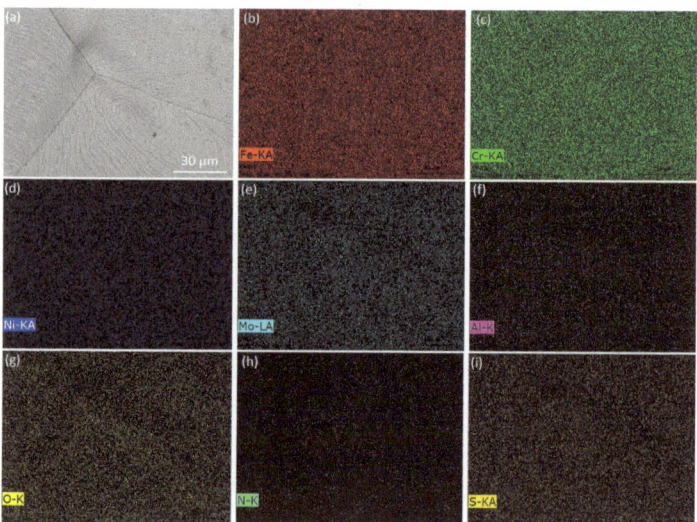

Figure 6. Typical morphology and chemical element maps of the S33207 steel after solidification with a cooling rate of 4 °C/min, (**a**) SEM images, and (**b–i**) chemical maps of the Fe, Cr, Ni, Mo, Mn, Al, O, N, and S elements, respectively.

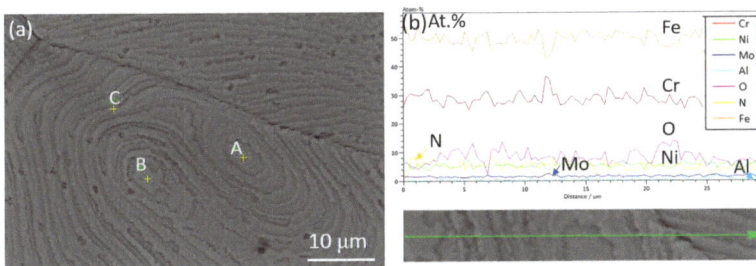

Figure 7. SEM image and line scan of the S33207 steel after solidification with a cooling rate of 4 °C/min, (**a**) morphology image, and (**b**) line scan of the Fe, Cr, Ni, Mo, Mn, Al, O, and N elements, respectively, A, B and C represent different selected locations for point analysis.

In order to view the detailed composition of all the elements presented as a mass %, the chemical content of positions A, B, and C in Figure 7a has been presented in Table 3. It is seen that the variation is rather small, since there is almost a single δ-ferrite phase observed after the solidification under a slow cooling rate. This finding is consistent with the above line scan and chemical mapping analysis.

Table 3. Chemical compositions of the different locations of the S33207 samples with a 4 °C/min cooling rate.

Point in Figure 7a	Chemical Elements (mass %)						
	O *	Al	Si	Cr	Ni	Mo	Fe
A	1.40	0.20	0.22	29.30	7.03	2.94	Bal.
B	1.65	0.04	0.54	30.85	6.02	3.08	Bal.
C	1.93	0.06	0.22	29.69	6.71	2.96	Bal.
Mean value	1.66	0.10	0.33	29.95	6.59	2.99	Bal.
Sigma	0.26	0.09	0.19	0.81	0.52	0.08	-
Sigma mean	0.15	0.05	0.11	0.47	0.30	0.04	-

* Light elements of O was measured using an EDS and was conducted as a quantitative analysis to confirm the identification of the phase including oxygen. The composition of the rest light elements e.g. N is not listed here.

In addition, many fine particles with a concave morphology on the solidified surface with a cooling rate of 4 °C/min were observed; the SEM-EDS line scan analysis result is shown in Figure 8. It is clearly seen that the particles are Al_2O_3; this is due to the re-oxidation reaction between the trace of Al in the matrix and the newly absorbed O on the sample surface. According to the HT-CLSM observation, it is not the non-metallic inclusion that formed in the liquid, and rather the formation could have been due to either surface re-oxidation or the undertaking of O to the sample surface.

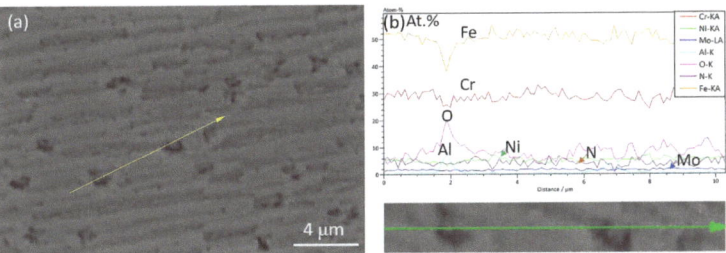

Figure 8. SEM image and the line scan of local particles on the S33207 steel surface with a cooling rate of 4 °C/min. (**a**) Morphology image. (**b**) Line scan of the Fe, Cr, Ni, Mo, Mn, Al, O, and N elements, respectively. Yellow and green arrows represent line scan area and direction.

3.2.2. Features on the Surface of the Specimen with a Fast Cooling Rate

The typical morphology of the steel surface after solidification with 150 °C/min has been presented in Figure 9. It is clearly seen that the dendrite microstructure can be observed. This microstructure was believed to be formed due to the fast cooling rate, and was not observed in the case with a slow cooling rate of 4 °C/min. In order to obtain the chemical composition of each location, SEM-EDS point analysis as well as a line scan was performed; the line scan result can be seen in Figure 10. It is seen that the variation of the γ-stabilized elements (e.g., Ni and N) and δ-stabilized elements (e.g., Cr) correspond with each other and were able to be distinguished between the austenite and ferrite phases. Figure 10c–j shows the chemical maps of different elements (i.e., Fe, Cr, Ni, Mo, Mn, N, Nb, and V). The concentration of the different elements in these maps was not so obvious, which can indicate that the difference in these elements at each location was not so large after solidification. Detailed EDS data of each point was summarized in Table 4; points B and C hold a relatively higher range of Cr and a lower range of Ni, which can indicative of the δ-ferrite phase. Alternatively, points A and D hold the higher Ni and lower Cr, which can be recognized as the γ austenite phase. Of note, in the solidified microstructure, the fraction of γ austenite was not really high, which has been confirmed by a separate work using electron backscatter diffraction (EBSD) [25]. This fact was deemed to be due to the loss in nitrogen during solidification, as well as lacking a post-heat treatment for austenite growth. Potential solution using a mixed Ar–N_2 gas instead of pure Ar and adding the isothermal ageing process at approximately 1000–1150 °C will favor the balance of the fractions of austenite and ferrite.

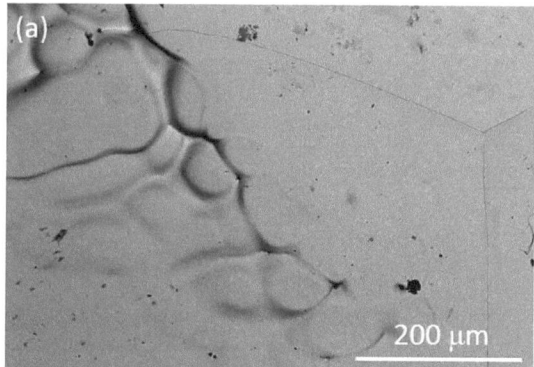 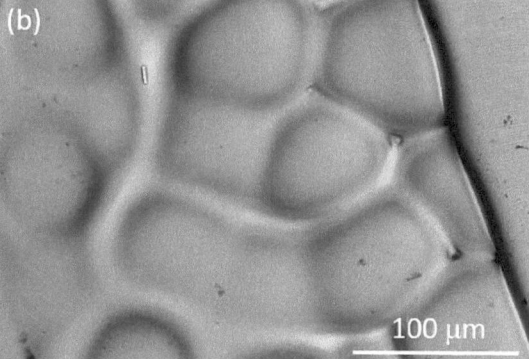

Figure 9. Typical morphology of S33207 after solidification with a cooling rate of 150 °C/min, with (**b**) displaying a magnification of a local place in (**a**).

Table 4. Chemical compositions (mass %) of the different locations of the S33207 samples after solidification.

Point in Figure 10a	Chemical Elements (mass %) *									
	Al	Si	Ti	V	Cr	Mn	Fe	Ni	Nb	Mo
A	0.02	0.27	0.02	0.03	32.51	0.34	Bal.	6.43	0.10	3.22
B	0.03	0.13	0.02	0.15	41.83	0.54	Bal.	5.59	0.06	3.99
C	0.07	0.24	0.01	0.02	35.55	0.46	Bal.	5.57	0.09	3.41
D	0.12	0.39	0.01	0.03	30.18	0.37	Bal.	6.93	0.16	3.00
Mean value	0.06	0.26	0.01	0.06	35.02	0.43	Bal.	6.13	0.10	3.41
Sigma	0.05	0.11	0.01	0.06	5.05	0.09	-	0.67	0.04	0.42
Sigma mean	0.02	0.05	0	0.03	2.52	0.05	-	0.33	0.02	0.21

* Light element of N cannot be detected quantitatively using an EDS, and so was not listed here.

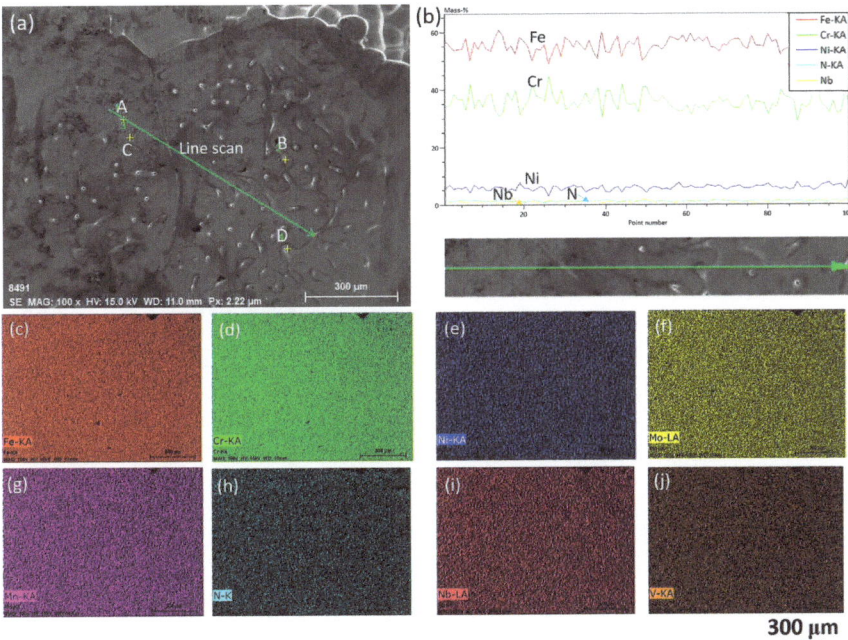

Figure 10. Typical morphology and chemical analysis of S33207 steel after solidification with a cooling rate of 150 °C/min. (**a**) SEM images. (**b**) Line scan result. (**c–j**) Chemical maps of the Fe, Cr, Ni, Mo, Mn, N, Nb, and V elements, respectively. The green arrow represents the line scan area and direction.

3.3. Comments on the Microstructure Evolution after Solidification

The nitrogen (N) content in the DSSs is one of the important elements used to determine the austenite formation temperature and fraction. Normally, the nitrogen was forced to be added into the DSSs in the liquid by e.g. pressure metallurgy. However, during the solidification process, and in the δ-ferrite phase, the solubility of N is much lower compared to the original amount, meaning the nitrogen will lose during in situ observations using Ar gas. In order to confirm this, the original DSS33207 alloy as well as the specimens after solidification with different rates are shown in Table 5; the data was measured by LECO analysis. It is seen that the N content after solidification was much lower than the original one (0.47%), which led to an unbalanced fraction between δ-ferrite and austenite. In fact, the loss of nitrogen during the melting and solidification of the DSSs is the actual limitation of the in situ observations of DSSs, and this process is inevitable. However, the obtained results are still quite useful and meaningful, since this simulates the actual case of DSS an as-cast ingot and heat-affected zone (HAZ) of welding, which also have a very low N content.

Table 5. Nitrogen content analysis in the specimens before and after solidification.

Condition	Original	Solidification with a CR of 4 °C/min	Solidification with a CR of 150 °C/min
N mass %	0.470	0.012	0.131

Thermodynamic calculations using Thermo-Calc 2023b with the TCFE12 database were used to calculate the phase diagram as well as the driving force of austenite in the DSS33207 with the increase in N content. The results are presented in Figures 11 and 12. In Figure 11, the stable phase region of austenite was much larger when N was relatively higher since N stabilizes the FCC phase. Furthermore, the normalized driving force of austenite also significantly increased the increasing N content, which can be seen in two typical temperatures of 1000 and 1200 °C, respectively, as shown in Figure 12. It is worth mentioning that the driving force value at 1000 °C is always higher than that at 1200 °C since lower temperatures always facilitate austenite formation.

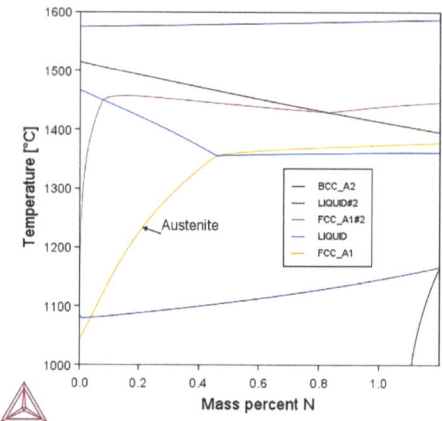

Figure 11. Phase diagram of DSS3207 with different N contents.

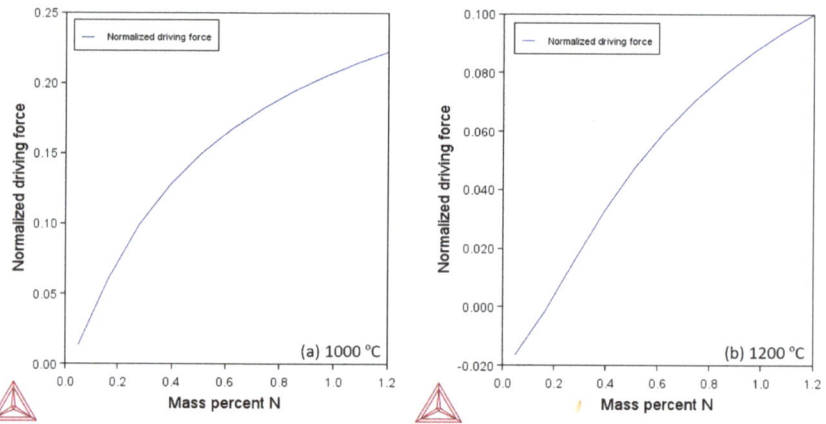

Figure 12. Driving force of austenite formation from the ferrite matrix at (**a**) 1000 °C and (**b**) 1200 °C, respectively.

Based on these calculations, it is known that the loss of N in the DSS3207 will lead to a decrease in the austenite fraction in the matrix. A typical microstructure of the as-received and solidified (fast cooling with a 150 °C/min) DSS33207 is displayed Figure 13. In Figure 13a, austenite (points A and B) presents the bright phase while ferrite (points C and D) presents as the dark matrix; the detailed composition of each phase is shown in Table 6. Chemical compositions of N and C are calculated by Thermo-Calc, and the rest was measured using an EDS. It is shown how N and Ni are obviously higher in the austenite phase than that in the ferrite phase, making it easy to identify the FCC structure. Also, a balanced mixture of each phase is shown in Figure 13a. In the solidified microstructure, it is almost the single ferrite phase since the polygonal morphology grain size can be seen, and the austenite can still be recognized due to the contrast difference. According to the unpublished work by the authors, the fraction of austenite in the vertical section of the DSS sample after solidification was around 5%, measured by EBSD. Points E and G represent the ferrite and austenite in the solidified sample. The composition of each phase was almost the same as the one in the as-received sample. Point F seems to be a mixture of both phases since the morphology of the newly formed austenite after solidification seems to resemble a tiny island morphology [25].

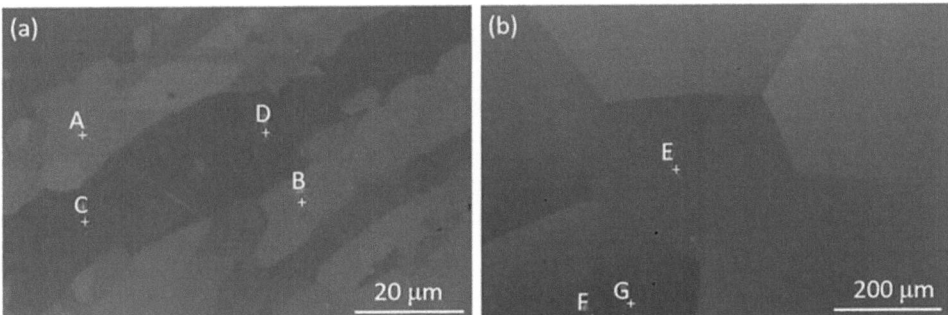

Figure 13. SEM images of the cross-section of the (**a**) as-received and (**b**) solidified DSS33207. Points A to G represent the different locations which are checked by point analysis.

Table 6. Chemical analysis of austenite and ferrite in the as-received and solidified DSS33207.

Points	Phase	Chemical Elements (mass %)						
		N *	C *	Si	Cr	Ni	Mo	Fe
A	FCC	0.831	0.023	0.28	30.49	7.93	2.54	Bal.
B	FCC			0.55	30.26	8.17	2.18	Bal.
Mean value	-	0.831	0.023	0.26	30.37	8.05	2.36	Bal.
Sigma	-	-	-	0.02	0.16	0.17	0.25	Bal.
Sigma mean	-	-	-	0.02	0.12	0.12	0.18	Bal.
C	BCC	0.105	0.008	0.31	31.02	5.91	3.60	Bal.
D	BCC			0.29	30.59	6.02	3.25	Bal.
Mean value	-	0.105	0.008	0.30	30.80	5.96	3.42	Bal.
Sigma	-	-	-	0.01	0.31	0.08	0.24	Bal.
Sigma mean	-	-	-	0.01	0.22	0.06	0.17	Bal.
E	BCC	0.105	0.008	0.26	31.45	6.22	2.66	Bal.
F	Mixed	-	-	0.32	30.30	7.05	3.34	Bal.
G	FCC	0.831	0.023	0.25	30.06	8.20	2.90	Bal.
Mean value	-	-	-	0.27	30.06	7.16	2.97	Bal.
Sigma	-	-	-	0.04	0.74	0.99	0.35	Bal.
Sigma mean	-	-	-	0.02	0.43	0.57	0.20	Bal.

* Content calculated by Thermo-Calc.

The obtained finding regarding the unbalanced microstructure is the normal case when using Ar as the protection gas during solidification. One potential solution to fix this is adding external thermal aging at the austenite temperature to trigger the second austenite formation [2]. Another solution is mixing N_2 with Ar to use in the HT-CLSM observation, which will supply the N loss; however, a reverse problem leading to N_2 being undertaken may occur. It has been reported that when using N_2 as the protection gas, the DSS changes to become an almost single FCC phase steel [32]. In this case, the ratio of mixing N_2 with Ar needs to be optimized in future work.

4. Conclusions

In this study, the effects of the cooling rate on the solidification microstructure of the hyper duplex stainless steel S33207 was assessed using a HT-CLSM. The solidification process of S33207 steel and the formation and growth characteristics of the δ-ferrite phase were clarified. The main conclusions are as follows: a higher cooling rate leads to a greater undercooling and lower solidification temperatures of ferrite. As the temperature decreased, the liquid phase first solidified into δ-ferrite, and after solidification is complete, the δ-ferrite would remain for a while and then partially transform into austenite. The area fraction of δ-ferrite during solidification of S33207 steels at a cooling rate of 4 °C/min can be expressed as $f_{\delta,\text{slow}} = 1 - \exp(-7.4 \times 10^{-8} \times t^{3.4})$, while a cooling rate of 150 °C/min can be expressed as $f_{\delta,\text{fast}} = 1 - \exp(-1.2 \times 10^{-3} \times t^{1.9})$, respectively. A slow cooling rate favored the growth of δ-ferrite, whereas a high cooling rate favored the nucleation of δ-ferrite. These findings could provide guidelines for the solidification control of advanced hyper duplex stainless steel production. In addition, the microstructure evolution after solidification was briefly mentioned, and suggestions to increase the fraction of austenite have been provided.

Author Contributions: Conceptualization, W.M.; data curation, Y.W.; methodology, investigation, preparation for the original draft, Y.W. and W.M.; manuscript review, revision and editing, funding acquisition, project administration and submission, W.M. All authors have read and agreed to the published version of the manuscript.

Funding: W.M. would like to acknowledge the Swedish Foundation for International Cooperation in Research and Higher Education (STINT, Project No. PT2017-7330 & IB2022-9228), VINNOVA (No. 2022-01216), SSF Strategic Mobility Grant (SM22-0039), the Swedish Steel Producers' Association (Jernkontoret), and in particular, Axel Ax:-son Johnsons forskningsfond, Prytziska fonden nr 2, and Gerhard von Hofstens Stiftelse för Metallurgisk forskning for the financial support. Y. Wang would like to acknowledge the National Natural Science Foundation of China (No. U21A20113 and 52074198) and the Japan Society for the Promotion of Science (JSPS) for supporting his research.

Data Availability Statement: The data presented in this study are available on request from the corresponding author.

Acknowledgments: Hiroyuki Shibata and Sohei Sukenaga from IMRAM, Tohoku University (Japan) are acknowledged for supporting the HT-CLSM measurement.

Conflicts of Interest: The authors declare no conflict of interest.

References

1. Qu, H.; Hou, H.; Li, P.; Li, S.; Ren, X. The effect of thermal cycling in superplastic diffusion bonding of heterogeneous duplex stainless steel. *Mater. Des.* **2016**, *96*, 499–505.
2. Pettersson, N.; Wessman, S.; Hertzman, S.; Studer, A. High-temperature phase equilibria of duplex stainless steels assessed with a novel in-situ neutron scattering approach. *Metall. Mater. Trans. A* **2017**, *48*, 1562–1571. [CrossRef]
3. Tucker, J.; Miller, M.K.; Young, G.A. Assessment of thermal embrittlement in duplex stainless steels 2003 and 2205 for nuclear power applications. *Acta Mater.* **2015**, *87*, 15–24.
4. Karahan, T.; Emre, H.E.; Tümer, M.; Kacar, R. Strengthening of AISI 2205 duplex stainless steel by strain ageing. *Mater. Des.* **2014**, *55*, 250–256. [CrossRef]
5. Gholami, M.; Hoseinpoor, M.; Moayed, M.H. A statistical study on the effect of annealing temperature on pitting corrosion resistance of 2205 duplex stainless steel. *Corros. Sci.* **2015**, *94*, 156–164.
6. Pan, J. Studying the passivity and breakdown of duplex stainless steels at micrometer and nanometer scales–the influence of microstructure. *Front. Mater.* **2020**, *7*, 133.
7. Gopal, M.; Gutema, E.M. Factors affecting and optimization methods used in machining duplex stainless steel-a critical review. *J. Eng. Sci. Technol. Rev.* **2021**, *14*, 119–135. [CrossRef]
8. Kazakov, A.A.; Zhitenev, A.I.; Fedorov, A.S.; Fomina, O.V. Development of duplex stainless steels Compositions. *CIS Iron Steel Rev.* **2019**, *18*, 20–26.
9. Xu, X.Q.; Zhao, M.; Feng, Y.R.; Li, F.G.; Zhang, X. A Comparative Study of Critical Pitting Temperature (CPT) of Super Duplex Stainless Steel S32707 in NaCl Solution. *Int. J. Electrochem. Sci* **2018**, *13*, 4298–4308.
10. Nilsson, J.O. Super duplex stainless steels. *Mater. Sci. Technol.* **1992**, *8*, 685–700.

11. Petrovič, D.S.; Pirnat, M.; Klančnik, G.; Mrvar, P.; Medved, J. The effect of cooling rate on the solidification and microstructure evolution in duplex stainless steel: A DSC study. *J. Therm. Anal. Calorim.* **2012**, *109*, 1185–1191. [CrossRef]
12. Chen, L.; Tan, H.; Wang, Z.; Li, J.; Jiang, Y. Influence of cooling rate on microstructure evolution and pitting corrosion resistance in the simulated heat-affected zone of 2304 duplex stainless steels. *Corros. Sci.* **2012**, *58*, 168–174. [CrossRef]
13. Cronemberger, M.E.R.; Nakamatsu, S.; Della Rovere, C.A.; Kuri, S.E.; Mariano, N.A. Effect of cooling rate on the corrosion behavior of as-cast SAF 2205 duplex stainless steel after solution annealing treatment. *Mater. Res.* **2015**, *18*, 138–142. [CrossRef]
14. Zhu, C.; Zeng, J.; Wang, W.; Chang, S.; Lu, C. Mechanism of $\delta \rightarrow \delta + \gamma$ phase transformation and hardening behavior of duplex stainless steel via sub-rapid solidification process. *Mater. Charact.* **2020**, *170*, 110679. [CrossRef]
15. Mu, W.; Mao, H.; Jönsson, P.G.; Nakajima, K. Effect of carbon content on the potency of the intragranular ferrite formation. *Steel Res. Int.* **2016**, *87*, 311–319. [CrossRef]
16. Shibata, H.; Arai, Y.; Suzuki, M.; Emi, T. Kinetics of peritectic reaction and transformation in Fe-C alloys. *Metall. Mater. Trans. B* **2000**, *31*, 981–991. [CrossRef]
17. Wang, Y.; Wang, Q.; Mu, W. In Situ Observation of Solidification and Crystallization of Low-Alloy Steels: A Review. *Metals* **2023**, *13*, 517.
18. Wu, C.; Li, S.; Zhang, C.; Wang, X. Microstructural evolution in 316LN austenitic stainless steel during solidification process under different cooling rates. *J. Mater. Sci.* **2016**, *51*, 2529–2539. [CrossRef]
19. Li, X.; Gao, F.; Jiao, J.; Cao, G.; Wang, Y.; Liu, Z. Influences of cooling rates on delta ferrite of nuclear power 316H austenitic stainless steel. *Mater. Charact.* **2021**, *174*, 111029. [CrossRef]
20. Li, Y.; Zou, D.; Chen, W.; Zhang, Y.; Zhang, W.; Xu, F. Effect of cooling rate on solidification and segregation characteristics of 904L super austenitic stainless steel. *Met. Mater. Int.* **2022**, *28*, 1907–1918. [CrossRef]
21. Sun, Y.; Zhao, Y.; Li, X.; Jiao, S. Effects of heating and cooling rates on $\delta \leftrightarrow \gamma$ phase transformations in duplex stainless steel by in situ observation. *Ironmak. Steelmak.* **2019**, *46*, 277–284. [CrossRef]
22. Wang, T. The Effects of Cooling Rate and Alloying Elements on the Solidification Behaviour of Continuously Cast Super-Austenitic and Duplex Stainless Steels. Master's Thesis, University of Wollongong, Wollongong, Australia, 2019.
23. Zhao, Y.; Sun, Y.; Li, X.; Song, F. In-situ observation of $\delta \leftrightarrow \gamma$ phase transformations in duplex stainless steel containing different nitrogen contents. *ISIJ Int.* **2017**, *57*, 1637–1644. [CrossRef]
24. Shin, B.H.; Park, J.; Jeon, J.; Heo, S.B.; Chung, W. Effect of cooling rate after heat treatment on pitting corrosion of super duplex stainless steel UNS S 32750. *Anti-Corros. Methods Mater.* **2018**, *65*, 492–498. [CrossRef]
25. Wang, Y.; Sukenaga, S.; Shibata, H.; Wang, Q.; Mu, W. Combination of in-situ confocal microscopy and calorimetry to investigate solidification of super and hyper duplex stainless steels. *Steel Res. Int.* **2023**, in press.
26. Mu, W.; Shibata, H.; Hedström, P.; Jönsson, P.G.; Nakajima, K. Combination of in situ microscopy and calorimetry to study austenite decomposition in inclusion engineered steels. *Steel Res. Int.* **2016**, *87*, 10–14. [CrossRef]
27. Shi, X.; Duan, S.C.; Yang, W.S.; Guo, H.J.; Guo, J. Effect of cooling rate on microsegregation during solidification of superalloy INCONEL 718 under slow-cooled conditions. *Metall. Mater. Trans. B* **2018**, *49*, 1883–1897. [CrossRef]
28. Liang, G.; Wan, C.; Wu, J.; Zhu, G.; Yu, Y.; Fang, Y. In situ observation of growth behavior and morphology of delta-ferrite as function of solidification rate in an AISI304 stainless steel. *Acta Metall. Sin.* **2006**, *19*, 441–448. [CrossRef]
29. Wang, T.; Wexler, D.; Guo, L.; Wang, Y.; Li, H. In situ observation and phase-field simulation framework of duplex stainless-steel slab during solidification. *Materials* **2022**, *15*, 5517. [CrossRef] [PubMed]
30. Christian, J.W. *The Theory of Transformations in Metals and Alloys*; Newnes: Oxford, UK, 2002.
31. Bruna, P.; Crespo, D.; González-Cinca, R.; Pineda, E. On the validity of Avrami formalism in primary crystallization. *J. Appl. Phys.* **2006**, *100*, 054907. [CrossRef]
32. Wang, T.; Phelan, D.; Wexler, D.; Qiu, Z.; Cui, S.; Franklin, M.; Guo, L.; Li, H. New insights of the nucleation and subsequent phase transformation in duplex stainless steel. *Mater. Char.* **2023**, *203*, 113115. [CrossRef]

Disclaimer/Publisher's Note: The statements, opinions and data contained in all publications are solely those of the individual author(s) and contributor(s) and not of MDPI and/or the editor(s). MDPI and/or the editor(s) disclaim responsibility for any injury to people or property resulting from any ideas, methods, instructions or products referred to in the content.

Article

Effects of Ag on High-Temperature Creep Behaviors of Peak-Aged Al-5Cu-0.8Mg-0.15Zr-0.2Sc(-0.5Ag)

Ying Wang [1,2], Ge Zhou [1,2], Xin Che [1,2], Feng Li [1,2] and Lijia Chen [1,2,*]

[1] School of Materials Science and Engineering, Shenyang University of Technology, Shenyang 110870, China
[2] Shenyang Key Laboratory of Advanced Structural Materials and Applications, Shenyang University of Technology, Shenyang 110870, China
* Correspondence: chenlj-sut@163.com

Abstract: The tensile creep of Al-5Cu-0.8Mg-0.15Zr-0.2Sc(-0.5Ag) was tested at 150–250 °C and 125–350 MPa, and the effect of Ag on the high-temperature creep of Al-Cu-Mg alloys was discussed. After the addition of Ag, the high-temperature creep performances of the alloy were significantly improved at 150 °C/300 MPa and 200 °C/(150 MPa, 175 MPa). Then, constitutive relational models of the alloy during high-temperature creep were built, and the activation energy was calculated to be 136.65 and 104.06 KJ/mol. Based on the thermal deformation mechanism maps, the high-temperature creep mechanism of the alloy was predicted. After the addition of Ag, the creep mechanism of the alloy at 150 °C transitioned from lattice diffusion control to grain boundary diffusion control. At 250 °C, the mechanism was still controlled by grain boundary slip, but as the stress index increased and after Ag was added, the alloy fractures lead to the formation of dimples, thus improving the high-temperature creep performance.

Keywords: Al-Cu-Mg-Ag alloy; high-temperature creep; ageing; constitutive relational model of creep; deformation mechanism map

Citation: Wang, Y.; Zhou, G.; Che, X.; Li, F.; Chen, L. Effects of Ag on High-Temperature Creep Behaviors of Peak-Aged Al-5Cu-0.8Mg-0.15Zr-0.2Sc(-0.5Ag). *Crystals* **2023**, *13*, 1096. https://doi.org/10.3390/cryst13071096

Academic Editor: Indrajit Charit

Received: 13 June 2023
Revised: 3 July 2023
Accepted: 6 July 2023
Published: 13 July 2023

Copyright: © 2023 by the authors. Licensee MDPI, Basel, Switzerland. This article is an open access article distributed under the terms and conditions of the Creative Commons Attribution (CC BY) license (https://creativecommons.org/licenses/by/4.0/).

1. Introduction

Heat-resistant 2xxx Al alloys (e.g., 2124, 2219 and 2618 alloys) are extensively applied in aerospace, due to their upper lightness and high heat resistance [1–4]. As the design requirements for aerospace and aircraft are progressively increased, the required operation temperature of Al alloys is increased accordingly. For this reason, researchers have devoted their energies to improving the heat resistance of Al-Cu-Mg Al alloys.

As for the design of material composition, the second phase in the balanced Al-Cu-Mg alloys mainly consists of θ, S and T phases [5]. The precipitate-phase precipitation series can be altered by regulating the Cu/Mg ratio of Al-Cu-Mg Al alloys and through ageing treatment. In other words, at a Cu/Mg ratio of >8, the main intensified phase is the θ' phase. At a Cu/Mg ratio of 4–8, the main intensified phases are the θ' and S' phases. The main intensified phase is the S' phase at a Cu/Mg ratio of 1.5–4. The coherent or semicoherent precipitate phases formed above can more effectively improve the alloy strength, and the S' phase can strengthen the heat resistance of alloys. However, when the operation temperature is above 150 °C, the mechanical properties of the alloys are significantly weakened, due to the coarsening of intensified phases, which are unable to meet the requirements for key components in aerospace aircraft. For this reason, Ag is added to high-Cu/Mg-ratio Al-Cu-Mg alloys, in order to alter the ageing precipitate series. As a result, the Ω phase—which is consistently below 200 °C and does not aggregate or grow upwards—in minimally sized and dispersed distribution is dispersed, thus improving the room-temperature or high-temperature strength of the alloys and increasing their thermal stability [6,7]. As for research on the creep behaviors of aged alloys, the existing findings on the Ag and Mg distributions of aged Al-Cu-Mg-Ag alloys are listed in Table 1 [8–13]. As has previously been reported, Al-Cu-Mg-Ag alloys exhibit outstanding creep performance at high temperature, and the steady-state creep rate at 125 °C/265 MPa is 1.6×10^{-5} h^{-1}, which is lower than the 2.8×10^{-5} h^{-1} of Al-Cu-Mg alloys at the same conditions [14–16]. The steady-state creep rate of Al-Cu-Mg alloys at 150 °C/265 MPa is 3.9×10^{-4} h^{-1}, which is higher than the 1.3×10^{-4} h^{-1} of Al-Cu-Mg-Ag alloys. The creep performances of Al-Cu-Mg-Ag alloys after different ageing treatments have been studied. The steady-state creep rate of under-aged alloys (185 °C, 2 h) was 3.5×10^{-10} s^{-1}, and that of peak-aged alloys (185 °C, 10 h) was 1.12×10^{-9} s^{-1}. The above results indicate that composition and ageing treatment critically affect the creep properties of alloys.

Table 1. Distribution of Mg and Ag atoms in Al-Cu-Mg-Ag alloys after ageing [8–13].

Alloy Composition, wt.%	Heat Treatment Condition	Ag or Mg Detected Inside Ω	Ag or Mg Detected at Ω/Matrix Interface
Al-4Cu-0.3Mg-0.4Ag	200 °C, 2 h, 10 h	(Ag)	(Ag)
Al-6Cu-0.45Mg-0.5Ag-0.5Mn-0.14Zr	Air cool from 500 °C	No	Ag, (Mg)
Al-4Cu-0.3Mg-0.4Ag	170 °C, 24 h	Mg, Ag	No evidence of pref. Seg
Al-4Cu-0.3Mg-0.4Ag	170 °C, 24 h	Mg, Ag	No evidence of pref. Seg
Al-6Cu-0.45Mg-0.5Ag-0.5Mn-0.14Zr	190 °C, 8 h	No	Ag, Mg
Al-4.3Cu-0.3Mg-0.8Ag	190 °C, 2 h, 8 h	No	Ag, Mg
Al-4Cu-0.5Mg-0.45Ag	250 °C, 6 min	Mg	Ag, Mg
Al-3.9Cu-0.3Mg-0.4Ag	200 °C, 1000 h	No	Ag, Mg
Al-4.3Cu-0.3Mg-0.8Ag	180 °C, 2 h, 10 h	No	Ag, Mg

Generally, the creep behaviors of metal materials under service conditions decide the service life of alloys [17], but the existing literature on the structural performance control of Al-Cu-Mg alloys has focused primarily on three aspects [18–26]: (1) effect of Ag on microstructures of alloys; (2) effect of ageing on room-temperature or high-temperature mechanical properties of alloys; (3) creep behaviors under deformation and thermal treatment cooperative control. However, thus far, no researcher has added rare earth elements, Sc or Ag, to Al-Cu-Mg ternary alloys to prepare peak-aged structures after ageing treatment, and studied the creep behaviors of this peak-aged alloy system. According to the analysis of existing research results, it is not sufficient to improve the high-temperature creep performance of an Al-Cu-Mg alloy by adjusting its composition alone and combining it with heat treatment methods. If a small amount of Zr, Sc and Ag are added on the basis of the alloy system, combined with aging treatment to give full reign to the peak-aging effect, the high-temperature mechanical properties and high-temperature creep properties of the alloy can be significantly improved. In this study, Al-5Cu-0.8Mg-0.15Zr-0.2Sc(-0.5Ag) alloy has been prepared, and the microstructures and mechanical properties of peak-aged alloys after ageing treatment have been obtained. In particular, the effects of Ag on the creep properties of peak-aged alloys at different temperatures and stresses were explored. Constitutive relational models of creep were also built, and the deformation mechanism map involving dislocation quantity was plotted. Together with microstructure characterization, the high-temperature creep fracture mechanism of Al-5Cu-0.8Mg-0.15Zr-0.2Sc(-0.5Ag) was uncovered. Thereby, the findings provide a reliable theoretical basis for the practical aerospace applications of this alloy system.

2. Materials and Methods

High-purity Al, Mg, Sc and Ag of industrial purity, and Al-Cu, Al-Mn and Al-Zr intermediate alloys were used to prepare Al-5Cu-0.8Mg-0.15Zr-0.2Sc(-0.5Ag) alloy cast ingots (wt%) via cast-ingot metallurgy. The composition of the alloy was listed in Table 2. Then, the cast ingots were homogenized at 500 °C for 24 h, and then extruded at 430 °C into bars in diameter of 20 mm. The bars were subjected to solution treatment and ageing in an SX-4-10-box-type resistance furnace. The solution treatment was conducted at 510 °C, with heat preservation for 2 h followed by water-cooling. The ageing treatment was conducted at 180 °C for 2–10 h, followed by air cooling. The peak-aged Al-5Cu-0.8Mg-0.2Sc-0.15Zr(-0.5Ag) alloy under extruded deformation was subjected to static tensile creep experiments under constant load in an SRD-100-microcomputer-controlled electronic creep tester. The experiment conditions were 150 °C/300 MPa, 150 °C/325 MPa, 150 °C/350 MPa; 200 °C/150 MPa, 200 °C/175 MPa, 200 °C/200 MPa; 250 °C/100 MPa, 250 °C/125 MPa, 250 °C/150 MPa. All static tensile creep experiments were conducted until the specimens fractured. The corresponding time of fracture upon creep was considered as the creep fracture life of the alloy. The microstructures of the Al alloys after different thermal treatments were observed and analyzed under an Axio Observer A1m optical microscope, and the corrosion reagent was 2.5%HNO$_3$ + 1.5%HCl + 0.5%HF. The microstructures of the alloys in the creep deformation zone after different thermal treatments were observed under a JEM-2100 transmission electron microscope (TEM). The electrolytic solution was 30%HNO$_3$ + 70%CH$_3$OH and controlled around −30 °C, and the voltage was 21 V. The fracture morphology of the creep-fractured samples was observed and analyzed with an S-3400N scanning electron microscope (SEM).

Table 2. Chemical composition of alloys (wt.%).

Alloys	Cu	Mg	Zr	Sc	Ag	Al
Al-Cu-Mg-Sc-Zr	5	0.8	0.15	0.2	0	Bal.
Al-Cu-Mg-Sc-Zr-Ag	5	0.8	0.15	0.2	0.5	Bal.

3. Results

3.1. Effects of Ageing on Mechanical Properties of Al-5Cu-0.8Mg-0.15Zr-0.2Sc(-0.5Ag) Alloy

Figure 1 shows the mechanical performance curves of Al-5Cu-0.8Mg-0.15Zr-0.2Sc(-0.5Ag) after certain duration of ageing. Clearly, after certain duration of ageing treatment, the room-temperature tensile strength of Al-5Cu-0.8Mg-0.15Zr-0.2Sc was enhanced after the addition of Ag (Figure 1a), indicating that Ag can effectively improve the ageing strengthening effect on alloys. After certain ageing time, the above alloys showed an evident ageing strengthening effect, which was mainly divided into three stages: under-ageing, peak ageing, and over-ageing. At the under-ageing stage, the tensile strength was enhanced with the prolonged ageing time, and peak ageing occurred at 8th h when the tensile strength was maximized to 468.2 MPa (Al-5Cu-0.8Mg-0.15Zr-0.2Sc-0.5Ag) and 455.2 MPa (Al-5Cu-0.8Mg-0.15Zr-0.2Sc). At the over-ageing stage, the tensile strength was slightly weakened with the prolonged ageing time. The elongation at the break of the under-aged Al-5Cu-0.8Mg-0.15Zr-0.2Sc(-0.5Ag) was substantial, but the elongation at the break of the over-aged alloy was significantly weakened (Figure 1b).

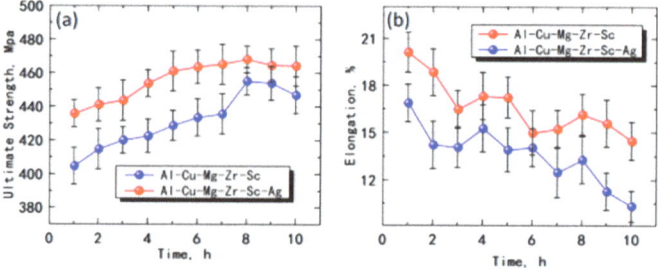

Figure 1. Effects of ageing time on room temperature mechanical properties of Al-5Cu-0.8Mg-0.15Zr-0.2Sc(-0.5Ag) (**a**) tensile strength; (**b**) elongation at break.

Figure 2 shows the microstructure of an Al-5Cu-0.8Mg-0.15Zr-0.2Sc(-0.5Ag) alloy in under-aged (4 h), peak-aged (8 h) and over-aged (10 h) states. It can be seen from the figure that the precipitate phase exists both inside the grain and at the grain boundaries of the Al-5Cu-0.8Mg-0.15Zr-0.2Sc(-0.5Ag) alloy in the solution and aging state. At the same aging time, the grains of the two alloys are equiaxed, and the grain size of Al-5Cu-0.8Mg-0.15Zr-0.2Sc-0.5Ag alloys is relatively small. The intercept method can be used to calculate the average grain size under different holding times in Figure 2, which is about 15 μm–22 μm. The grain size first increases then decreases with the increase in temperature, and the amount of precipitate phase in the alloy is significantly higher than that of Al-5Cu-0.8Mg-0.15Zr-0.2Sc alloy. At peak aging (8 h), the precipitate phase quantity of both alloys showed an increasing trend, and at over-aging (10 h), the precipitate phase quantity of both alloys showed a decreasing trend. The TEM electron microscopic analysis was performed on the precipitate phase in Figure 2, as shown in Figure 3.

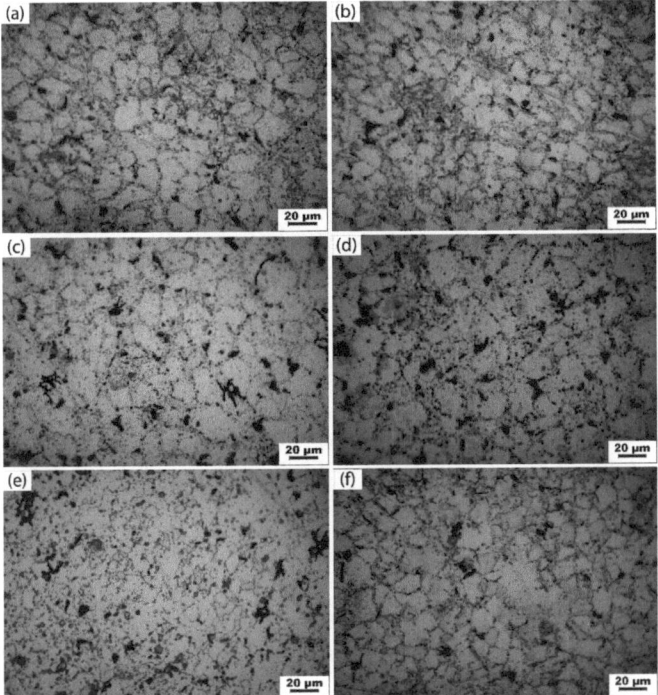

Figure 2. Microstructures of Al-5Cu-0.8Mg-0.15Zr-0.2Sc(-0.5Ag) alloys subjected to aging treatment for various durations (**a,c,e**): Al-5Cu-0.8Mg-0.15Zr-0.2Sc at 4, 8, 10 h; (**b,d,f**) Al-5Cu-0.8Mg-0.15Zr-0.2Sc-0.5Ag at 4, 8, 10 h.

It can be seen from the figure that a large amount of precipitate phase was produced in the grain of the peak-aged Al-5Cu-0.8Mg-0.15Zr-0.2Sc(-0.5Ag) alloy. According to the characterization results of Figure 3a,b, the precipitate phase with a rod-like shape is θ phase ($CuAl_2$). The precipitate phase is an Al3Sc phase, and no S phase ($CuMgAl_2$) was found. The Al_2Cu phase preferentially aggregates at the grain boundaries, which is consistent with the results obtained in the studies of Al-Cu-Mg-Ag series alloys [24–28]. During the aging process, in addition to the formation of Al_2Cu precipitates inside the grains, there are also intermittent distributions of precipitates at the grain boundaries, and precipitation-free zones (PFZ) appear near the grain boundaries, as shown in Figure 3c,d. The addition of the Ag element slightly increased the size of the precipitate phase at the grain boundaries of Al-5Cu-0.8Mg-0.15Zr-0.2Sc alloy, and the width of the precipitated band at the grain boundary decrease. Therefore, the peak aging state Al-5Cu-0.8Mg-0.15Zr-0.2Sc(-0.5Ag) alloy with good comprehensive mechanical properties was selected to study the creep behavior at high temperatures.

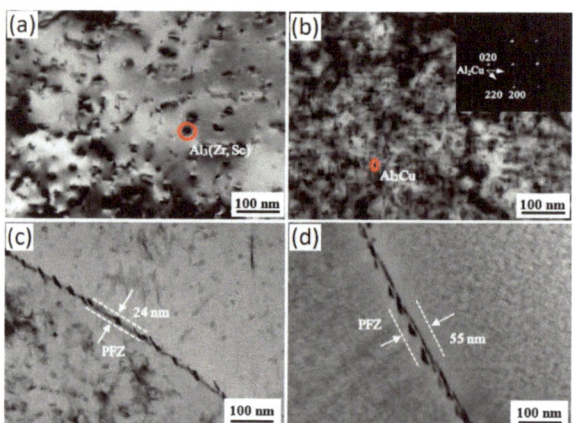

Figure 3. TEM microstructure images of Al-5Cu-0.8Mg-0.15Zr-0.2Sc(-0.5Ag) after 8 h of ageing. (**a**,**b**) Intracrystalline precipitate phase of Al-5Cu-0.8Mg-0.15Zr-0.2Sc(-0.5Ag); (**c**,**d**) no precipitate band on crystal boundary of Al-5Cu-0.8Mg-0.15Zr-0.2Sc(-0.5Ag).

3.2. Effects of Ag on Tensile Creep Performance of Peak-Aged Al-5Cu-0.8Mg-0.15Zr-0.2Sc(-0.5Ag)

Figure 4 demonstrates the tensile creep curves of peak-aged Al-5Cu-0.8Mg-0.2Sc-0.15Zr(-0.5Ag) at 150, 200, 250 °C and external stresses of 300, 325, 350 MPa. After the addition of Ag, the creep life of the alloys was prolonged (Figure 4a) and in particular, the creep life was significantly extended at the external stress of 300 or 325 MPa. This was because after Ag was added into Al-5Cu-0.8Mg-0.2Sc-0.15Zr, the content of the Ω phase (Al_2Cu) rose in the peak ageing stage achieved after long-time ageing, and the alloy showed high thermal stability. The second-phase strengthening effect was significant during the creep, but hindered dislocation motion, forming a dislocation pile-up cluster that prevented creep. As a result, the creep speed was decelerated, and the stable creep stage was extended, thus prolonging the creep life of the alloys. When the external stress was increased to 350 MPa, which was large, the dislocation motion was aggravated, and the dislocations were easily removed from the constraint by the second phase. As a result, the creep rate gradually increased and the steady-state creep stage was significantly shortened, so the creep life was not significantly prolonged. Figure 4b demonstrates the tensile creep curves of peak-aged Al-5Cu-0.8Mg-0.2Sc-0.15Zr(-0.5Ag) at 200 °C and different external stresses. Clearly, the creep life was significantly prolonged after the addition of Ag, and the creep life of Al-5Cu-0.8Mg-0.2Sc-0.15Zr-0.5Ag at 200 °C/200MPa was longer than that of Al-5Cu-0.8Mg-0.2Sc-0.15Zr at 200 °C/150 MPa. This was because with the temperature rise, the atom and cavity diffusion speeds were accelerated, which facilitated dislocation motion and creep deformation. With the presence of Ag, the peak-aged alloys separated out abundant Ω phase, which pinned and inhibited the dislocations. At the early creep stage, the dislocation motion was intense, and the Ω phase hindered dislocation motion, thus decelerating creep speed. At the steady-state creep stage, the Ω phase grew upwards, but this process depended on the diffusion and redistribution of Cu, Mg and Ag. At 200 °C, however, the above atom diffusion and redistribution were limited, so the Ω phase growing speed was slow, which thereby pinned the dislocation and crystal boundaries and inhibited the increase in steady-state creep speed. Figure 4c shows the tensile creep curves of peak-aged Al-5Cu-0.8Mg-0.2Sc-0.15Zr(-0.5Ag) at 250 °C and external stresses of 100, 125 and 150 MPa. Clearly, the presence of Ag prolonged the steady-state creep stage and extended the creep life of the alloys to an extent. In comparison with 150 or

200 °C, however, the creep life of alloys was significantly shortened after the addition of Ag. The reason for this was that although the newly added Ag formed abundant thermally stable Ω phase, the Ω phase grew upwards after the temperature rise, so the second phase intensifying was weakened. Moreover, dislocation motion was intensified. These changes contributed to creep. Hence, at this temperature, the effect of Ag on prolonging the creep life of alloys was limited.

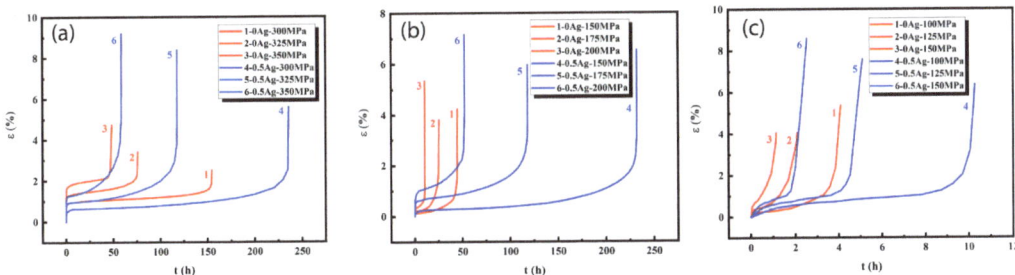

Figure 4. Tensile creep curves of peak-aged Al-5Cu-0.8Mg-0.15Zr-0.2Sc(-0.5Ag) at different temperatures (**a**) 150 °C, (**b**) 200 °C, (**c**) 250 °C.

The high-temperature tensile creep curves of peak-aged Al-5Cu-0.8Mg-0.2Sc-0.15Zr(-0.5Ag) under different conditions at the steady-state creep stage were linearly fitted, and the steady-state creep speeds can be calculated thereby (Table 3).

Table 3. Steady-state creep rates of peak-aged Al-5Cu-0.8Mg-0.15Zr-0.2Sc(-0.5Ag) under the same conditions.

	Al-5Cu-0.8Mg-0.15Zr-0.2Sc		
Temperatures/°C	Stress/MPa		
	300	325	350
150	1.85×10^{-8} s^{-1}	4.61×10^{-8} s^{-1}	8.31×10^{-8} s^{-1}
Temperatures/°C	Stress/MPa		
	150	175	200
200	4.85×10^{-8} s^{-1}	9.89×10^{-8} s^{-1}	1.67×10^{-7} s^{-1}
Temperatures/°C	Stress/MPa		
	100	100	100
250	1.60×10^{-6} s^{-1}	3.63×10^{-6} s^{-1}	6.18×10^{-6} s^{-1}
	Al-5Cu-0.8Mg-0.15Zr-0.2Sc-0.5Ag		
Temperatures/°C	Stress/MPa		
	300	325	350
150	2.52×10^{-8} s^{-1}	6.19×10^{-8} s^{-1}	1.59×10^{-7} s^{-1}
Temperatures/°C	Stress/MPa		
	150	175	200
200	2.16×10^{-8} s^{-1}	2.87×10^{-8} s^{-1}	1.24×10^{-7} s^{-1}
Temperatures/°C	Stress/MPa		
	100	100	100
250	6.39×10^{-7} s^{-1}	2.22×10^{-6} s^{-1}	1.72×10^{-6} s^{-1}

At the same temperature, the steady-state creep speed of Al-5Cu-0.8Mg-0.15Zr-0.2Sc(-0.5Ag) rose in tandem with external stress, but that of the alloys increased alongside the creep temperature and the drop of external stress, and the alloys were very sensitive to temperature (Table 2). The creep properties of alloys before and after the addition of 0.5 Ag were compared. Clearly, the steady-state creep speed at 150 °C increased after the addition of Ag, but the increasing rate was significantly correlated with the increase in external stress. It was apparent that the creep properties of the alloys with added Ag were weakened by the decrease in external stress at 150 °C. The steady-state creep rate decreased at both 200 and 250 °C, but did not change significantly with the increment of applied stress. It may be that the addition of 0.5 Ag enhances the high-temperature creep properties of Al-5Cu-0.8Mg-0.2Sc-0.15Zr to an extent.

In all, Ag substantially prolonged the tensile creep life of peak-aged Al-5Cu-0.8Mg-0.2Sc-0.15Zr(-0.5Ag), and this effect was significant under low-temperature and high-stress conditions (150 °C/300 MPa), as well as medium-temperature and low-stress conditions (200 °C/(150MPa, 175 MPa)).

3.3. Constitutive Relational Model of Alloy Creep

The peak-aged Al-5Cu-0.8Mg-0.2Sc-0.15Zr(-0.5Ag) had a substantial high-temperature creep resistance (Figure 4). Reportedly, steady-state creep speed is associated with creep temperature and stress [29]:

$$\dot{\varepsilon} = A\sigma^n \exp(-Q/RT), \tag{1}$$

where $\dot{\varepsilon}$ is steady-state creep rate, T is creep temperature, σ is stress, A is constant, n is stress index, Q is activation energy, and R is mol gas constant. Equation (1) is mathematically transformed:

$$\ln \dot{\varepsilon} = \ln A + n\ln\sigma - \frac{Q}{R} \cdot \frac{1}{T} \tag{2}$$

Based on the above equation, when the stress is invariable, the slope of the and 1/T curve is $-Q/R$. When the temperature is invariable, the slope on the curve of against $\ln\sigma$ is n and the intercept is $\ln A - (Q/(RT))$. On this basis, the stress index n and constant A can be determined. Thus, the data on the high-temperature creep curve of peak-aged Al-5Cu-0.8Mg-0.2Sc-0.15Zr under different conditions in Figure 4 were chosen and fitted to obtain $\ln \dot{\varepsilon} - 1/T$ and $-\ln\sigma$ curves (Figure 5). Furthermore, we determined that n = 5.97, A = 1.89×10^{-5}, and Q = 136.65 kJ/mol.

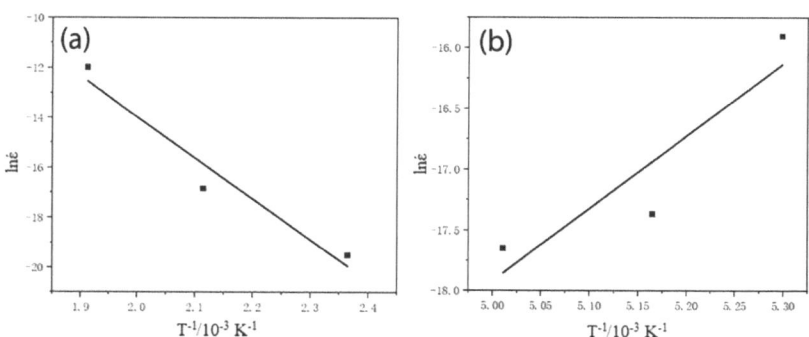

Figure 5. Relation curve of high-temperature creep of Al-5Cu-0.8Mg-0.2Sc-0.15Zr (a) and 1/T curve; (b) $-\ln\sigma$ curve.

With the above method, we determined that n = 4.81, A = 4.27×10^{-5} and Q = 104.06 kJ/mol during the creep of Al-5Cu-0.8Mg-0.2Sc-0.15Zr-0.5Ag. Thus, at the steady-state creep stage of Al-5Cu-0.8Mg-0.2Sc-0.15Zr(-0.5Ag) at 150–250 °C, and 100–350 MPa, the relationships of creep speed with temperature and external stress can be obtained:

Constitutive relational model of Al-5Cu-0.8Mg-0.2Sc-0.15Zr:

$$\dot{\varepsilon} = 1.89 \times 10^{-5} \sigma^{5.97} \exp(-136{,}650/RT), \tag{3}$$

Constitutive relational model of Al-5Cu-0.8Mg-0.2Sc-0.15Zr-0.5Ag:

$$\dot{\varepsilon} = 4.27 \times 10^{-5}\, \sigma^{4.81} \exp(-104{,}060/RT), \tag{4}$$

3.4. Alloy Creep Life Equation and Prediction

Figure 6 shows the double-logarithm curves of steady-state creep rate and creep fracture life of Al-5Cu-0.8Mg-0.2Sc-0.15Zr(-0.5Ag) at 150, 200 and 250 °C. Clearly, the double-logarithm curves of creep fracture time and steady-state creep speed are linear, so the creep life meets the Monkman-Grant relation [30]:

$$(\dot{\varepsilon})^{\beta} t_f = C_{MG}, \tag{5}$$

where $\dot{\varepsilon}$ is steady-state creep speed, t_f is time of fracture, C_{MG} is constant (related to the material and testing temperature), β is the slope of the double-logarithm curve between creep life and steady-state creep speed. Then, the β and C_{MG} were calculated under different creep conditions (Table 4). Based on the Monkman–Grant relational models, the creep life of Al-5Cu-0.8Mg-0.2Sc-0.15Zr(-0.5Ag) was predicted and compared with the testing results (Figure 7).

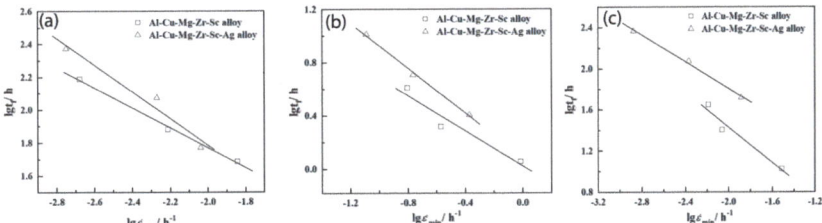

Figure 6. Double-logarithm curve of creep life and steady-state creep rate of Al-5Cu-0.8Mg-0.2Sc-0.15Zr(-0.5Ag) (**a**) 150 °C; (**b**) 200 °C; (**c**) 250 °C.

Table 4. The β, and C_{MG} of Al-5Cu-0.8Mg-0.2Sc-0.15Zr(-0.5Ag) under different conditions.

Alloy	Temperature/°C	β	C_{MG}
Al-5Cu-0.8Mg-0.2Sc-0.15Zr	150	0.61	1.34×10^{-2}
	200	0.6	1.29×10^{-2}
	250	0.62	1.57×10^{-2}
Al-5Cu-0.8Mg-0.2Sc-0.15Zr-0.5Ag	150	0.88	1.96×10^{-2}
	200	0.77	1.88×10^{-2}
	250	0.8	1.73×10^{-2}

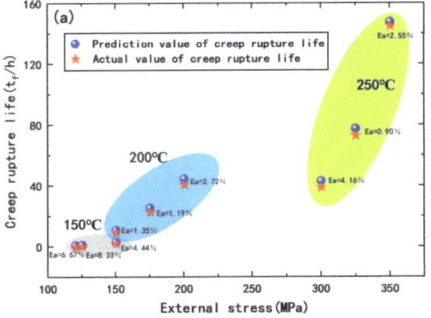
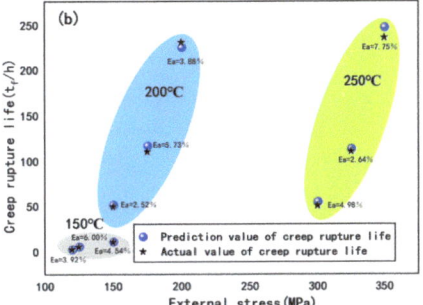

Figure 7. Comparison between predicted value and actual value of creep life of (**a**) Al-5Cu-0.8Mg-0.2Sc-0.15Zr; (**b**) Al-5Cu-0.8Mg-0.2Sc-0.15Zr-0.5Ag.

The Monkman–Grant models can accurately predict the creep life of Al-5Cu-0.8Mg-0.2Sc-0.15Zr(-0.5Ag) at 150, 200 or 250 °C and under various external stresses (Figure 7). The maximum and minimum relative errors of the predicted results are 8.33% and 0.90%, respectively, indicating that the creep life models are applicable.

4. Discussion

Deformation mechanism maps, also called the phase maps of metal hot deformation, are theoretically based on the constitutive relational models based on diffusion control, the constitutive relational models based on grain boundary slide, and the constitutive relational models based on dislocation slide control. The high-temperature deformation mechanisms of metal materials all obey a velocity control equation:

$$\dot{\varepsilon}_i = A_i \left(\frac{b}{d_i}\right)^p \cdot \frac{D}{K \cdot T \cdot b^2} \cdot \left(\frac{\sigma_i}{E}\right)^n, \qquad (6)$$

where $\dot{\varepsilon}_i$ is the steady-state strain rate; σ_i is stress; E is Young's modulus; d_i is grain size, b is the Burgers vector; k is Boltzmann's constant; D is the diffusion coefficient (including the grain boundary diffusion coefficient D_{gb} and the lattice diffusion coefficient D_L); and A_i, n and P are material constants. The main parameters used in the deformation mechanism map plotting are listed in Table 5. Constitutive relational models ①–⑤, ②–⑤, ⑤–⑥, ⑥–⑦, ⑥–②, ②–③, ②–①, ②–④, ⑥–③, ⑦–③, ①–④ and ③–④ in Table 4 were all solved, and the coordinates $(\sigma/E, d/b)_i$ of the key points in the deformation mechanism maps were determined. The internal dislocation root count of single crystal grains can be computed as follows [31]:

$$n_i = 2[(1-v)\pi d_i \tau_i]/(Gb), \qquad (7)$$

where n_i is the number of internal dislocation roots in the crystal grains, v is Poisson's ratio and τ_i is the shear stress (MPa), $\tau_i = 0.5\sigma_i$. The n_i was solved and marked at the intersection point on each deformation mechanism map. The above constitutive equations were solved. Then, RWS deformation mechanism maps were plotted on a plotting software with modulus compensation stress as X-axis, and grain size with Burgers vector compensation as Y-axis (Figure 8).

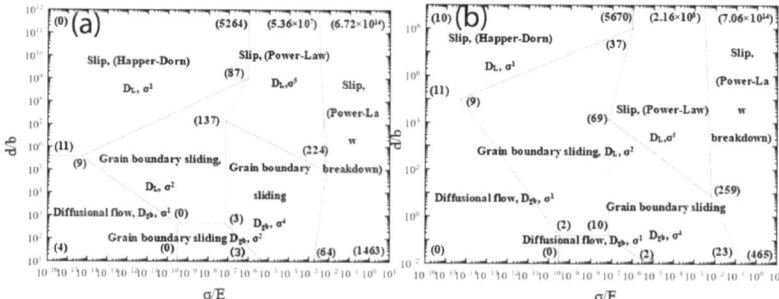

Figure 8. Steady-state controlling deformation mechanism maps for aluminum alloy at different temperatures (**a**) 150 °C; (**b**) 250 °C.

The experimental data of creep of Al-5Cu-0.8Mg-0.2Sc-0.15Zr(-0.5Ag) under various conditions were analyzed to determine the grain size with Burgers vector compensation and to determine the stress with modulus compensation under different temperatures and different strain rates (Table 6).

Table 5. The main parameters of the deformation mechanism map for Al-5Cu-0.8Mg-0.2Sc-0.15Zr(-0.5Ag).

	Creep Process	Equation
◆	Diffusional creep	
①	Coble [32]	$\dot{\varepsilon} = 50\left(D_{gb}b/d^3\right)\left(Eb^3/kT\right)(\sigma/E)$
◆	Grain boundary sliding(GBS)	
②	Lattice-diffusion controlled [33]	$\dot{\varepsilon} = 6.4 \times 10^9 \left(D_L/d^2\right)(\sigma/E)^2$
③	Pipe-diffusion controlled [34]	$\dot{\varepsilon} = 3.2 \times 10^{11}\alpha\left(D_p/d^2\right)(\sigma/E)^4$
④	Grain boundary diffusion controlled [35]	$\dot{\varepsilon} = 5.6 \times 10^8 \left(D_{gb}b/d^3\right)(\sigma/E)^2$
◆	Slip creep	
⑤	Harper–Dorn [36]	$\dot{\varepsilon} = 1.7 \times 10^{-11}\left(D_L/b^2\right)\left(Eb^3/kT\right)(\sigma/E)$
⑥	Lattice-diffusion controlled (dislocation climb creep) [37]	$\dot{\varepsilon} = 10^{11}\left(D_L/b^2\right)(\sigma/E)^5$
⑦	Pipe-diffusion controlled (dislocation climb creep) [38]	$\dot{\varepsilon} = 5 \times 10^{12}\left(D_p/b^2\right)(\sigma/E)^7$

The material constants used for Al-5Cu-0.8Mg-0.15Zr-0.2Sc(-0.5Ag) alloy

$\gamma = 0.33$	$E = 7.2 \times 10^5$ MPa
$b = 0.255$ nm [39]	$D_{W,L} = \sum_i X_i D^*_{i,L} = 1.58 \times 10^{-4} \exp\left(\frac{-278,300}{RT}\right)$ m^2/s
$k = 1.38 \times 10^{-23}$ J/K [36–38]	$D_{W,gb} = D_p = \sum_i X_i D^*_{i,gb} = 1.38 \times 10^{-2} \exp\left(\frac{-178,800}{RT}\right)$ m^2/s

Table 6. Experimentally calculated data on creep of Al-5Cu-0.8Mg-0.2Sc-0.15Zr(-0.5Ag).

Alloy	Temperature/°C	Normalized Grain Size with Burgers Vector Compensation (d/b) × 10^{-7}	Normalized Flow Stress of Modulus Compensation (σ/E) × 10^4	Strain Rate $\dot{\varepsilon}/(10^{-8} \cdot s^{-1})$
Al-5Cu-0.8Mg-0.2Sc-0.15Zr	150	6.7~9.7	11.2~43.9	1.85~618
Al-5Cu-0.8Mg-0.2Sc-0.15Zr	250	8.9~10.9	2.6~17.8	1.85~618
Al-5Cu-0.8Mg-0.2Sc-0.15Zr-0.5Ag	150	9.7~12.7	13.2~49.9	1.85~618
Al-5Cu-0.8Mg-0.2Sc-0.15Zr-0.5Ag	250	10.4~13.9	4.69~19.0	1.85~618

The deformation mechanism of the creep of 5Cu-0.8Mg-0.2Sc-0.15Zr(-0.5Ag) was deduced from Table 5 and Figure 8. At the deformation temperature of 150 °C, the high-temperature creep of Al-5Cu-0.8Mg-0.2Sc-0.15Zr fell within the dislocation polygon (137)(224)(64)(13)(3)(3), indicating that the creep mechanism of this alloy was a dislocation glide mechanism with a stress index of 4 and was controlled by grain boundary slip. At the deformation temperature of 250 °C, during deformation, this alloy fell within the dislocation polygon (2)(10)(2)(0), indicating that the creep mechanism at this temperature was a grain boundary slip system with a stress index of 1 and controlled by diffusion. At the deformation temperature of 150 °C, the high-temperature creep of Al-5Cu-0.8Mg-0.2Sc-0.15Zr-0.5Ag fell within the dislocation polygon (9)(87)(137)(3)(0), indicating that the creep mechanism of this alloy was a dislocation glide mechanism with a stress index of 2 and controlled by grain boundary sliding. At the deformation temperature of 250 °C, during deformation, this alloy fell within the dislocation polygon (10)(12)(23)(259)(69), indicating that the creep mechanism at this temperature was a grain boundary slip mechanism with a stress index of 4 and controlled by diffusion. It was observed that after the addition of Ag, the creep mechanism at 150 °C transitioned from lattice diffusion control to grain boundary diffusion control, but at 250 °C, the mechanism was controlled by grain boundary slip, and the stress index was observed to increase. It is apparent that the addition of Ag enhances the high-temperature creep properties of Al-5Cu-0.8Mg-0.2Sc-0.15Zr.

Figure 9 shows the creep fracture morphology of Al-5Cu-0.8Mg-0.2Sc-0.15Zr at 150 °C and at stresses of 300 or 350 MPa. The creep fractures of the alloy at 150 °C/300 MPa contained certain amounts of dimples and torn edges as well as a few shallow pores on the crystal boundary (Figure 9a,b). As the stress increased further to 350 MPa, the number of cavities at the crystal boundary of creep fractures also increased significantly, and the pores were aggregated. This was because, during creeping of aged alloys, as the stress was increased, the dislocations further reduced the constraint of the second-phase particles, so that the crystal boundaries were weakened and slip and torsion occurred more easily, leading to the formation of hollows. As a result, the number of hollows was increased. At the same time, as the creep proceeded, the small hollows gradually grew upwards, aggregated and expanded to form long strips and even fractures. Nevertheless, the fractures of this alloy under the above creep conditions were clearly ductile.

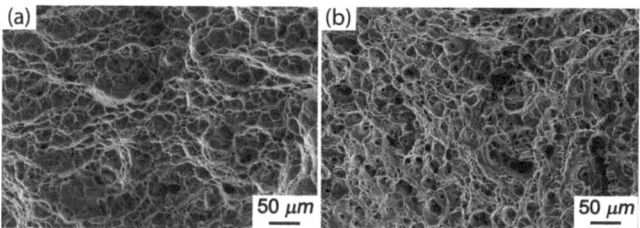

Figure 9. Creep fracture morphology of Al-5Cu-0.8Mg-0.2Sc-0.15Zr under different stresses (150 °C): (**a**) 300 MPa; (**b**) 350 MPa.

Figure 10 shows the creep fracture morphology of Al-5Cu-0.8Mg-0.2Sc-0.15Zr-0.5Ag at 150 °C and under stress conditions of either 300 or 350 MPa. Clearly, the creep fractures of the alloy with added 0.5Ag were still ductile, and the number of dimples increased. When the external stress was at 350 MPa, the speed of pore aggregation, growth and expansion was reduced, which inhibited the expansion of holes near crystal boundaries. It was observed that the addition of 0.5Ag enhanced the creep property of alloys.

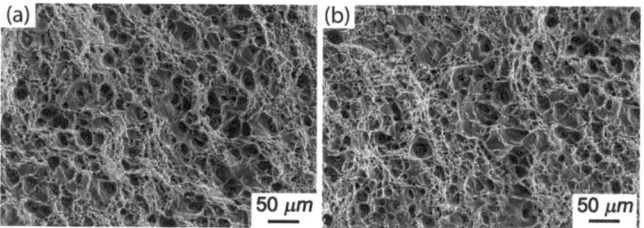

Figure 10. Creep fracture morphology of Al-5Cu-0.8Mg-0.2Sc-0.15Zr-0.5Ag under different stresses (150 °C): (**a**) 300 MPa; (**b**) 350 MPa.

Figure 11 shows the creep fracture morphology of Al-5Cu-0.8Mg-0.2Sc-0.15Zr(-0.5Ag) at 200 °C and under stress conditions of 150 MPa. Clearly, after the addition of 0.5Ag, dimples increased notably in both number and depth, and were found to coexist with torn edges. This was because more precipitate phase was formed after the addition of Ag, which caused the dimples to form and be distributed evenly. Hence, the creep life of Al-5Cu-0.8Mg-0.2Sc-0.15Zr-0.5Ag was significantly prolonged under the above creep conditions.

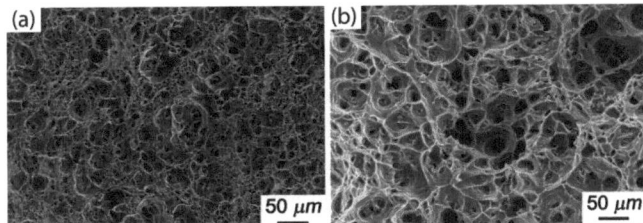

Figure 11. Creep fracture morphology of Al-5Cu-0.8Mg-0.2Sc-0.15Zr-0.5Ag under different stresses (200 °C): (**a**) Al-5Cu-0.8Mg-0.2Sc-0.15Zr; (**b**) Al-5Cu-0.8Mg-0.2Sc-0.15Zr-0.5Ag.

5. Conclusions

The peak-aged Al-5Cu-0.8Mg-0.15Zr-0.2Sc(-0.5Ag) alloy, which had outstanding comprehensive mechanical properties, was tested under different conditions in terms of its high-temperature creep performance.

(1) Ag prolonged the high-temperature creep life of peak-aged Al-5Cu-0.8Mg-0.2Sc-0.15Zr, and this effect was significant under conditions of 150 °C/300 MPa or 200 °C/(150 MPa, 175 MPa).

(2) The stress indices of peak-aged Al-5Cu-0.8Mg-0.15Zr-0.2Sc(-0.5Ag) during high-temperature tensile creep were 5.97 and 4.81, and the activation energy was 136.65 and 104.06 KJ/mol. Accordingly, constitutive relational models between creep speed and creep temperature or stress were built:

Constitutive relational model of Al-5Cu-0.8Mg-0.2Sc-0.15Zr:

$$\dot{\varepsilon} = 1.89 \times 10^{-5} \sigma^{5.97} \exp(-136,650/RT), \tag{8}$$

Constitutive relational model of Al-5Cu-0.8Mg-0.2Sc-0.15Zr-0.5Ag:

$$\dot{\varepsilon} = 4.27 \times 10^{-5} \sigma^{4.81} \exp(-104,060/RT), \tag{9}$$

(3) Monkman–Grant creep life models of Al-5Cu-0.8Mg-0.2Sc-0.15Zr(-0.5Ag) were built and utilized in order to predict the creep life of the alloys at temperatures of 150, 200 and 250 °C, and under different stresses. The maximum and minimum relative errors of the predicted results are 8.33% and 0.90%, respectively, indicating that the models are applicable.

(4) RWS deformation mechanism maps involving dislocation quantity were plotted, and the high-temperature creep mechanism of Al-5Cu-0.8Mg-0.15Zr-0.2Sc(-0.5Ag) was predicted. At 150 °C, the creep mechanism, after the addition of Ag, transitioned from lattice diffusion control to grain boundary diffusion control, but at 250 °C, the mechanism was still controlled by grain boundary slip, and the stress index was observed to increase. After Ag was added, dimples were formed upon fracturing and tended to be evenly distributed, which effectively enhanced the high-temperature creep properties of the alloys.

Author Contributions: Investigation, Y.W. and G.Z.; conceptualization, F.L., X.C. and L.C.; formal analysis, Y.W., L.C. and G.Z.; writing—original draft preparation, Y.W. and L.C. All authors have read and agreed to the published version of the manuscript.

Funding: This research was funded by the Liaoning Provincial Science and Technology Department Applied Basic Research Program Project and the grant number [No. 2022JH2/101300078].

Data Availability Statement: No new data were created or analyzed in this study. Data sharing is not applicable to this article.

Conflicts of Interest: The authors declare no conflict of interest.

References

1. Sukumaran, K.; Ravikumar, K.K.; Pillai, S.G.K.; Rajan, T.P.D.; Ravi, M.; Pillai, R.M.; Pai, B.C. Studies on squeeze casting of Al 2124 alloy and 2124-10% SiCp metal matrix composite. *Mater. Sci. Eng. A* **2008**, *490*, 235–241. [CrossRef]
2. Raju, P.N.; Rao, K.S.; Reddy, G.M.; Kamaraj, M.; Rao, K.P. Microstructure and high temperature stability of age hardenable AA2219 aluminium alloy modified by Sc, Mg and Zr additions. *Mater. Sci. Eng. A* **2007**, *464*, 192–201. [CrossRef]
3. Wang, J.; Yi, D.; Su, X.; Yin, F. Influence of deformation ageing treatment on microstructure and properties of aluminum alloy 2618. *Mater. Charact.* **2008**, *59*, 965–968. [CrossRef]
4. Yu, K.; Li, W.; Li, S.; Zhao, J. Mechanical properties and microstructure of aluminum alloy 2618 with Al3 (Sc, Zr) phases. *Mater. Sci. Eng. A* **2004**, *368*, 88–93. [CrossRef]
5. Villars, P. *Handbook of Ternary Alloy Phase Diagrams*; ASM International: Detroit, MI, USA, 1995; Volume 7, pp. 8754–8755.
6. Chang, C.H.; Lee, S.L.; Lin, J.C.; Yeh, M.S.; Jeng, R.R. Effect of Ag content and heat treatment on the stress corrosion cracking of Al–4.6 Cu–0.3 Mg alloy. *Mater. Chem. Phys.* **2005**, *91*, 454–462. [CrossRef]
7. Xiao, D.H.; Wang, J.N.; Ding, D.Y.; Chen, S.P. Effect of Cu content on the mechanical properties of an Al–Cu–Mg–Ag alloy. *J. Alloy. Compd.* **2002**, *343*, 77–81. [CrossRef]
8. Kerry, S.; Scott, V.D. Structure and orientation relationship of precipitates formed in Al-Cu-Mg-Ag alloys. *Met. Sci.* **1984**, *18*, 289–294. [CrossRef]
9. Scott, V.D.; Kerry, S.; Trumper, R.L. Nucleation and growth of precipitates in Al–Cu–Mg–Ag alloys. *Mater. Sci. Technol.* **1987**, *3*, 827–835. [CrossRef]
10. Rainforth, W.M.; Rylands, L.M.; Jones, H. Nano-beam analysis of {Omega} precipitates in a Al-Cu-Mg-Ag alloy. *Scr. Mater.* **1996**, *35*, 261–265. [CrossRef]
11. Auld, J.H.; Cousland, S.M. On the structure of the M'phase in Al–Zn–Mg alloys. *J. Appl. Crystallogr.* **1985**, *18*, 47–48. [CrossRef]
12. Sano, N.; Hono, K.; Sakurai, T.; Hirano, K. Atom-probe analysis of Ω and θ′ phases in an Al-Cu-Mg-Ag alloy. *Scr. Metall. Et Mater.* **1991**, *25*, 491–496. [CrossRef]
13. Hutchinson, C.R.; Fan, X.; Pennycook, S.J.; Shiflet, G.J. On the origin of the high coarsening resistance of Ω plates in Al–Cu–Mg–Ag alloys. *Acta Mater.* **2001**, *49*, 2827–2841. [CrossRef]
14. Bakavos, D.; Prangnell, P.B.; Bes, B.; Eberl, F. The effect of silver on microstructural evolution in two 2xxx series Al-alloys with a high Cu: Mg ratio during ageing to a T8 temper. *Mater. Sci. Eng. A* **2008**, *491*, 214–223. [CrossRef]
15. Reddy, A.S. Fatigue and creep deformed microstructures of aged alloys based on Al–4% Cu–0.3% Mg. *Mater. Des.* **2008**, *29*, 763–768. [CrossRef]
16. Lumley, R.N.; Morton, A.J.; Polmear, I.J. Enhanced creep performance in an Al–Cu–Mg–Ag alloy through underageing. *Acta Mater.* **2002**, *50*, 3597–3608. [CrossRef]
17. Liu, M.T.; Tian, Y.; Wang, Y.; Wang, K.L.; Zhang, K.M.; Lu, S.Q. Critical Conditions for Dynamic Recrystallization of S280 Ultra-High-Strength Stainless Steel Based on Work Hardening Rate. *Metals* **2022**, *12*, 1122. [CrossRef]
18. Lin, Y.C.; Jiang, Y.Q.; Xia, Y.C.; Zhang, X.C.; Zhou, H.M.; Deng, J. Effects of creep-aging processing on the corrosion resistance and mechanical properties of an Al–Cu–Mg alloy. *Mater. Sci. Eng. A* **2014**, *605*, 192–202. [CrossRef]
19. Knipling, K.E.; Dunand, D.C. Creep resistance of cast and aged Al–0.1 Zr and Al–0.1 Zr–0.1 Ti (at.%) alloys at 300–400 °C. *Scr. Mater.* **2008**, *59*, 387–390. [CrossRef]
20. Xu, Y.; Zhan, L.; Xu, L.; Huang, M. Experimental research on creep aging behavior of Al-Cu-Mg alloy with tensile and compressive stresses. *Mater. Sci. Eng. A* **2017**, *682*, 54–62. [CrossRef]
21. Xu, F.S.; Zhang, J.; Deng, Y.L.; Zhang, X.M. Precipitation orientation effect of 2124 aluminum alloy in creep aging. *Trans. Nonferrous Met. Soc. China* **2014**, *24*, 2067–2071. [CrossRef]
22. Zhan, L.H.; Li, Y.G.; Huang, M.H. Effects of process parameters on mechanical properties and microstructures of creep aged 2124 aluminum alloy. *Trans. Nonferrous Met. Soc. China* **2014**, *24*, 2232–2238. [CrossRef]
23. Farkoosh, A.R.; Pekguleryuz, M. The effects of manganese on the T-phase and creep resistance in Al–Si–Cu–Mg–Ni alloys. *Mater. Sci. Eng. A* **2013**, *582*, 248–256. [CrossRef]
24. Erdeniz, D.; Nasim, W.; Malik, J.; Yost, A.R.; Park, S.; De Luca, A.; Dunand, D.C. Effect of vanadium micro-alloying on the microstructural evolution and creep behavior of Al-Er-Sc-Zr-Si alloys. *Acta Mater.* **2017**, *124*, 501–512. [CrossRef]
25. Song, Y.; Pan, Q.; Cao, S.; Wang, Y.; Li, C. Effects of Ag Content on Microstructures and Properties of Al-Cu-Mg Alloy. *J. Aeronaut. Mater.* **2013**, *33*, 7–13.
26. Ringer, S.P.; Hono, K.; Polmear, I.J.; Sakurai, T. Nucleation of precipitates in aged AlCuMg (Ag) alloys with high Cu: Mg ratios. *Acta Mater.* **1996**, *44*, 1883–1898. [CrossRef]
27. Gazizov, M.; Kaibyshev, R. High cyclic fatigue performance of Al–Cu–Mg–Ag alloy under T6 and T840 conditions. *Trans. Nonferrous Met. Soc. China* **2017**, *27*, 1215–1223. [CrossRef]
28. Li, J.; An, Z.; Hage, F.S.; Wang, H.; Xie, P.; Jin, S.; Sha, G. Solute clustering and precipitation in an Al–Cu–Mg–Ag–Si model alloy. *Mater. Sci. Eng. A* **2019**, *760*, 366–376. [CrossRef]
29. Yin, X.; Zhan, L.; Zhan, J. Establishment of steady creep constitutive equation of 2219 aluminum alloy. *Chin. J. Nonferrous Met.* **2014**, *24*, 7.
30. Monkman, F.C. An empirical relationship between rupture life and minimum creep rate in creep-rupture tests. *Proc. ASTM* **1956**, *56*, 593–620.

31. Liu, C.; Wang, X.; Zhou, G.; Li, F.; Zhang, S.; Zhang, H.; Liu, H. Dislocation-controlled low-temperature superplastic deformation of Ti-6Al-4V alloy. *Front. Mater.* **2021**, *7*, 606092. [CrossRef]
32. Coble, R.L. A model for boundary diffusion controlled creep in polycrystalline materials. *J. Appl. Phys.* **1963**, *34*, 1679–1682. [CrossRef]
33. Lüthy, H.; White, R.A.; Sherby, O.D. Grain boundary sliding and deformation mechanism maps. *Mater. Sci. Eng.* **1979**, *39*, 211–216. [CrossRef]
34. Ruano, O.A.; Miller, A.K.; Sherby, O.D. The influence of pipe diffusion on the creep of fine-grained materials. *Mater. Sci. Eng.* **1981**, *51*, 9–16. [CrossRef]
35. Harper, J.; Dorn, J.E. Viscous creep of aluminum near its melting temperature. *Acta Metall.* **1957**, *5*, 654–665. [CrossRef]
36. Robinson, S.L.; Sherby, O.D. Mechanical behavior of polycrystalline tungsten at elevated temperature. *Acta Metall.* **1969**, *17*, 109–125. [CrossRef]
37. Vaidya, M.; Pradeep, K.G.; Murty, B.S.; Wilde, G.; Divinski, S.V. Bulk tracer diffusion in CoCrFeNi and CoCrFeMnNi high entropy alloys. *Acta Mater.* **2018**, *146*, 211–224. [CrossRef]
38. Fuentes-Samaniego, R.; Nix, W.D. Appropriate diffusion coefficients for describing creep processes in solid solution alloys. *Scr. Metall.* **1981**, *15*, 15–20. [CrossRef]
39. Hasaka, M.; Morimura, T.; Uchiyama, Y.; Kondo, S.; Furuse, T.; Watanabe, T.; Hisatsune, K. Diffusion of copper, aluminum and boron in nickel. *Scr. Metall. Mater. USA* **1993**, *29*, 959–962. [CrossRef]

Disclaimer/Publisher's Note: The statements, opinions and data contained in all publications are solely those of the individual author(s) and contributor(s) and not of MDPI and/or the editor(s). MDPI and/or the editor(s) disclaim responsibility for any injury to people or property resulting from any ideas, methods, instructions or products referred to in the content.

Article

Microstructure, Mechanical Properties and Thermal Stability of Ni-Based Single Crystal Superalloys with Low Specific Weight

Dengyu Liu [1], Qingqing Ding [1,*], Qian Zhou [1], Dingxin Zhou [1], Xiao Wei [1], Xinbao Zhao [1], Ze Zhang [1,2,*] and Hongbin Bei [1,*]

[1] School of Materials Science and Engineering, Zhejiang University, Hangzhou 310027, China; mseweixiao@zju.edu.cn (X.W.)
[2] State Key Laboratory of Silicon Materials, Zhejiang University, Hangzhou 310027, China
* Correspondence: qq_ding@zju.edu.cn (Q.D.); zezhang@zju.edu.cn (Z.Z.); hbei2018@zju.edu.cn (H.B.)

Abstract: Ni-based single crystal (SX) superalloy with low specific weight is vital for developing aero engines with a high strength-to-weight ratio. Based on an alloy system with 3 wt.% Re but without W, namely Ni-Co-Cr-Mo-Ta-Re-Al-Ti, a specific weight below 8.4 g/cm^3 has been achieved. To reveal the relationship among the composition, mechanical properties, and thermal stability of Ni-based SX superalloys, SXs with desirable microstructures are fabricated. Tensile tests revealed that the SX alloys have comparable strength to commercial second-generation SX CMSX-4 (3 wt.% Re and 6 wt.% W) and Rene' N5 alloys (3 wt.% Re and 5 wt.% W) above 800 °C. Moreover, the elongation to fracture (EF) below 850 °C (>20%) is better than that of those two commercial SX superalloys. During thermal exposure at 1050 °C for up to 500 h, the topological close-packed (TCP) phase does not appear, indicating excellent phase stability. Decreasing Al concentration increases the resistance of γ' rafting and replacing 1 wt.% Ti with 3 wt.% Ta is beneficial to the stability of the shape and size of γ' phase during thermal exposure. The current work might provide scientific insights for developing Ni-based SX superalloys with low specific weight.

Keywords: Ni-based single crystal superalloy; specific weight; mechanical properties; thermal stability

1. Introduction

Ni-based SX superalloys are widely employed in turbine blades of aero engines because of their unique combination of mechanical properties and corrosion resistance at high temperatures [1–4]. Developing aero engines with high strength/weight ratio, improved temperature capacity, and reduced specific weight are still major research directions for the development of Ni-based SX superalloys [5–7]. Improving temperature capacity can directly increase the turbine inlet temperature and, therefore, the performance of aero engines [8]. Current advanced Ni-based SX superalloys have a temperature capacity as high as 1100 °C [9], but the hottest spots in turbine blades occasionally approach 1200 °C, which is already ~90% of the melting point of the alloy [10–12]. Therefore, further increase in the temperature capability by alloy design is intrinsically limited by the onset of melting/phase stability [13,14]. Low specific weight can directly reduce the weight of Ni-based SX superalloys used as turbine blades and indirectly reduce the weight of the entire rotor (disk, hub, and shaft) as well as non-rotating support structures [6].

To reduce the specific weight in first-generation Ni-based SX superalloys, increasing the concentration of light elements (Ti + Al) is adopted in the 1980s [15]. For example, Wortmann et al. [15] developed CMSX-6 with a specific weight of less than 8.0 g/cm^3, which still has comparable creep strength to first-generation SX CMSX-2 and CMSX-3 superalloys, by increasing the concentration of Al + Ti to 9.5 wt.%. With Ni-based SX superalloys developing into higher generations, the research direction toward low specific weight mainly focuses on balancing the concentration of heavy elements (Re + Ta + W +

Mo + Ru). Among those heavy elements, Re, Ta, and W have the greatest influence on increasing the specific weight of the alloys [16–18]. Macky et al. [6]. reduced the specific weight of Ni-based SX superalloy to less than 8.8 g/cm^3 by using Mo replacement of W and Re, and the creep rupture strength is higher than that of second-generation Rene' N5 superalloy. Helmer et al. [19] removed Re and obtained the SX ERBO/15 alloy (specific weight is 8.4 g/cm^3), whose creep properties at stress below 150 MPa are still equivalent to that of the second-generation CMSX-4 superalloy. In addition to Re, DD16 alloy further removes W, making its specific weight less than 8.0 g/cm^3, but it has a yield strength (YS) equivalent to first-generation DD3 superalloy (with ~5.4 wt.% W) [20]. In addition, Du et al. [21] confirmed that the SX superalloy with a specific weight of less than 8.2 g/cm^3 could be obtained by controlling the concentration of heavy elements (Mo + W + Ta) to 7.9 wt.%. However, the Ni-based SX superalloys with low specific weight still need further systematic development.

Recently, through numerical models on specific weight, γ/γ' phase volume fraction, elemental partitioning behavior, heavy elements are carefully balanced, and several compositions (with Re but without W) were designed, and their microstructures validated in their polycrystalline form [16]. After solution and aging treatments, all alloys have a typical γ/γ' two-phase microstructure, and the specific weights are all lower than 8.4 g/cm^3. However, SX growth has not been performed [16]. Thus, all properties and principles of alloy development based on SXs, such as mechanical properties, thermal stability, and dendrite segregation, have not been reported. Here, in order to reveal the relationship among the composition, mechanical properties, and thermal stability of newly developed compositions [16], the entire processing circle, including casting, SX growth, and heat treatment, is employed. Tensile tests at temperatures ranging from room temperature (RT) to 1050 °C are used to understand mechanical behaviors. Thermal exposure experiments at 1050 °C for up to 500 h are used to investigate the thermal/phase stability of the SX superalloys. With the help of advanced microscopy, the fracture behavior in tensile tests and microstructure evolution during thermal exposure can be investigated in multi-scales. The current work might provide scientific insights for the development of Ni-based SX superalloys with low specific weight.

2. Materials and Methods

2.1. Alloy Preparation and SX Growth

Four Ni-based SX superalloys are fabricated to reveal the relationship among the composition, mechanical properties, and thermal stability. The nominal and measured compositions of those alloys are given in Table 1. Master alloys are prepared by arc melting of pure Ni, Co, Mo, Re, Ti, Ta, Al, and Cr (purity > 99.9 wt.%) in an argon atmosphere, and the details of arc melting and drop-casting can be found elsewhere [16,22]. Subsequently, the as-cast polycrystalline rod is placed in an alumina crucible with an SX seed with a [001] direction and directionally solidified to produce SX in a Bridgman furnace [13], as shown schematically in Figure 1. During SX growth, the withdrawal rate has been tried from 17 to 55 μm/s. When the withdrawal rate is 33 μm/s, the dendrite spacing is about 320 μm, which is similar to the dendrite spacing in a second-generation SX superalloy (~350 μm) [23]. Thus the withdrawal rate of 33 μm/s is selected in this study. The diameter of our experimental SX rods is about 12.5 mm.

Table 1. The nominal and measured compositions (wt.%) of Ni-based SX superalloys. The measured compositions are listed in parentheses. The alloys are named by the concentration of Ta, Al, and Ti. For example, if the alloy contains 3 wt.% Ta, 6.4 wt.% Al, and 1 wt.% Ti, the alloy is called 3Ta6.4Al1Ti alloy for convenience.

Alloys	Cr	Co	Mo	Ta	Re	Al	Ti	Ni
3Ta6.4Al1Ti	5 (4.9)	15 (15.1)	6 (5.8)	3 (3.1)	3 (2.2)	6.4 (6.4)	1 (1.1)	Bal.
0Ta6.4Al2Ti	5 (4.9)	15 (15.2)	6 (6.0)	-	3 (2.4)	6.4 (6.5)	2 (2.0)	Bal.

Table 1. Cont.

Alloys	Cr	Co	Mo	Ta	Re	Al	Ti	Ni
3Ta6Al1Ti	5 (4.9)	15 (14.9)	6 (6.0)	3 (3.1)	3 (2.3)	6 (6.1)	1 (1.1)	Bal.
0Ta6Al2Ti	5 (5.0)	15 (15.1)	6 (6.1)	-	3 (2.4)	6 (6.0)	2 (2.0)	Bal.

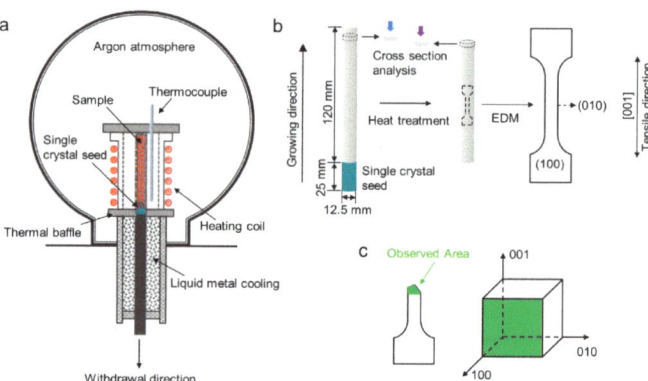

Figure 1. Schematic diagram showing SX growth and sample preparation procedures. (**a**) SX growth in a Bridgeman furnace; (**b**) The preparation and the orientation of tensile samples; (**c**) Observation position for deformed microstructure. The viewing direction is normal to (100) plane.

2.2. Heat Treatment

To eliminate the compositional inhomogeneities after SX growth and obtain a typical γ/γ' two-phase microstructure, the as-grown SX rods need solid solution and aging treatments. Similar to heat treatment procedures adopted previously [16], a stepwise process [1290 °C/2 h + 1305 °C/2 h + 1320 °C/4 h + 1330 °C/4 h/AC (air cooling)] is used to eliminate the dendritic segregation in the as-grown SXs and two-step aging treatment (1080 °C/4 h/AC + 870 °C/4 h/AC) is carried out to obtain typical γ/γ' two-phase microstructure. All solution and aging heat treatments are conducted in an argon-protected tube furnace to prevent surface oxidation of alloys.

2.3. Mechanical Testing

Dog bone-shaped tensile specimens with a gauge section of $1.9 \times 1.0 \times 9.5$ mm^3 are cut by using an electro-discharge machining (EDM) from the SX rods after the two-step aging treatment (as shown in Figure 1b). The axial direction of the tensile specimen is [001], and the other two sides of the gauge section are (100) and (010), respectively. All samples are carefully polished with 600 grit SiC paper to eliminate micro-cracks and oxide layer caused by EDM.

Tensile tests at a temperature range from RT to 1050 °C are carried out on a screw-driven mechanical testing machine equipped with an induction heater [24]. Specimens are heated to the test temperature and retained for 15 min to ensure temperature uniformity in the sample before loading. All samples are tested at a constant displacement rate of 0.57 mm/min, corresponding to an engineering strain rate of 1×10^{-3} s^{-1}. YS is calculated by using the 0.2% offset method.

2.4. Thermal Stability

Specimens for thermal stability tests are disks 12.5 mm in diameter and 4 mm in height, which are cut from the SX rods after the two-step aging treatment using EDM (as shown in Figure 1b). Before thermal exposure, all disks are ground down to 600 grit SiC papers to ensure a flat and shiny surface. Thermal exposure up to 500 h is carried out at 1050 °C in the laboratory atmosphere. A thermocouple is placed at a 10 mm distance from the sample to ensure temperature accuracy in the thermal stability experiments.

2.5. Microstructure Characterization

Samples for optical microscopy (OM) and scanning electron microscopy (SEM) are prepared using standard metallographic procedures, including grinding and mechanical polishing. To better reveal the microstructures, some samples are electrochemically etched in 90% H_2O + 10% H_3PO_4 solution at 6 V at RT.

Secondary electron (SE) images, energy dispersive spectroscopy (EDS) results, and electron backscattered diffraction (EBSD) patterns are all captured in an FEI Quanta 650 SEM (FEI Company, Eindhoven, The Netherlands) equipped with Oxford EDS and EBSD detectors, using an acceleration voltage of 15 kV and a working distance of 12 mm. Crystal orientations of the SX rods are determined by EBSD. Image J software (V1.8.0) is used to measure the γ' size.

For the transmission electron microscopy (TEM) investigations, thin disks were first mechanically ground to a thickness of about 50 μm using SiC paper and then punched into 3 mm diameter disks. TEM samples are prepared using a twin-jet electropolisher in an alcoholic solution containing 5 vol.% perchloric acid at −30 °C. The composition of two phases in the four alloys is analyzed by EDS equipped on an FEI Tecnai G^2 F20 TEM operated at 200 kV.

3. Results and Discussion

3.1. Microstructure

SX rods for all four compositions are successfully grown using the Bridgman method and exhibit similar microstructure. An example of an as-grown SX rod before solution and aging treatments is shown in Figure 2a. A cross-sectional optical micrograph of the as-grown rod (Figure 2b) confirms that no high-angle grain boundaries are observed on the entire surface, indicating the good quality of the SX. EBSD is used to determine the orientation of the SX, as shown in the inverse pole figure (inset of Figure 2b). The growth direction is within 10° of the [001] direction. The as-grown microstructures are typical dendrite structures with an average primary dendrite spacing of ~320 μm. The interdendritic region mainly consists of coarse γ' particles with an average size of 370 (±68) nm (Figure 2c). In contrast, the γ' particles in dendritic regions are relatively fine, and the average size is 220 (±30) nm (Figure 2d).

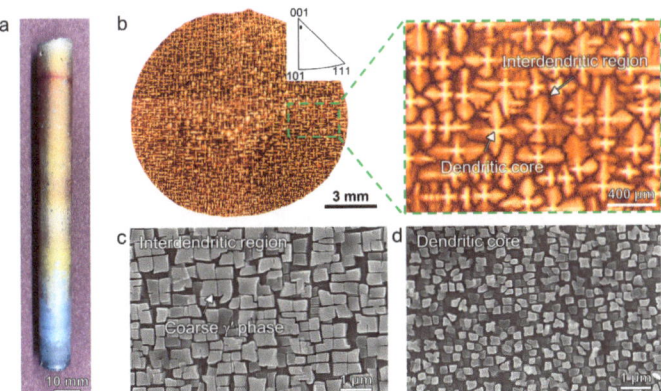

Figure 2. Photograph and cross-sectional microstructures of an as-grown Ni-based SX rod (3Ta6Al1Ti). (**a**) An SX rod; (**b**) Optical micrograph and inverse pole figure (inset) of the cross-section of the SX alloy, indicating that growth direction is along the [001] direction. SE images show that the γ' phase in the interdendritic region (**c**) is larger than that in the dendritic core (**d**).

Microstructures of the four SX alloys after solution and aging treatments are presented in Figure 3. Typical γ/γ' two-phase microstructure of Ni-based SX superalloy is obtained for all four alloys, and the γ' particles are square and uniformly distributed in the γ phase. Moreover, the inhomogeneity between the dendritic and interdendritic region are eliminated after heat treatment. Based on the measurement of more than 100 γ' particles in each alloy, average sizes of γ' particles in 3Ta6.4Al1Ti, 0Ta6.4Al2Ti, 3Ta6Al1Ti, and 0Ta6Al2Ti SX superalloys are 260 ($\pm$61) nm, 270 ($\pm$59) nm, 270 ($\pm$59) nm, and 290 ($\pm$66) nm, respectively.

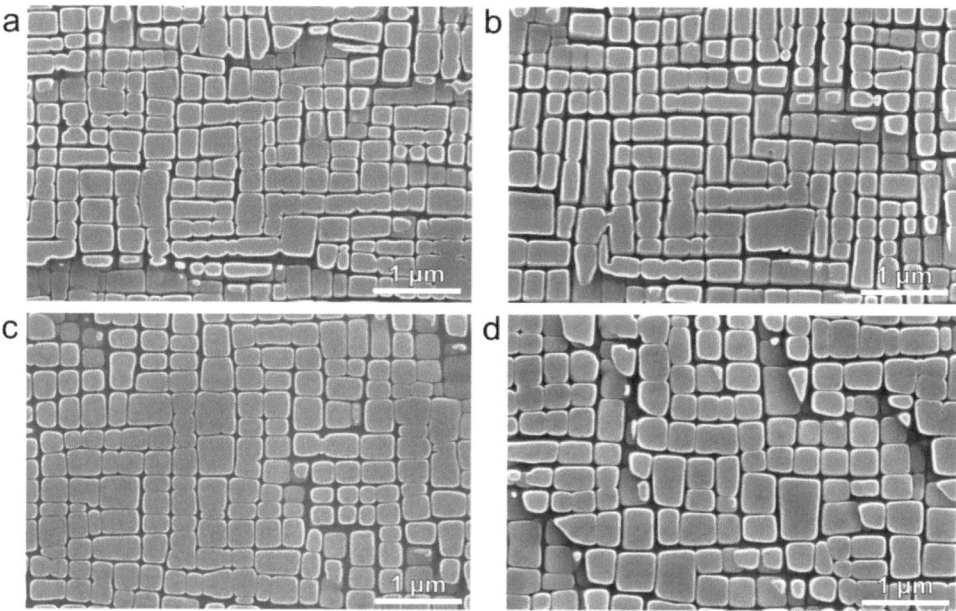

Figure 3. SEM images of Ni-based SX superalloys after solution and aging treatments: (**a**) 3Ta6.4Al1Ti; (**b**) 0Ta6.4Al2Ti; (**c**) 3Ta6Al1Ti; (**d**) 0Ta6Al2Ti. Four SX superalloys all have typical γ/γ' two-phase microstructure.

3.2. Tensile Properties

Tensile tests at the temperature range from RT to 1050 °C are conducted to investigate the mechanical properties of SX superalloys with typical γ/γ' two-phase microstructure (shown in Figure 3). Four alloys have similar plastic deformation behavior. Therefore, stress-strain curves for one alloy (3Ta6.4Al1Ti alloy) are shown in Figure 4a as a representative. Similar to other Ni-based SX superalloys [25–27], at RT-750 °C, the samples fail after continuous work hardening, which might be related to the dislocation slip in the matrix and superdislocation shearing of γ' precipitates [28]. At 850–950 °C, the flow stress decreases and then increases after yielding, which is the so-called yield drop phenomenon. The yield drop phenomena have been observed in other Ni-based SX superalloys as well, such as 3.8Cr-8.5Co-7W-5.2Al-6Ta-1.6Re-1.5Mo-Ni [29] and 5.8Al-8Co-2Mo-18(W + Cr + Ta)-3Re-Ni [30] alloys, which is often companied by the observation of Kear-Wilsdorf lock (KWL) during deformation in this temperature range [30]. At 950–1050 °C, only slight work hardening is observed after the yield point, then the flow stress gradually decreases until fracture. The decrease in flow stress may be related to the low strength of γ and γ' phases at high temperatures because the movement of dislocations is fast and cannot provide effective obstacles to dislocation movements [23,31].

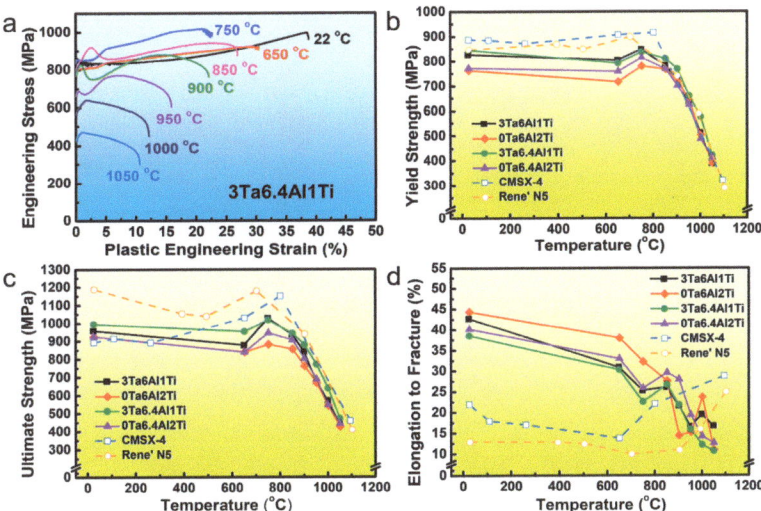

Figure 4. Tensile properties of Ni-based SX superalloys at a temperature range from RT to 1050 °C. (**a**) Stress-strain curves of Ni-based SX superalloy (3Ta6.4Al1Ti). (**b**) YS, (**c**) UTS, and (**d**) EF of four experimental alloys. For comparison, YS, UTS, and EF of two second-generation commercial alloys, CMSX-4 [32] and Rene′ N5 [33], are also included in (**b**–**d**).

Figure 4b–d summarizes the temperature dependence of YS, ultimate tensile strength (UTS), and elongation to fracture (EF) of four experimental SX superalloys. The newly developed SX superalloys contain 3 wt.%Re but without W for a low specific weight. For comparison, data for two commercial Ni-based single crystal superalloys CMSX-4 and Rene′ N5 [32,33], which also have 3 wt.% Re but contain W, is also included in the figures. As Figure 4b,c shows, both YS and UTS of four experimental alloys slightly increase to the maximum value from RT to 750 °C, then drop quickly with increased temperature, which is consistent with those two commercial alloys and many other Ni-based SX superalloys [32,33]. At temperatures below 800 °C, the four experimental alloys have slightly lower YS and UTS than CMSX-4 and Rene′ N5 alloys. However, service temperatures of Ni-based SX superalloys as turbine blades are normally in the range from 780 to 1050 °C [1]. The YS and UTS of the four alloys at temperatures above 800 °C are close to those of the two commercial superalloys. Moreover, the specific weight of the four alloys is lower than 8.4 g/cm^3. Therefore, the newly developed SX superalloys might have potential applications at high temperatures. The comparison of the EF is shown in Figure 4d. The EF of all four alloys below 850 °C exceeds 20%, which is significantly better than that of CMSX-4 and Rene′ N5 alloys. This is consistent with the previously alloy design principle [16]. The high concentration of Co (about 15 wt.%) in the four alloys may effectively reduce the stacking fault energy (SFE) of the face centered cubic (FCC) structured γ matrix. This might be responsible for the good ductility below 850 °C and relatively low ductility above 850 °C [25,34–36]. However, at temperatures above 850 °C, the alloys still exhibit reasonable ductility with EF exceeding 10%.

3.3. Fracture Analysis

Microstructural analysis after tensile tests is examined to reveal the fracture behavior of the alloys. In consistence with the stress-strain behavior, the temperature dependence of fracture modes can also be divided into three groups according to the deformation temperature for all four alloys: low temperatures (RT-750 °C), intermediate temperatures (850–900 °C), and high temperatures (950–1050 °C). The microstructures for one alloy (3Ta6.4Al1Ti) fractured at RT, 850 °C, and 1050 °C are shown as representatives in Figures 5–7.

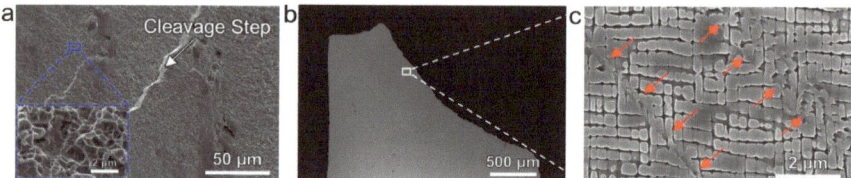

Figure 5. Fracture analysis after the tensile tests at RT. (**a**) Fracture surface of the sample showing the cleavage step (white arrow) and ductile dimples (blue rectangle). The loading direction is out of the plane. (**b**) Side view of the fractured sample. (**c**) Magnified image of the white rectangle in (**b**). The red arrows in (**c**) indicate the shear traces near the fracture surface. The tensile direction in (**b**,**c**) is vertical.

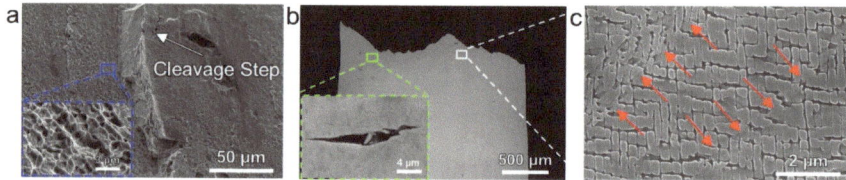

Figure 6. Fracture analysis after the tensile tests at 850 °C. (**a**) Fracture surface of the sample showing the cleavage step (white arrow) and ductile dimples (blue rectangle). The loading direction is out of plane. (**b**) Side view of the fractured sample showing the microvoids (green rectangle) in the interdendritic region. (**c**) Magnified image of the white rectangle in (**b**). The red arrows in (**c**) indicate shear traces near the fracture surface. The tensile direction in (**b**,**c**) is vertical.

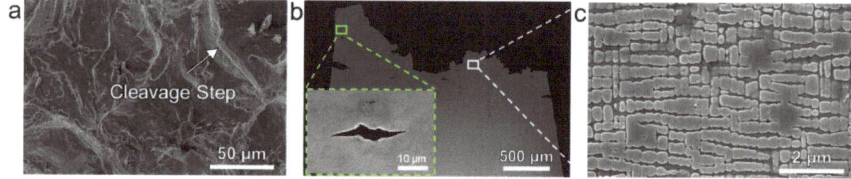

Figure 7. Fracture analysis after the tensile tests at 1050 °C. (**a**) Fracture surface of the sample showing the cleavage step (white arrow). The loading direction is out of the plane. (**b**) Side view of the fractured sample showing the microvoids (green rectangle) in the interdendritic region. (**c**) Magnified image of the white rectangle in (**b**), showing rafting microstructures. The tensile direction in (**b**) and (**c**) is vertical.

As shown in Figure 5a, the alloy fractures in a ductile manner at RT. A large number of cleavage steps and dimples appear on the fracture surface, indicating that the alloy has undergone a large amount of plastic deformation before fracture, which is consistent with the large EF of 35%. The deformation mechanism can be preliminarily determined in Figure 5b,c. There are some shear traces along 45° of the tensile direction passing through the γ/γ' interface (red arrows in Figure 5c), indicating that some of the γ' precipitates are already sheared off. Therefore, the deformation of the alloy at RT is dominated by the localized shear of both phases.

As shown in Figure 6a, similar to the RT samples, the cleavage step and dimples on the fracture surface at 850 °C also show a ductile nature, corresponding to the high EF (>20%). The shear traces on the side image also indicate that the failure of the alloy is dominated by the localized shear of both phases (Figure 6c). One significant difference between the 850 °C and RT samples is that many microvoids (inset in Figure 6b) are observed on the side surface of the tensile specimens after fracturing at 850 °C. Careful examination reveals that those microvoids are located in the interdendritic region, indicating that the interdendritic region is weak during deformation at this temperature range.

As the testing temperature increases to 1050 °C, the fracture surface and deformation mechanism of the samples are changed. The fracture surface is dominated by cleavage steps and without dimples (Figure 7a), which is consistent with the relatively low EF of ~10% (shown in Figure 4d). Moreover, no shear traces in both phases have been observed at 1050 °C (Figure 7c), indicating that deformation might be diffusion dominated rather than dislocation slip [23]. The γ' phase appears to be stretched along the tensile direction of [001] and starts to connect in the horizontal direction indicating that γ phase has been squeezed due to Poisson's effect. In addition, similar to the sample fractured at 850 °C, the weak region is still the interdendritic region, where microvoids are observed (Figure 7b).

3.4. Microstructural Evolution during Thermal Exposure at 1050 °C

The microstructures of the four SX superalloys after thermal exposure at 1050 °C for up to 500 h are shown in Figure 8. No TCP phase formation is observed in all four alloys, indicating excellent phase stability. The γ' morphology evolution during thermal exposure of those SX superalloys shows that the coarsening behavior depends on the composition. During 500 h thermal exposure, the γ' phase of alloys with 6 wt.% Al (3Ta6Al1Ti and 0Ta6Al2Ti) coarsens slower than that with 6.4 wt.% Al. Moreover, γ' particles in the alloys with 6 wt.% Al become irregular, but they are still separated by the γ phase after 500 h exposure at 1050 °C. In comparison, γ' particles in alloys with 6.4 wt.% Al start connection to form rafting microstructure after 100 h thermal exposure, which continuously coarsens as thermal exposure time increases. It is clear that decreasing Al concentration appears to increase the resistance to rafting in the four alloys.

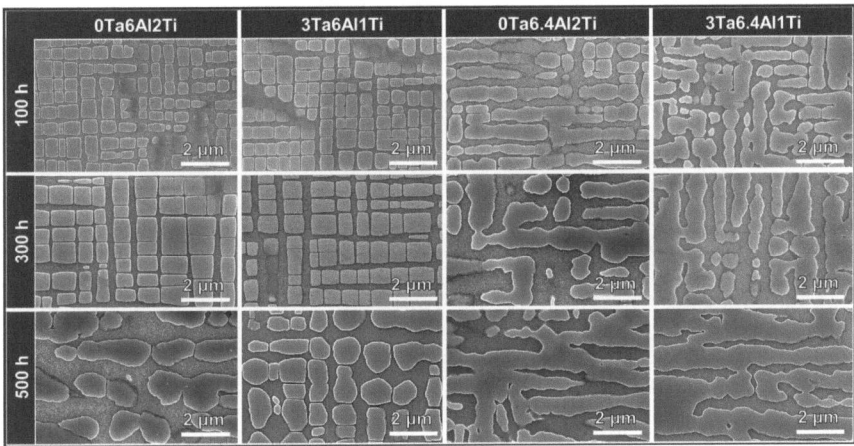

Figure 8. Microstructure evolution of 0Ta6Al2Ti, 3Ta6Al1Ti, Ta6.4Al2Ti, and 3Ta6.4Al1Ti alloys after thermal exposure at 1050 °C for up to 500 h. The microscopic observations identify no TCP phase formation at different exposure times in all alloys.

Normally, the γ' phase coarsening behavior is closely related to lattice misfit (δ) between γ' and γ phases [37,38], which is defined in ref. [1] as:

$$\delta = 2 \times \frac{a^{\gamma'} - a^{\gamma}}{a^{\gamma'} + a^{\gamma}} \quad (1)$$

where a^γ and $a^{\gamma'}$ are lattice parameters of the γ and γ' phases, which can be calculated by using the Caron model [37,39,40] according to the elemental concentrations in γ and γ' phases:

$$a_{RT}^\gamma (\text{Å}) = 3.524 + 0.11 C_{Cr}^\gamma + 0.0196 C_{Co}^\gamma + 0.478 C_{Mo}^\gamma + 0.444 C_W^\gamma + 0.179 C_{Al}^\gamma + 0.422 C_{Ti}^\gamma + 0.7 C_{Ta}^\gamma + 1.03 C_{Hf}^\gamma + 0.15 C_{Fe}^\gamma + 0.441 C_{Re}^\gamma + 0.3125 C_{Ru}^\gamma + 0.7 C_{Nb}^\gamma + 5.741 \times 10^{-5} \tfrac{\text{Å}}{K} \times T - 1.010 \times 10^{-9} \tfrac{\text{Å}}{K^2} \times T^2 \quad (2)$$

$$a_{RT}^{\gamma'} (\text{Å}) = 3.57 - 0.004 C_{Cr}^{\gamma'} - 0.0042 C_{Co}^{\gamma'} + 0.208 C_{Mo}^{\gamma'} + 0.194 C_W^{\gamma'} + 0.258 C_{Ti}^{\gamma'} + 0.5 C_{Ta}^{\gamma'} + 0.78 C_{Hf}^{\gamma'} - 0.004 C_{Fe}^{\gamma'} + 0.262 C_{Re}^{\gamma'} + 0.1335 C_{Ru}^{\gamma'} + 0.46 C_{Nb}^{\gamma'} + 6.162 \times 10^{-5} \tfrac{\text{Å}}{K} \times T - 1.132 \times 10^{-8} \tfrac{\text{Å}}{K^2} \times T^2 \quad (3)$$

where C_i^γ and $C_i^{\gamma'}$ are the atomic percentages of the element i in the γ and γ' phases, and T is the temperature of Kelvin.

In order to calculate the lattice parameters by using the Caron model, compositions of both phases after solid solution and aging treatments are obtained by using EDS quantification analysis. TEM-EDS maps (Figure 9) show the elemental distribution of the 3Ta6.4Al1Ti alloy and locations for quantification in γ and γ' phases. To guarantee data accuracy, more than six positions are selected for each phase in each alloy. The averaged value of phase composition is calculated and listed in Table 2. Accordingly, the lattice misfits of four SX superalloys are calculated using Equations (1)–(3), and the results are also listed in Table 2.

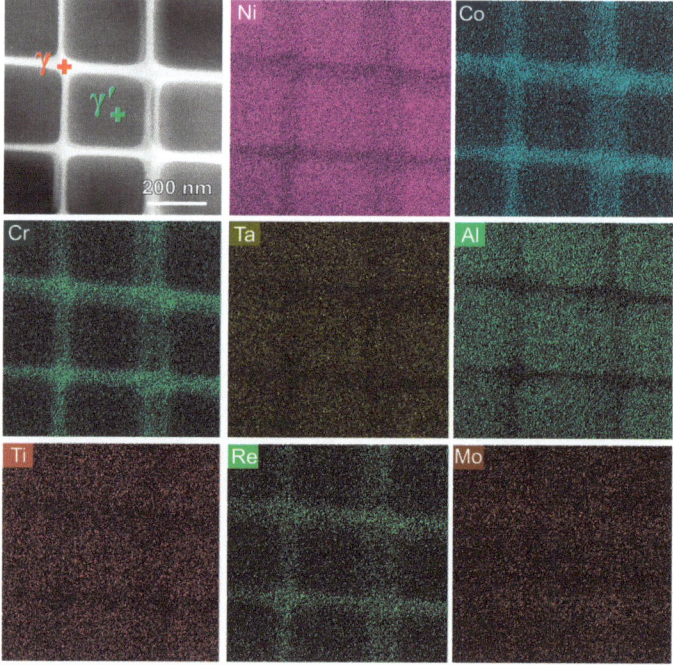

Figure 9. TEM-EDS maps showing the elemental distribution in γ and γ' phases of the 3Ta6.4Al1Ti alloy.

The calculated δ of 3Ta6.4Al1Ti, 0Ta6.4Al2Ti, 3Ta6Al1Ti, and 0Ta6Al2Ti alloys at 1050 °C are -0.37%, -0.3%, -0.27%, and -0.26%, respectively. Generally, high absolute values of the lattice misfit between the γ and γ' phases indicate relatively high interphase boundary energy, which is responsible for quick coarsening/rafting during thermal exposure [38,41,42]. In the four alloys, both 6 wt.% Al alloys have lower absolute values of lattice misfit (~0.26%) than that in the 6.4 wt.% Al alloys, which is consistent with the experimental observation of relatively slow coarsening/rafting.

Table 2. Two-phase compositions (at.%) of four SX superalloys after the solid solution and two-step aging treatments and their lattice misfit at 1050 °C.

Alloy		Cr	Co	Mo	Ta	Re	Al	Ti	Ni	Lattice Misfit (%) at 1050 °C
3Ta6.4Al1Ti	γ	14.6	26.4	5.4	0.3	3.2	4.5	0.2	45.4	−0.37
	γ'	2.3	10.3	2.7	1.8	0.2	16.0	1.6	65.2	
0Ta6.4Al2Ti	γ	14.3	26.6	4.9	-	3.0	3.3	0.4	47.5	−0.3
	γ'	3.2	11.2	2.9	-	0.4	12.6	3.0	66.8	
3Ta6Al1Ti	γ	13.5	25.2	5.2	0.3	3.4	4.7	0.2	47.5	−0.27
	γ'	2.1	9.7	2.6	2.2	0.2	16.1	1.8	66.2	
0Ta6Al2Ti	γ	13.3	25.3	4.8	-	2.9	4.0	0.5	49.2	−0.26
	γ'	2.4	9.8	2.6	-	0.3	15.1	3.6	66.3	

Although 3Ta6Al1Ti and 0Ta6Al2Ti alloys have similar lattice misfits (about −0.26%), their γ' phase coarsening behavior shows a slight difference. After 500 h thermal exposure, the γ' phase of 3Ta6Al1Ti alloy still maintains a certain cubic shape, while the γ' phase in 0Ta6Al2Ti alloy has changed completely into an irregular shape and started to join together (Figure 8). Moreover, the γ' size of 3Ta6Al1Ti alloy is lower than that of 0Ta6Al2Ti alloy. Coarsening is not only influenced by the lattice misfit of the alloy but also requires elemental diffusion in the γ phase, which might be slowed down by the addition of Ta. To maintain consistency with the experimental observation that replacing 1 wt.% Ti with 3 wt.% Ta is beneficial to the stability of the shape and size of γ' phase during thermal exposure, replacing 1 wt.% Ti with 3 wt.% Ta increases the Re and Mo partitioning into the γ phase, which leads to a slow diffusion rate of the γ phase due to both Re and Mo being slow diffusers [42,43].

Collectively, with carefully balancing heavy elements Mo, Ta, Re, and W, four low specific weight SX superalloys with 3 wt.% Re but without W have been fabricated. Tensile tests revealed that the SX alloys have comparable strength to commercial second-generation SX CMSX-4 (3 wt.% Re and 6 wt.% W) and Rene' N5 alloys (3 wt.% Re and 5 wt.% W) above 800 °C. Moreover, the EF below 850 °C (>20%) is better than that of those two commercial SX superalloys. The TCP phase does not appear during thermal exposure at 1050 °C for up to 500 h, indicating excellent phase stability of the alloys. The current work might provide scientific insights for developing Ni-based SX superalloys with low specific weight. However, current work focuses on 3 wt.% Re (typically in second-generation superalloys) without W, how the heavy elements type (Mo, Ta, W, and Re) and concentration combinatorially affect the microstructure, mechanical properties, thermal stability as well as specific weight are complicated and need further systematic exploration.

4. Conclusions

In order to reveal the relationship among the composition, mechanical properties, and thermal stability of SX Ni-based superalloys with low specific weight, SXs are grown. After solution and aging heat treatments, typical γ/γ' two-phase microstructures are obtained to conduct tensile tests at a wide temperature range between RT and 1050 °C and thermal exposure tests at 1050 °C. With the help of advanced microscopy, the following conclusions can be drawn:

(1) All four alloys can be grown into SXs form using the Bridgman method, and the as-grown SXs have typical dendritic microstructure. After solution and aging treatments, alloys all have typical γ/γ' two-phase microstructure, with γ' size of about 260–290 nm.
(2) Tensile tests revealed that the yield strength and ultimate tensile strength of the newly developed SX superalloys are similar to those of typical commercial second-

generation SX CMSX-4 and Rene' N5 superalloys at a temperature above 800 °C. Moreover, the ductility below 850 °C is greater than 20 %, better than that of those two commercial alloys.

(3) Four alloys show similar plastic deformation and fracture behaviors. At RT to 850 °C, the deformation is dominated by localized shear of both phases. Above 950 °C, the deformation appears to be diffusion-dominated and rafting-like microstructures are observed.

(4) During thermal exposure at 1050 °C for up to 500 h, the topological close-packed phase does not appear, indicating excellent phase stability.

(5) The evolution of γ' phase during thermal exposure at 1050 °C is related to the concentration of Al, Ti, and Ta. Decreasing Al concentration appears to increase the resistance of rafting in current experimental alloys and replacing 1 wt.% Ti with 3 wt.% Ta is beneficial to the stability of the shape and size of γ' phase during thermal exposure.

Author Contributions: D.L.: formal analysis, investigation, writing—original draft, writing—review and editing. Q.D.: conceptualization, methodology, formal analysis, investigation, writing—original draft, writing—review and editing. Q.Z.: investigation, writing—review, and editing. D.Z.: investigation, writing—review, and editing. X.W.: resources, writing—review and editing. X.Z.: resources, writing—review and editing. Z.Z.: conceptualization, formal analysis, resources, writing—review and editing, supervision. H.B.: conceptualization, methodology, formal analysis, investigation, resources, writing—original draft, writing—review and editing, supervision. All authors have read and agreed to the published version of the manuscript.

Funding: This work was supported by Basic Science Center Program for Multiphase Media Evolution in Hypergravity of the National Natural Science Foundation of China (No. 51988101), the Key R & D Project of Zhejiang Province (No. 2020C01002), National Science and Technology Major Project of China (J2019-III-0008-0051), National Natural Science Foundation of China (No. 52201027 & 91960201).

Data Availability Statement: The data presented in this study are available on request from the corresponding author.

Conflicts of Interest: The authors declare no conflict of interest.

References

1. Reed, R.C. *The Superalloys Fundamentals and Applications*; Cambridge University Press: Cambridge, UK, 2008; pp. 1–28.
2. Yao, X.; Ding, Q.; Zhao, X.; Wei, X.; Wang, J.; Zhang, Z.; Bei, H. Microstructural rejuvenation in a Ni-based single crystal superalloy. *Mater. Today Nano* **2021**, *17*, 100152. [CrossRef]
3. Wang, M.; Cheng, X.; Jiang, W.; Cao, T.; Liu, X.; Lu, J.; Zhang, Y.; Zhang, Z. The effect of amorphous coating on high temperature oxidation resistance of Ni-based single crystal superalloy. *Corros. Sci.* **2023**, *213*, 111000. [CrossRef]
4. Giamei, A.F. Development of Single Crystal Superalloys: A Brief History. *Adv. Mater. Process* **2013**, *171*, 26–30.
5. Li, J.R.; Zhong, Z.G.; Liu, S.Z.; Tang, D.Z.; Han, M. A Low-Cost Second Generation Single Crystal Superalloy DD6. In *Superalloys 2000, Proceedings of the Ninth International Symposium on Superalloys, Seven Springs, PA, USA, 17–21 September 2000*; Pollock, T.M., Ed.; The Minerals, Metals & Materials Society: Warrendale, PA, USA, 2000; pp. 777–783.
6. Mackay, R.A.; Gabb, T.P.; Smialek, J.L.; Nathal, M.V. Alloy Design Challenge: Development of Low Density Superalloys for Turbine Blade Applications. Available online: https://ntrs.nasa.gov/citations/20100011899 (accessed on 30 September 2020).
7. Xia, W.; Zhao, X.; Yue, L.; Zhang, Z. A review of composition evolution in Ni-based single crystal superalloys. *J. Mate. Sci. Technol.* **2020**, *44*, 76–95. [CrossRef]
8. Yokokawa, T.; Harada, H.; Kawagishi, K.; Kobayashi, T.; Yuyama, M.; Takata, Y. Advanced Alloy Design Program and Improvement of Sixth-Generation Ni-Base Single Crystal Superalloy TMS-238. In Proceedings of the Fourteenth International Symposium on Superalloys, Seven Springs, PA, USA, 12–16 September 2021; Sammy, T., Ed.; The Minerals, Metals & Materials Society: Warrendale, PA, USA, 2020; pp. 122–130.
9. Yao, X.; Ding, Q.; Wei, X.; Wang, J.; Zhang, Z.; Bei, H. The effects of key elements Re and Ru on the phase morphologies and microstructure in Ni-based single crystal superalloys. *J. Alloys Compd.* **2022**, *926*, 166835. [CrossRef]
10. Bewlay, B.P.; Jackson, M.R.; Zhao, J.C. A Review of Very-High-Temperature Nb-Suicide-Based Composites. *Metall. Mater. Trans. A* **2003**, *34A*, 2043–2052. [CrossRef]
11. Lemberg, J.A.; Ritchie, R.O. Mo-Si-B alloys for ultrahigh-temperature structural applications. *Adv. Mater.* **2012**, *24*, 3445–3480. [CrossRef]

12. Luo, L.; Ru, Y.; Qin, L.; Pei, Y.; Ma, Y.; Li, S.; Zhao, X.; Gong, S. Effects of Alloyed Aluminum and Tantalum on the Topological Inversion Behavior of Ni-Based Single Crystal Superalloys at High Temperature. *Adv. Eng. Mater.* **2019**, *21*, 1800793. [CrossRef]
13. Pollock, T.M. Alloy design for aircraft engines. *Nat. Mater.* **2016**, *15*, 809–815. [CrossRef]
14. Xia, W.; Zhao, X.; Yue, L.; Zhang, Z. Microstructural evolution and creep mechanisms in Ni-based single crystal superalloys: A review. *J. Alloys Compd.* **2020**, *819*, 152954. [CrossRef]
15. Harris, K.; Erickson, G.; Wortmann, J.; Froschhammer, D. Development of low density single crystal superalloy CMSX-6. *TMS (Metall. Soc.) Pap. Sel. (USA)* **1984**, *56*, CONF-840909.
16. Liu, D.; Ding, Q.; Yao, X.; Wei, X.; Zhao, X.; Zhang, Z.; Bei, H. Composition design and microstructure of Ni-based single crystal superalloy with low specific weight—Numerical modeling and experimental validation. *J. Mater. Res.* **2022**, *37*, 3773–3783. [CrossRef]
17. Fleischmann, E.; Miller, M.K.; Affeldt, E.; Glatzel, U. Quantitative experimental determination of the solid solution hardening potential of rhenium, tungsten and molybdenum in single-crystal nickel-based superalloys. *Acta Mater.* **2015**, *87*, 350–356. [CrossRef]
18. Horst, O.M.; Adler, D.; Git, P.; Wang, H.; Streitberger, J.; Holtkamp, M.; Jöns, N.; Singer, R.F.; Körner, C.; Eggeler, G. Exploring the fundamentals of Ni-based superalloy single crystal (SX) alloy design: Chemical composition vs. microstructure. *Mater. Des.* **2020**, *195*, 108976. [CrossRef]
19. Helmer, H.; Matuszewski, K.; Müller, A.; Rettig, R.; Ritter, N.; Singer, R. Development of a Low-Density Rhenium-Free Single Crystal Nickel-Based Superalloy by Application of Numerical Multi-Criteria Optimization Using Thermodynamic Calculations. In Proceedings of the Thirteenth International Symposium on Superalloys, Seven Springs, PA, USA, 11–15 September 2016; Mark, H., Ed.; The Minerals, Metals & Materials Society: Warrendale, PA, USA, 2016; pp. 35–44.
20. Zhang, J.; Luo, Y.; Zhao, Y.; Yang, S.; Jia, Y.; Cui, L.; Xu, J.; Tang, D. The research of microstructure and property of a low density nickel base single crystal superalloy. *J. Aeron. Mater.* **2011**, *31*, 90–93.
21. Du, Y.; Tan, Z.; Yang, Y.; Wang, X.; Zhou, Y.; Li, J.; Sun, X. Creep Properties of a Nickel-Based Single Crystal Superalloy with Low Density. *Met. Mater. Int.* **2021**, *27*, 5173–5178. [CrossRef]
22. Ding, Q.; Bei, H.; Wei, X.; Gao, Y.F.; Zhang, Z. Nano-twin-induced exceptionally superior cryogenic mechanical properties of a Ni-based GH3536 (Hastelloy X) superalloy. *Mater. Today Nano* **2021**, *14*, 100110. [CrossRef]
23. Ding, Q.; Bei, H.; Zhao, X.; Gao, Y.; Zhang, Z. Processing, Microstructures and Mechanical Properties of a Ni-Based Single Crystal Superalloy. *Crystals* **2020**, *10*, 572. [CrossRef]
24. Ding, Q.; Bei, H.; Li, L.; Ouyang, J.; Zhao, X.; Wei, X.; Zhang, Z. The dependence of stress and strain rate on the deformation behavior of a Ni-based single crystal superalloy at 1050 °C. *Int. J. Mech. Syst. Dyn.* **2021**, *1*, 121–131. [CrossRef]
25. Song, W.; Wang, X.G.; Li, J.G.; Meng, J.; Yang, Y.H.; Liu, J.L.; Liu, J.D.; Zhou, Y.Z.; Sun, X.F. Effect of Ru on tensile behavior and deformation mechanism of a nickel-based single crystal superalloy. *Mater. Sci. Eng. A* **2021**, *802*, 140430. [CrossRef]
26. Wang, G.L.; Liu, J.L.; Liu, J.D.; Wang, X.G.; Zhou, Y.Z.; Sun, X.D.; Zhang, H.F.; Jin, T. Temperature dependence of tensile behavior and deformation microstructure of a Re-containing Ni-base single crystal superalloy. *Mater. Des.* **2017**, *130*, 131–139. [CrossRef]
27. Xiong, X.; Quan, D.; Dai, P.; Wang, Z.; Zhang, Q.; Yue, Z. Tensile behavior of nickel-base single-crystal superalloy DD6. *Mater. Sci. Eng. A* **2015**, *636*, 608–612. [CrossRef]
28. Ding, Q.; Bei, H.; Yao, X.; Zhao, X.; Wei, X.; Wang, J.; Zhang, Z. Temperature effects on deformation substructures and mechanisms of a Ni-based single crystal superalloy. *Appl. Mater. Today* **2021**, *23*, 101061. [CrossRef]
29. Yin, Q.; Wen, Z.; Wang, J.; Lian, Y.; Lu, G.; Zhang, C.; Yue, Z. Microstructure characterization and damage coupled constitutive modeling of nickel-based single-crystal alloy with different orientations. *Mater. Sci. Eng. A* **2022**, *853*, 143761. [CrossRef]
30. Tan, Z.H.; Wang, X.G.; Du, Y.L.; Duan, T.F.; Yang, Y.H.; Liu, J.L.; Liu, J.D.; Yang, L.; Li, J.G.; Zhou, Y.Z.; et al. Temperature dependence on tensile deformation mechanisms in a novel Nickel-based single crystal superalloy. *Mater. Sci. Eng. A* **2020**, *776*, 138997. [CrossRef]
31. Dieter, G.E. *Mechanical Metallurgy*; McGraw-Hill Book Company: New York, NY, USA, 1961; pp. 110–117.
32. Sengupta, A.; Putatunda, S.K.; Bartosiewicz, L.; Hangas, J.; Nailos, P.J.; Peputapeck, M.; Alberts, F.E. Tensile behavior of a new single-crystal nickel-based superalloy (CMSX-4) at room and elevated temperatures. *J. Mater. Eng. Perform.* **1994**, *3*, 73–81. [CrossRef]
33. Corrigan, J.; Launsbach, M.G.; Mihalisin, J.R. Nickel Base Superalloy and Single Crystal Castings. U.S. Patent 8241560 B2, 14 August 2012.
34. Yang, W.; Qu, P.; Sun, J.; Yue, Q.; Su, H.; Zhang, J.; Liu, L. Effect of alloying elements on stacking fault energies of γ and γ' phases in Ni-based superalloy calculated by first principles. *Vacuum* **2020**, *181*, 109682. [CrossRef]
35. Yang, W.; Qu, P.; Liu, C.; Cao, K.; Qin, J.; Su, H.; Zhang, J.; Ren, C.; Liu, L. Temperature dependence of compressive behavior and deformation microstructure of a Ni-based single crystal superalloy with low stacking fault energy. *Trans. Nonferrous Met. Soc. China* **2023**, *33*, 157–167. [CrossRef]
36. Tian, C.; Han, G.; Cui, C.; Sun, X. Effects of stacking fault energy on the creep behaviors of Ni-base superalloy. *Mater. Des.* **2014**, *64*, 316–323. [CrossRef]
37. Long, H.; Wei, H.; Liu, Y.; Mao, S.; Zhang, J.; Xiang, S.; Chen, Y.; Gui, W.; Li, Q.; Zhang, Z.; et al. Effect of lattice misfit on the evolution of the dislocation structure in Ni-based single crystal superalloys during thermal exposure. *Acta Mater.* **2016**, *120*, 95–107. [CrossRef]

38. Zhuang, X.; Antonov, S.; Li, L.; Feng, Q. Effect of alloying elements on the coarsening rate of γ' precipitates in multi-component CoNi-based superalloys with high Cr content. *Scr. Mater.* **2021**, *202*, 114004. [CrossRef]
39. Reed, R.C.; Tao, T.; Warnken, N. Alloys-By-Design: Application to nickel-based single crystal superalloys. *Acta Mater.* **2009**, *57*, 5898–5913. [CrossRef]
40. Wei, B.; Lin, Y.; Huang, Z.; Huang, L.; Zhou, K.; Zhang, L.; Zhang, L. A novel Re-free Ni-based single-crystal superalloy with enhanced creep resistance and microstructure stability. *Acta Mater.* **2022**, *240*, 118336. [CrossRef]
41. Harada, H.; Murakami, H. Design of Ni-base superalloys. In *Computational Materials Design*; Saito, T., Ed.; Springer: Heidelberg, Germany, 1999; Volume 34, pp. 39–70.
42. Pan, Q.; Zhao, X.; Cheng, Y.; Yue, Q.; Gu, Y.; Bei, H.; Zhang, Z. Effects of Co on microstructure evolution of a 4th generation nickel-based single crystal superalloys. *Intermetallics* **2023**, *153*, 107798. [CrossRef]
43. Lu, F.; Antonov, S.; Lu, S.; Zhang, J.; Li, L.; Wang, D.; Zhang, J.; Feng, Q. Unveiling the Re effect on long-term coarsening behaviors of γ' precipitates in Ni-based single crystal superalloys. *Acta Mater.* **2022**, *233*, 117979. [CrossRef]

Disclaimer/Publisher's Note: The statements, opinions and data contained in all publications are solely those of the individual author(s) and contributor(s) and not of MDPI and/or the editor(s). MDPI and/or the editor(s) disclaim responsibility for any injury to people or property resulting from any ideas, methods, instructions or products referred to in the content.

Article

The Effect of Interatomic Potentials on the Nature of Nanohole Propagation in Single-Crystal Nickel: A Molecular Dynamics Simulation Study

Xinmao Qin [1,2,3,4], Yilong Liang [1,3,4,*], Jiabao Gu [1,3,4] and Guigui Peng [1,3,4]

1. College of Materials Science and Metallurgical Engineering, Guizhou University, Guiyang 550025, China
2. School of Electronic and Information Engineering, Anshun University, Anshun 561000, China
3. Guizhou Key Laboratory for Mechanical Behavior and Microstructure of Materials, Guiyang 550025, China
4. National & Local Joint Engineering Laboratory for High-Performance Metal Structure Material and Advanced Manufacturing Technology, Guiyang 550025, China
* Correspondence: ylliang@gzu.edu.cn

Abstract: Based on a molecular dynamics (MD) simulation, we investigated the nanohole propagation behaviors of single-crystal nickel (Ni) under different styles of Ni–Ni interatomic potentials. The results show that the MEAM (the modified embedded atom method potential) potential is best suited to describe the brittle propagation behavior of nanoholes in single-crystal Ni. The EAM/FS (embedded atom method potential developed by Finnis and Sinclair) potential, meanwhile, is effective at characterizing the plastic growth behavior of nanoholes in single-crystal Ni. Furthermore, the results show the difference between the different styles of interatomic potentials in characterizing nanohole propagation in single-crystal Ni and provide a theoretical basis for the selection of interatomic potentials in the MD simulation of Ni crystals.

Keywords: nanohole propagation; interatomic potentials; dislocation; single-crystal Ni

Citation: Qin, X.; Liang, Y.; Gu, J.; Peng, G. The Effect of Interatomic Potentials on the Nature of Nanohole Propagation in Single-Crystal Nickel: A Molecular Dynamics Simulation Study. *Crystals* **2023**, *13*, 585. https://doi.org/10.3390/cryst13040585

Academic Editors: Wangzhong Mu and Chao Chen

Received: 1 March 2023
Revised: 22 March 2023
Accepted: 22 March 2023
Published: 29 March 2023

Copyright: © 2023 by the authors. Licensee MDPI, Basel, Switzerland. This article is an open access article distributed under the terms and conditions of the Creative Commons Attribution (CC BY) license (https:// creativecommons.org/licenses/by/ 4.0/).

1. Introduction

Fracture is a widespread and complex process of crack initiation, propagation, and coalescence, spanning a range of scales from the macroscale, mesoscale, and microscale to the atomic scale. The fracture of components is a critical issue that determines integrity and safety. At the macroscale [1–8], a variety of continuum fracture mechanics theories and empirical formulas have been established to analyze the macroscale failure behavior of materials and components. The traditional continuum fracture mechanics theory, however, is unsuitable for describing the basic physical mechanism of the failure process at the atomic scale. At the atomic scale, the essence of the fracture process of materials is the breaking of bonds that bind atoms of materials during crack initiation and propagation. Hence, atomic-scale modeling and simulations are required. The molecular dynamics (MD) simulation is a useful tool that can be used to explore the physical and mechanical properties of materials at the atomic level [9].

MD simulation is a powerful tool that can be used to study the microstructural evolution (involving dislocations, stacking faults, and twins) of plastic deformation and the fracture processes of materials [10–13]. To more systematically obtain the mechanism of deformation and fracture, crack initiation, propagation, and coalescence have been investigated based on the MD simulation. The main factors determining the crack propagation behavior are initiated crack length, crack distribution, temperature, strain rate, and the stress state of the crack tip [14–20].

In addition, nickel (Ni)-based single-crystal superalloys have been used in high-performance applications, such as turbine disks and blades, due to their good performance in creep resistance and fatigue resistance [21–24]. Therefore, it has been necessary to investigate the deformation, crack nucleation, and propagation mechanisms of the FCC γ phase (matrix) in the Ni-based single-crystal superalloy. Yang et al. [25] explored the effects of grain boundary structures on crack nucleation during the deformation process in a Ni-nano-laminated structure. Yao [26] studied the microstructure evolution and stress distribution of pre-crack single-crystal Ni at different temperatures. The effects of temperature, strain rate, and orientation on the crack propagation of single-crystal Ni were demonstrated by Chen [27]. Furthermore, the effects of three-dimensional defects on crack growth were investigated [28]. The crack propagation mechanisms and behaviors of crystalline Ni-based materials have been studied based on the MD simulation using different styles of interatomic potentials [29–35]. However, no systematic investigation has been conducted to examine the nanohole propagation behaviors and mechanisms for the different styles of Ni–Ni interatomic potentials.

In this study, we used the large-scale atomic/molecular massively parallel simulator (LAMMPS) software based on the MD simulation to investigate the nanohole propagation behaviors of single-crystal Ni at different styles of Ni–Ni interatomic potentials. We systematically compared the nanohole propagation behaviors of the three styles of Ni–Ni interatomic potentials and investigated the differences between them in characterizing the nanohole propagation of single-crystal Ni. The results offer a theoretical basis for the selection of interatomic potentials in the MD simulation of Ni crystals.

2. Simulation Conditions

2.1. Initial Conditions

In this work, we investigated the nanohole propagation behaviors and mechanisms of single-crystal Ni according to uniaxial tensile deformation along the Y [010] direction of MD simulation models, as shown in Figure 1. The single-crystal Ni was in the cubic orientations of X—[100], Y—[010], and Z—[001]. The size of the model was 50 a × 50 a × 5 a (176 Å × 176 Å × 17.6 Å), where a = 3.52 Å is the lattice parameter of Ni crystal (Figure 1). By deleting specified Ni atoms at the central region of the deformation system, we created a model of a cylindrical nanohole with a specified size. The diameter and thickness of the nanohole were 20 Å and 17.6 Å, respectively.

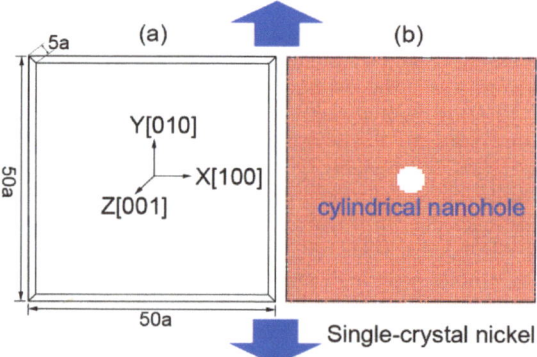

Figure 1. The MD model of FCC single-crystal Ni with a central cylindrical nanohole: (**a**) the size and orientation of simulated region and (**b**) single-crystal Ni with cylindrical nanohole.

We applied periodic boundary conditions in all directions. Using the conjugate gradient (CG) algorithm, we performed energy minimization by iteratively adjusting the coordinates of the Ni atoms of single-crystal Ni. Before tensile deformation, using an isothermal–isobaric ensemble (NPT) [36–38], we relaxed the tensile model at 20 K and 0 bar pressure for 10 ps (Tdamp = 0.01 ps and Pdamp =1 ps).Then, the tensile deformation of the system was performed at a constant temperature of 20 K, which was realized using a canonical ensemble (NVT)(Tdamp = 0.001 ps).The application of NVT ensembles means that the lateral dimensions (X and Z directions) are not allowed to relax. Uniaxial tensile deformation with a strain rate of 0.001 ps^{-1} was applied to the Y direction of single-crystal Ni. In the simulation, the simulation timestep was 0.001 ps. To analyze the nanohole propagation behaviors of single-crystal Ni, we visualized the atomic configurations and stress distributions of Ni atoms using the Open Visualization Tool (OVITO) [39].

To obtain the nanohole propagation behaviors, we calculated the atomic stress definitions of the front of the nanohole during tensile deformation and the average atomic stress $\sigma_{\alpha\beta}(i)$ [40–42] as follows:

$$\sigma_{\alpha\beta}(i) = -\frac{1}{2\Omega_i} \sum_{j \neq i}^{N} f_\alpha(i,j) r_\beta(i,j) \tag{1}$$

where α and β represent x, y, or z; N is the number of the atoms in a region around i within a potential cutoff distance; $f_\alpha(i,j)$ is the vector component form of the interaction force exerted by atom j on atom i; $r_\beta(i,j)$ is the vector component form of the relative position form of atom j on atom i; and Ω_i is the volume of atom i given by the calculation of the Voronoi tessellation of the atom i in the simulation box.

In addition, the microstructure evolution of the tensile system was analyzed using common neighbor analysis (CNA) [43,44] and the dislocation extraction algorithm (DXA) of the model.

2.2. Potential between Atoms

We applied the three styles of potentials in our MD simulation—namely, the modified embedded atom method potential (hereinafter referred to as the MEAM potential) [45], the embedded atom method potential developed by Finnis and Sinclair (hereinafter referred to as the EAM/FS potential) [46], and the embedded atom method potential developed by Foiles and Baskes (hereinafter referred to as the EAM potential) [47]. Furthermore, the relevant parameters of the MEAM, EAM/FS, and EAM potentials are included in Supplementary Materials.

3. Simulation Results and Discussion

3.1. Stress–Strain Behavior

Figure 2 shows the stress–strain behavior of single-crystal Ni at various interatomic potentials, comprising the (a) MEAM potential, (b) EAM/FS potential, and (c) EAM potential. For the different styles of Ni–Ni interatomic potentials, during the initial stage of tensile deformation, the single-crystal Ni exhibited different stress–strain behaviors. ε denotes the tensile strain, and ε_e, ε_p, and ε_t denote the elastic, plastic, and total strain, respectively. When $\varepsilon < \varepsilon_e$, the tensile process was in the elastic deformation stage. It was when the tensile stress of these models reached the peak stress that the tensile process began to enter the plastic deformation stage (the peak stress denoted the yield stress). For the MEAM potential, after the tensile stress reached its peak, it decreased quickly to zero with an increase in strain (Figure 2a). The accumulated plastic strain, which was defined as the total strain (ε_t) at the fracture minus the elastic strain (ε_e) [48,49], was only 6% (as shown in Figure 3). Conversely, in the process of the plastic deformation of single-crystal Ni at the EAM/FS potential, the flow stress dropped slowly to non-zero followed by a jerky flow and gradual decrease. This feature indicated the representative ductile nature of single-crystal Ni (Figure 2b), and this ductile nature was further demonstrated by the accumulated plastic strain of 13% (Figure 3). As shown in Figures 2 and 3, the stress–strain behavior of single-crystal Ni at the EAM potential also had the stress–strain characteristics of single-crystal Ni at the MEAM potential and EAM/FS potential. The accumulated plastic strain was 10% (between 6% of the MEAM potential and 13% of the EAM/FS potential).

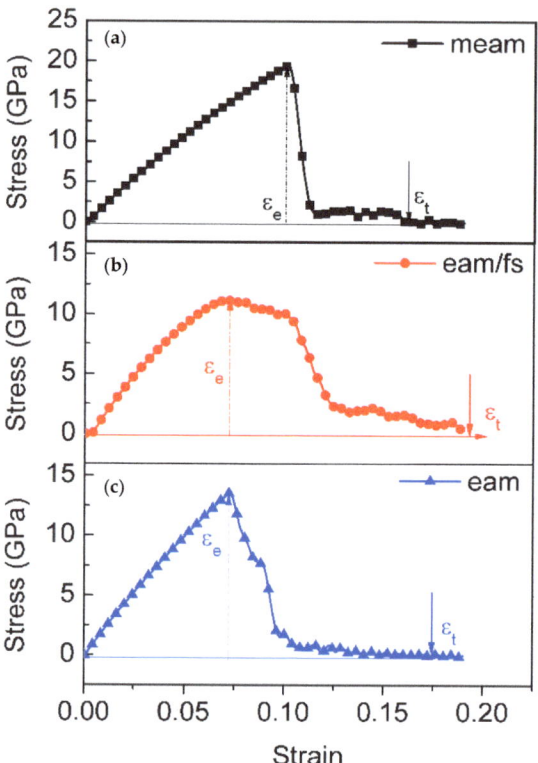

Figure 2. The stress–strain behavior of single-crystal Ni under the (**a**) MEAM potential, (**b**) EAM/FS potential, and (**c**) EAM potential. The failure location is marked by the solid arrow.

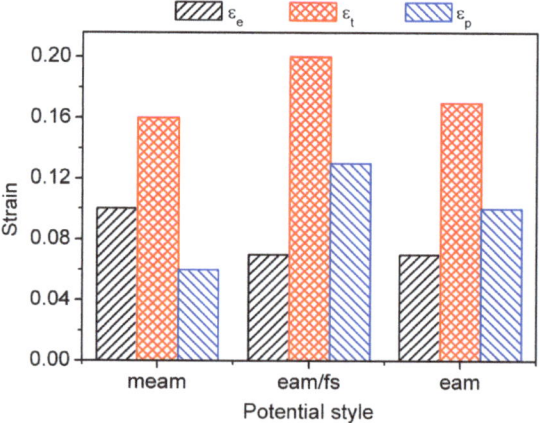

Figure 3. The elastic strain ε_e, total strain ε_t and accumulated plastic strain ε_p of single-crystal Ni under different styles of interatomic potentials.

3.2. Nanohole Propagation Behavior

As the MEAM potential was used to study the nanohole propagation process, we found that the central nanohole propagated first using a fast brittle propagation model that included the process of formation and coalescence of nanopores at the front of the nanohole, as shown in Figure 4. In the process of uniaxial tensile along the y direction, the stress concentration was present at the left and right sides of the region of the central nanohole (see (a1) inset in Figure 4). When ε increased from 0% to 10.1%, the no. 1 nanopore formed at the left-bottom corner of the central nanohole because the relevant atoms of this region had a maximum tensile stress (about σ_{yy} = 32.2 GPa). As the ε value increased, the no. 1 nanohole gradually grew and coalesced with the main nanohole. At the same time, the no. 2 nanopore formed at the right-bottom corner of the main nanohole due to the stress concentration (ε = 10.2%, σ_{yy} = 31 GPa; see Figure 4(c1)). Then, the no. 2 nanopore gradually grew and coalesced with the main nanohole, and the left region of the main nanohole also produced two nanopores (no. 3 and no. 4 nanopores). As shown in Figure 4d, the plastic deformation occurred in the upper local area of the right nanopore. When ε = 10.7%, the new no. 3 and no. 4 nanopores continued to grow, and the misorientation between the tensile direction and the nanohole growth direction was 45°, indicating that the crack mainly propagated along the (110) plane of single-crystal Ni (see Figure 4g). Meanwhile, the stress concentration was present in the region of the front of the right-bottom corner of the propagated nanohole (Figure 4g; σ_{yy} = 34 GPa), which gave rise to the new no. 5 nanopore initiation (Figure 4h). As ε = 15.9%, the nanohole propagated across the whole single-crystal Ni (Figure 4i). When the ε value was below 10.4%, the nanohole was propagated using a fast brittle propagation model that included the process of formation and the coalescence of nanopores at the front of the nanohole with almost no emission of dislocations from the nanohole. With the strain increasing from 10.4% to 10.9%, however, the process of nanohole propagation was accompanied by the emission and slip of dislocations.

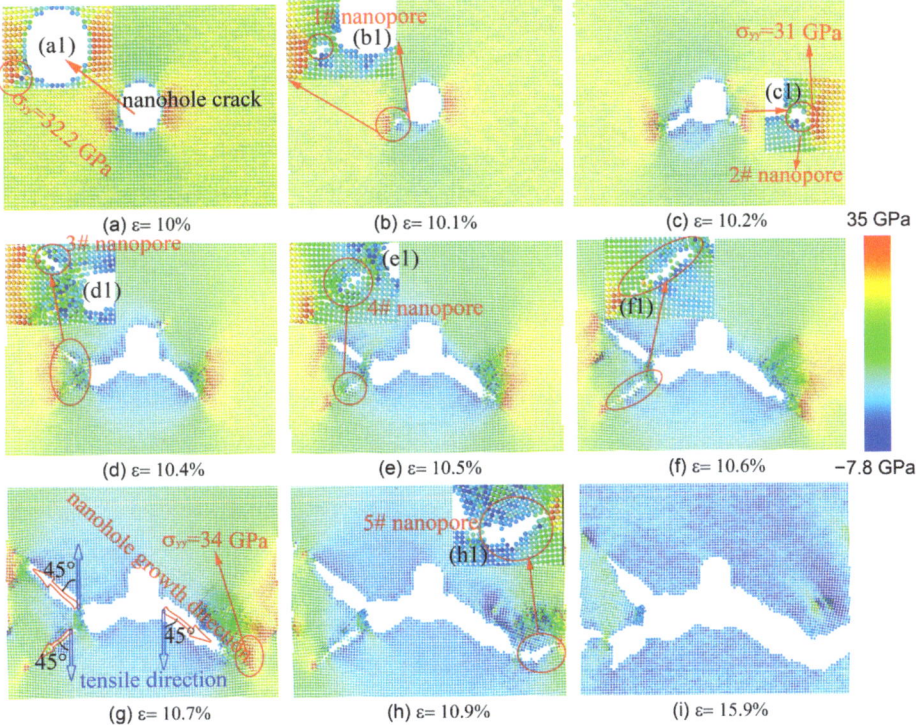

Figure 4. The contour plots of the atomic tensile stress field and nanohole growth states at different tensile strains (MEAM potential).

Figure 5 shows the process of the uniaxial tensile of single-crystal Ni with the EAM/FS potential. To conveniently analyze the structural change in the nanohole local region during the nanohole propagation process, we performed a CNA and deleted the perfect atoms of the FCC structure, as shown in Figure 5(b1–f1), in which the gray and red atoms denote the amorphous atoms and the HCP structure atoms, respectively. The insets of Figure 5(b2,c2) show the change in dislocation type of different tensile strains. From Figure 5, we found that the front of the nanohole first presented stress concentration, and then the stress concentration level increased with the tensile strain. To decrease the stress concentration of the local region of the front of the nanohole, the nanohole propagation was carried out by changing the shape from a cylindrical nanohole to a rectangular nanohole (Figure 5a,b). When $\varepsilon = 8.3\%$, the stress concentration of the nanohole resulted in the lattice structure transformation of the local region of the front of the nanohole (from a perfect FCC structure to amorphous atoms (gray atoms) and an HCP structure (red atoms); Figure 5b1). We also found that the stair-rod dislocations with a Burgess vector of $\frac{a}{6}[\bar{1}01]$ appeared at the boundary between the region of amorphous atoms and the perfect FCC structure (see Figure 5(b2), the magenta dislocation line). The stair-rod dislocation with a Burgess vector of $\frac{a}{6}[\bar{1}01]$ was formed through the dislocation reaction of $\frac{a}{6}[112] + \frac{a}{6}[\bar{2}\bar{1}\bar{1}] \rightarrow \frac{a}{6}[\bar{1}01]$. The dislocations of $\vec{b} = \frac{a}{6}[112]$ and $\vec{b} = \frac{a}{6}[\bar{2}\bar{1}\bar{1}]$ were Shockley partial dislocations. The stair-rod dislocation (also called the Lomer–Cottrell lock) further impeded the advance of the slip and resulted in a pile-up of the dislocation. Consequently, as the strain increased, the nanohole growth of the left-upper corner was hindered by the Lomer–Cottrell lock (as shown in Figure 6; see the red platform of the left nanohole length–strain curve). For the right-bottom corner of the nanohole, the nanohole was blunted throughout the local region atom's amorphization to release a stress concentration, and a Lomer–Cottrell lock did not form. Therefore, the nanohole growth of the right-bottom corner was not hindered. When $\varepsilon = 9.5\%$, the Lomer–Cottrell lock of $\vec{b} = \frac{a}{6}[\bar{1}01]$ disappeared from the left-upper corner of the nanohole via the relative motion of atoms in the local region. Therefore, the effect of the pile-up of the dislocation of the Lomer–Cottrell lock was removed. Two Shockley partial dislocations of $\vec{b} = \frac{a}{6}[\bar{1}21]$ and $\vec{b} = \frac{a}{6}[1\bar{2}\bar{1}]$ also formed (see the green line in Figure 5c2). Then, the nanohole propagated in the way of the local region crystal structure transformation and the dislocations slip (Figures 5d–f and 6).

Figure 7 shows the process of the uniaxial tensile test of single-crystal Ni for the use of EAM potential. When $\varepsilon = 7.1\%$, the stress concentration was present at the left and right regions of the nanohole (Figure 7a). Then, with an increase in tensile strain, the stress concentration level of the nanohole local region increased gradually, resulting in the formation of an amorphous structure in this region (Figure 7(b1,c1)), and the dislocations started to nucleate at the boundary between the region of the amorphous structure and the perfect FCC structure. The dislocation slips of the front of the nanohole resulted in nanohole propagation (Figure 7a–c). However, when the tensile strain was 8.7%, a stair-rod dislocation with a Burgess vector of $\frac{a}{6}[\bar{1}01]$ appeared at 20 Å from the front of the nanohole (Figure 7(d1,d2); see the magenta dislocation line). The stair-rod dislocation was a fixed dislocation, halting the right-side growth of the nanohole. These immobile high-density dislocations caused a maximum tensile stress of about $\sigma_{yy} = 26$ GPa at the right-side local region of the nanohole (Figure 7(d1,d2)). Further increased strain led to the formation of a new nanopore to release the stress concentration level (Figure 7(e1,e2)). Finally, through the process of dislocation slip and the formation and coalescence of the nanopore, the tensile model was completely fractured.

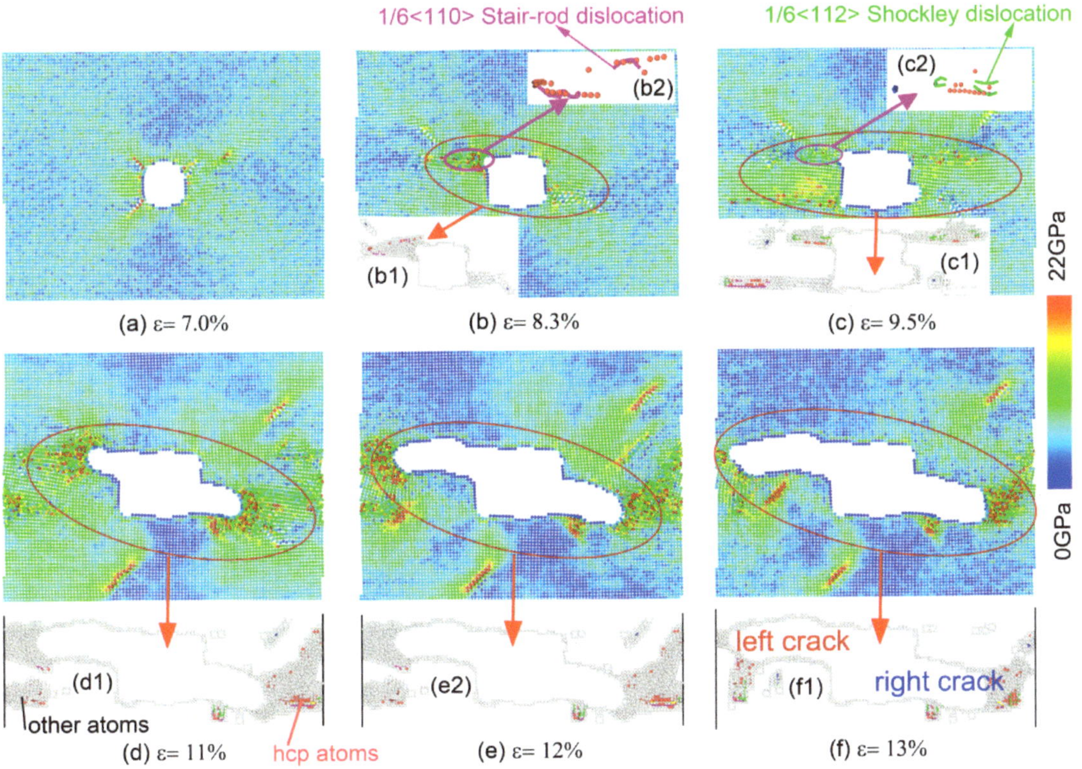

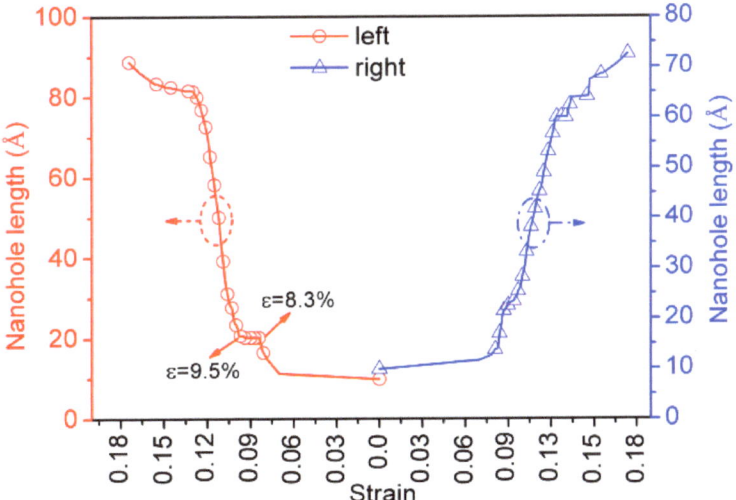

Figure 5. The contour plots of the atomic tensile stress field and crack growth states at different tensile strains (EAM/FS potential).

Figure 6. The nanohole length–strain curve of the nanohole propagation process (EAM/FS potential).

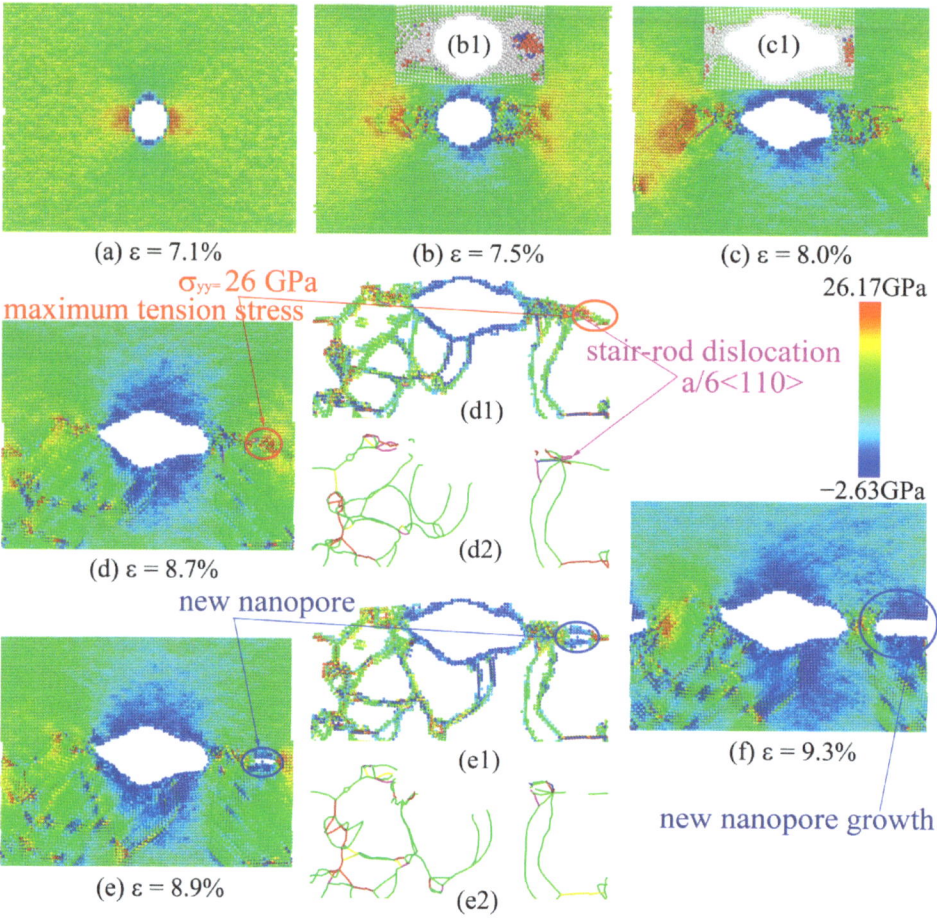

Figure 7. The contour plots of the atomic tensile field and crack growth states at different tensile strains (EAM potential).

3.3. Relationship between Crack Length and Tensile Strain

Generally, the propagation rate of nanoholes and the flow stress or tensile strains are closely related [50]. Figure 8 shows the relationship between tensile strain and nanohole length for the tensile process of single-crystal Ni at the MEAM, EAM/FS, and EAM potentials. For the MEAM potential, the center nanohole was propagated when the tensile strain was (ε) 10%. After that, the nanohole growth entered a rapid stage—for example, when the tensile strain increased from 10% to 11%, the total nanohole length increased rapidly from 20 Å to 180 Å. Then, the total nanohole length increased slowly. When the tensile strain was (ε_t)16%, the nanohole growth extended across the single-crystal Ni along the x-direction (the nanohole length was about 190 Å). The relationship between the nanohole length and strain further confirmed the nature of the nanohole propagation of single-crystal Ni at the MEAM potential. For the single-crystal Ni tensile model under the EAM/FS potential, the central nanohole began to propagate when the tensile strain was (ε) 7% (the single-crystal Ni tensile model under the EAM potential has the same behavior). Then, the tensile models under the EAM/FS and EAM potentials entered the rapid propagation stage. At the strain rates of 10% (for the EAM/FS potential) and 13% (for the EAM potential), the propagation rate of nanoholes decreased. However, in this rapid propagation stage of nanoholes, the nanohole propagation rate of the tensile model under the EAM/FS potential was relatively slow compared to that of the tensile model under the EAM potential. This slow rate was due to the nature of the dislocation (emission and slip) of the tensile model under the EAM/FS potential. Afterward, as the strain increased, the nanohole propagation was conducted in the form of a dislocation moving from the front of the nanohole to the edge of the tensile sample. The relationship between the crack length and strain confirmed the ductile crack propagation of single-crystal Ni at the EAM/FS potential.

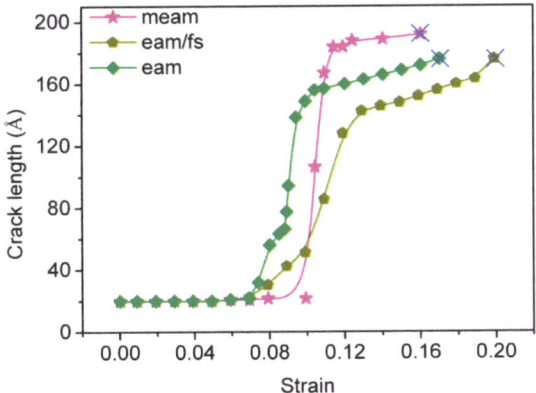

Figure 8. The crack length–strain curve of single-crystal Ni at different styles of potentials. The symbol '×' denotes the fracture point of the tensile model.

3.4. Discussion

Figures 2 and 3 show that the EAM/FS potential was effective in describing the Ni–Ni interaction, which showed good plastic deformation ability and a good maximum cumulative plastic strain (ε_p = 13%). In addition, the cylindrical nanohole first was transformed into a square nanohole due to its good plastic deformation ability before the nanohole propagation. Then, the nanohole was propagated forward due to the local region passivation of the front of the nanohole and the dislocation emission, which showed clear plastic propagation behavior. Therefore, for the single-crystal Ni tensile model under the EAM/FS potential, crack propagation showed the obvious plasticity behavior. For the condition of the MEAM potential to describe the Ni–Ni interaction, the single-crystal Ni showed the worst plastic deformation capacity, and the cumulative plastic strain at the main stage of crack propagation was only about 1% (Figure 4a–h, at which point the crack propagation was almost throughout the entire cross-section of the tensile model). It should be noted that, although the cumulative plastic strain corresponding to this potential was 6% (Figure 3), this was mainly due to the consumption of 5% plastic strain work in the final stage of crack propagation (Figure 4h–g). Hence, when the MEAM potential was used to describe single-crystal Ni, crack propagation showed obvious brittle behavior. For the EAM potential, the single-crystal Ni tensile model exhibited both plastic crack propagation related to dislocations and brittle crack propagation related to the micropore formation (Figure 7e,f). Furthermore, the above analysis can be further confirmed by the results of the crack length and the tensile strain curve in Figure 8.

To analyze the reason for the above difference in nanohole propagation behavior, we further compared the surface energy and stacking fault energy of single-crystal Ni for the MEAM, EAM/FS, and EAM potentials. The surface energy and stacking fault energy are shown in Table 1. It can be found that the model of single-crystal Ni described by the MEAM potential exhibited the maximum surface energy and stacking fault energy. Furthermore, the EAM/FS potential gave the minimum surface energy and stacking fault energy of single-crystal Ni. These differences in surface energy and stacking fault energy of single-crystal Ni at different styles of potentials eventually led to the difference in nanohole propagation behaviors.

Table 1. The computed properties of single-crystal Ni for the different styles of potentials.

		MEAM	EAM/FS	EAM
Surface energy (erg/cm^2)	(100) plane	1943	1444	1580
	(110) plane	2057	1548	1730
	(111) plane	1606	1153	1450
Stacking fault energy (erg/cm^2)		125	33	–

4. Conclusions

In this study, based on the MD simulation, we investigated the nanohole propagation behaviors of single-crystal Ni under different styles of potentials (MEAM potential, EAM/FS potential, and EAM potential). The simulation results revealed that the behaviors of nanohole propagation for the different styles of potentials were quite different. According to the experimental results, the following conclusions can be drawn:

(1) The MEAM potential is best suited to describe the brittle propagation behavior of nanoholes in single-crystal Ni.
(2) The EAM/FS potential is effective in characterizing the plastic growth behavior of nanoholes in single-crystal Ni.

The results showed the differences between different styles of potentials in characterizing nanohole propagation in single-crystal Ni. Furthermore, the results offer a theoretical basis for the selection of interatomic potentials in the MD simulation of Ni crystals.

However, the current results were obtained under special conditions (for example, a temperature of 20 K and a strain rate of $0.001\ ps^{-1}$). The microstructure evolution and nanohole propagation process in the single-crystal Ni can be different as the simulation conditions change. In the future, we will systematically consider the effects of temperature, strain rate, crack shape, and potential function on crack propagation in single-crystal Ni.

Supplementary Materials: The following supporting information can be downloaded at: https://www.mdpi.com/article/10.3390/cryst13040585/s1, Table S1. The relevant parameters of the Ni–Ni interatomic meam potential; Table S2. The relevant parameters of the Ni–Ni interatomic eam/fs potential and eam potential;

Author Contributions: Writing—original draft, X.Q.; methodology, Y.L.; visualization, J.G.; software, G.P. All authors have read and agreed to the published version of the manuscript.

Funding: This research was funded by the Engineering Technology Research Center (grant number [2019]5303) and the central government guide's local science and technology development (grant number [2019]4011).

Data Availability Statement: Not applicable.

Acknowledgments: This work was supported by the Engineering Technology Research Center (Grant NO. [2019]5303) and the central government guide's local science and technology development (Grant NO. [2019]4011).

Conflicts of Interest: The authors declare no conflict of interest.

References

1. Proudhon, H.; Li, J.; Wang, F.; Roos, A.; Chiaruttini, V.; Forest, S. 3D simulation of short fatigue crack propagation by finite element crystal plasticity and remeshing. *Int. J. Fatigue* **2016**, *82*, 238–246. [CrossRef]
2. Lin, B.; Zhao, L.G.; Tong, J. A crystal plasticity study of cyclic constitutive behavior, crack-tip deformation and crack-growth path for a polycrystalline nickel-based superalloy. *Eng. Fract. Mech.* **2011**, *78*, 2174–2192. [CrossRef]
3. Li, L.; Shen, L.; Proust, G. Fatigue crack initiation life prediction for aluminum alloy7075 using crystal plasticity finite element simulations. *Mech. Mater.* **2015**, *81*, 84–93. [CrossRef]

4. Yang, S.; Ma, G.; Ren, X.; Ren, F. Cover refinement of numerical manifold method for crack propagation simulation. *Eng. Anal. Bound. Elem.* **2014**, *43*, 37–49. [CrossRef]
5. Özden, U.A.; Mingard, K.P.; Zivcec, M.; Bezold, A.; Broeckmann, C. Mesoscopical finite element simulation of fatigue crack propagation in WC/Co-hard metal. *Int. J. Refract. Met. Hard Mater.* **2015**, *49*, 261–267. [CrossRef]
6. Dewang, Y.; Hora, M.S.; Panthi, S.K. Prediction of crack location and propagation in stretch flanging process of aluminum alloy AA-5052 sheet using FEM simulation. *Trans. Nonferrous Met. Soc. China* **2015**, *25*, 2308–2320. [CrossRef]
7. Özden, U.A.; Bezold, A.; Broeckmann, C. Numerical simulation of fatigue crack propagation in WC/Co based on a continuum damage mechanics approach. *Prog. Mater. Sci.* **2014**, *3*, 1518–1523. [CrossRef]
8. Keyhani, A.; Goudarzi, M.; Mohammadi, S.; Roumina, R. XFEM–dislocation dynamics multi-scale modeling of plasticity and fracture. *Comput. Mater. Sci.* **2015**, *104*, 98–107. [CrossRef]
9. Calvo, F.; Yurtsever, E. The quantum structure of anionic hydrogen clusters. *J. Chem. Phys.* **2018**, *148*, 102305. [CrossRef]
10. Hou, Y.; Wang, L.; Wang, D.; Qu, X.; Wu, J. Using a molecular dynamics simulation to investigate asphalt nano-cracking under external loading conditions. *Appl. Sci.* **2017**, *7*, 770. [CrossRef]
11. Ramezani, M.G.; Golchinfar, B. Mechanical properties of cellulose nanocrystal (CNC) bundles: Coarse-grained molecular dynamic simulation. *J. Compos. Sci.* **2019**, *3*, 57. [CrossRef]
12. Liu, C.; Yao, Y. Study of crack-propagation mechanism of $Al_{0.1}$CoCrFeNi high-entropy alloy by molecular dynamics method. *Crystals* **2023**, *13*, 11. [CrossRef]
13. Lee, S.; Kang, H.; Bae, D. Molecular dynamics study on crack propagation in Al containing Mg–Si clusters formed during natural aging. *Materials* **2023**, *16*, 883. [CrossRef] [PubMed]
14. Komanduri, R.; Chandrasekaran, N.; Raff, L.M. Molecular dynamics (MD) simulation of uniaxial tensile of some single-crystal cubic metals at nanolevel. *Int. J. Mech. Sci.* **2001**, *43*, 2237–2260. [CrossRef]
15. Xu, S.; Deng, X. Nanoscale void nucleation and growth and crack tip stress evolution ahead of a growing crack in a single crystal. *Nanotechnology* **2008**, *19*, 115705. [CrossRef]
16. Cui, C.B.; Beom, H.G. Molecular dynamics simulations of edge cracks in copper and aluminum single crystals. *Mater. Sci. Eng. A* **2014**, *15*, 102–109. [CrossRef]
17. Zhuo, X.R.; Kim, J.H.; Gyu Beom, H. Atomistic investigation of crack growth resistance in a single-crystal Al-nanoplate. *J. Mater. Res.* **2016**, *9*, 1185–1192. [CrossRef]
18. Ding, J.; Wang, L.-S.; Song, K.; Liu, B.; Huang, X. Molecular dynamics simulation of crack propagation in single-crystal Aluminum plate with central cracks. *J. Nanomater.* **2017**, *2017*, 5181206. [CrossRef]
19. Mikelani, M.; Panjepour, M.; Taherizadeh, A. Investigation on mechanical properties of nanofoam aluminum single crystal: Using the method of molecular dynamics simulation. *Appl. Phys. A Mater. Sci. Process.* **2020**, *126*, 921. [CrossRef]
20. Ji, H.; Ren, K.; Ding, L.; Wang, T.; Li, J.-M.; Yang, J. Molecular dynamics simulation of the interaction between cracks in single crystal Aluminum. *Mater. Today Commun.* **2022**, *30*, 103020. [CrossRef]
21. Yu, J.; Zhang, Q.; Liu, R.; Yue, Z.; Tang, M.; Li, X. Molecular dynamics simulation of crack propagation behaviors at the Ni/Ni_3Al grain boundary. *RSC Adv.* **2014**, *4*, 32749. [CrossRef]
22. Hou, N.X.; Wen, Z.X.; Yue, Z.F. Creep behavior of single crystal superalloy specimen under temperature gradient condition. *Mater. Sci. Eng. A* **2009**, *510–511*, 42–45. [CrossRef]
23. Mao, H.; Wen, Z.; Yue, Z.; Wang, B. The evolution of plasticity for nickel-base single crystal cooled blade with film cooling holes. *Mater. Sci. Eng. A* **2013**, *587*, 79–84. [CrossRef]
24. Kim, J.; Suh, C.; Amanov, A.; Kim, H.; Pyun, Y. Rotary bending fatigue properties of Inconel 718 alloys by ultrasonic nanocrystal surface modification technique. *J. Eng.* **2015**, *13*, 133–137. [CrossRef]
25. Yang, X.F.; He, C.Y.; Yuan, G.J.; Chen, H.; Wang, R.Z.; Jia, Y.F.; Tu, S.T. The effects of grain boundary structures on crack nucleation in nickel nanolaminsted structure: A molecular dynamics study. *Comput. Mater. Sci.* **2021**, *186*, 110019. [CrossRef]
26. Mishin, Y.; Farkas, D.; Mehl, M.J.; Papaconstantopoulos, D.A. Interatomic potentials for monatomic metals from experimental data and ab initio calculations. *Phys. Rev. B* **1999**, *59*, 3393–3407. [CrossRef]
27. Wu, W.-P.; Yao, Z.-Z. Molecular dynamics simulation of stress distribution and microstructure evolution ahead of a growing crack in single crystal nickel. *Theor. Appl. Fract. Mech.* **2012**, *62*, 67–75. [CrossRef]
28. Sung, P.-H.; Chen, T.-C. Studies of crack growth and propagation of single-crystal nickel by molecular dynamics. *Comput. Mater. Sci.* **2015**, *102*, 151–158. [CrossRef]
29. Ma, L.; Xiao, S.; Deng, H.; Hu, W. Atomistic simulation of mechanical properties and crack propagation on irradiated nickel. *Comput. Mater. Sci.* **2016**, *120*, 21–28. [CrossRef]
30. Zhang, Y.; Jiang, S. Molecular dynamics simulation of crack propagation in nanoscale polycrystal nickel based on different strain rate. *Metal* **2017**, *7*, 432. [CrossRef]
31. Zhang, Y.; Jiang, S.; Zhu, X.; Zhao, Y. Mechanisms of crack propagation in nanoscale single crystal, bicrystal and tricrystal nickels based on the molecular dynamics simulation. *Results Phys.* **2017**, *7*, 1722–1733. [CrossRef]
32. Zhang, Y.; Jiang, S. Investigation on dislocation-based mechanisms of void growth and coalescence on single and nanotwinned nickels by molecular dynamics simulation. *Philos. Mag.* **2017**, *97*, 2772–2794. [CrossRef]
33. Zhang, Y.; Jiang, S.; Zhu, X.; Zhao, Y. A molecular dynamics study of intercrystalline crack propagation in nano-nickel bicrystal films with (010) twist boundary. *Eng. Fract. Mech.* **2016**, *168*, 147–159. [CrossRef]

34. Zhang, Y.; Jiang, S.; Zhu, X.; Zhao, Y. Influence of twist angle on crack propagation of nanoscale bicrystal nickel film based on molecular dynamics simulation. *Phys. E Low-Dimens. Syst. Nanostruct.* **2017**, *87*, 281–294. [CrossRef]
35. Zhang, J.; Ghosh, S. Molecular dynamics based study and characterization of deformation mechanisms near a crack in a crystalline material. *J. Mech. Phys. Solids* **2013**, *61*, 1670–1690. [CrossRef]
36. Glenn, J.; Martyna, D.J.; Tobias, M.L. Klein. Constant pressure molecular dynamics algorithms. *J. Chem. Phys.* **1994**, *101*, 4177–4189.
37. Parrinello, M.; Rahman, A. Polymorphic transitions in single crystal: A new molecular dynamics method. *J. Appl. Phys.* **1981**, *52*, 7182–7190. [CrossRef]
38. Tuckerman, M.E.; Alejandre, J.; López-Rendón, R.; Jochim, A.L.; Martyna, G.J. A Liouville-operator derived measure-preserving integrator for molecular dynamics simulations in the isothermal–isobaric ensemble. *J. Phys. A Math. Gen.* **2006**, *39*, 5629–5651. [CrossRef]
39. Stukowski, A. Visualization and analysis of atomistic simulation data with OVITO—The open visualization tool. *Model. Simul. Mater. Sci. Eng.* **2010**, *18*, 015012. [CrossRef]
40. Heyes, D.M. Pressure tensor of partial-charge and point-dipole lattices with bulk and surface geometries. *Phys. Rev. B* **1994**, *49*, 755–764. [CrossRef]
41. Sirk, T.W.; Moore, S.; Brown, E.F. Characteristics of thermal conductivity in classical water models. *J. Chem. Phys.* **2013**, *138*, 064505. [CrossRef] [PubMed]
42. Aidan, P.; Thompson, S.J.; Plimpton, W.M. General formation of pressure and stress tensor for arbitrary many-body interaction potentials under periodic boundary conditions. *J. Chem. Phys.* **2009**, *131*, 154107.
43. Honeycutt, J.D.; Andersen, H.C. Andersen. Molecular dynamics study of melting and freezing of small Lennard-Jones clusters. *J. Phys. Chem.* **1987**, *91*, 4950–4963. [CrossRef]
44. Faken, D.; Jónsson, H. Systematic analysis of local atomic structure combined with 3D computer graphics. *Comput. Mater. Sci.* **1994**, *2*, 279–286. [CrossRef]
45. Lee, B.-J.; Shim, J.-H.; Baskes, M.I. Semiempirical atomic potentials for the fcc metals Cu, Ag, Au, Ni, Pd, Pt, Al, and Pb based on first and second nearest-neighbor modified embedded atom method. *Phys. Rev. B* **2003**, *68*, 144112. [CrossRef]
46. Ackland, G.J.; Tichy, G.; Vitek, V.; Finnis, M.W. Simple N-body potentials for the noble metals and nickel. *Philos. Mag. A* **1987**, *56*, 735–756. [CrossRef]
47. Foiles, S.M.; Baskes, M.I.; Daw, M.S. Embedded-atom-method functions for the fcc metals Cu, Ag, Au, Ni, Pd, Pt, and their alloys. *Phys. Rev. B* **1986**, *33*, 7983–7991. [CrossRef]
48. Sainath, G.; Choudhary, B.K. Atomistic simulations on ductile-brittle transition in <111> BCC Fe nanowires. *J. Appl. Phys.* **2017**, *122*, 095101.
49. Gordon, P.A.; Neeraj, T.; Luton, M.J.; Farkas, D. Crack-tip deformation mechanisms in α-Fe and binary Fe alloys: An atomistic study on single crystals. *Metall. Mater. Trans. A* **2007**, *38A*, 2191–2202. [CrossRef]
50. Sainath, G.; Nagesha, A. Atomistic simulations of twin boundary effect on the crack growth behavior in BCC Fe. *Trans. Indian Natl. Acad. Eng.* **2022**, *7*, 433–439. [CrossRef]

Disclaimer/Publisher's Note: The statements, opinions and data contained in all publications are solely those of the individual author(s) and contributor(s) and not of MDPI and/or the editor(s). MDPI and/or the editor(s) disclaim responsibility for any injury to people or property resulting from any ideas, methods, instructions or products referred to in the content.

Article

High Temperature Deformation Behavior of Near-β Titanium Alloy Ti-3Al-6Cr-5V-5Mo at α + β and β Phase Fields

Haoyu Zhang *, Shuo Zhang, Shuai Zhang, Xuejia Liu, Xiaoxi Wu, Siqian Zhang and Ge Zhou

School of Materials Science and Engineering, Shenyang University of Technology, Shenyang 110870, China
* Correspondence: zhanghaoyu@sut.edu.cn; Tel.: +86-024-25496301

Abstract: Most near-β titanium alloy structural components should be plastically deformed at high temperatures. Inappropriate high-temperature deformed processes can lead to macro-defects and abnormally coarse grains. Ti-3Al-6Cr-5V-5Mo alloy is a near-β titanium alloy with the potential application. The available information on the high-temperature deformation behavior of the alloy is limited. To provide guidance for the actual hot working of the alloy, the flow stress behavior and processing map at α + β phase field and β phase field were studied, respectively. Based on the experimental data obtained from hot compressing simulations at the range of temperature from 700 °C to 820 °C and at the range of strain rate from 0.001 s^{-1} to 10 s^{-1}, the constitutive models, as well as the processing map, were obtained. For the constitutive models at the α + β phase field and β phase field, the correlated coefficients between actual stress and predicted stress are 0.986 and 0.983, and the predictive mean relative errors are 2.7% and 4.1%. The verification of constitutive models demonstrates that constitutive equations can predict flow stress well. An instability region in the range of temperature from 700 °C to 780 °C and the range of strain rates from 0.08 s^{-1} to 10 s^{-1}, as well as a suitable region for thermomechanical processing in the range of temperature from 790 °C to 800 °C and the range of strain rates from 0.001 s^{-1} to 0.007 s^{-1}, was predicted by the processing map and confirmed by the hot-deformed microstructural verification. After the deformation at 790 °C/0.001 s^{-1}, the maximum number of dynamic recrystallization grains and the minimum average grain size of 17 μm were obtained, which is consistent with the high power-dissipation coefficient region predicted by the processing map.

Keywords: near-β titanium alloy; constitutive mode; thermomechanical processing

Citation: Zhang, H.; Zhang, S.; Zhang, S.; Liu, X.; Wu, X.; Zhang, S.; Zhou, G. High Temperature Deformation Behavior of Near-β Titanium Alloy Ti-3Al-6Cr-5V-5Mo at α + β and β Phase Fields. *Crystals* **2023**, *13*, 371. https://doi.org/10.3390/cryst13030371

Academic Editors: Wangzhong Mu and Chao Chen

Received: 4 February 2023
Revised: 16 February 2023
Accepted: 18 February 2023
Published: 21 February 2023

Copyright: © 2023 by the authors. Licensee MDPI, Basel, Switzerland. This article is an open access article distributed under the terms and conditions of the Creative Commons Attribution (CC BY) license (https://creativecommons.org/licenses/by/4.0/).

1. Introduction

Near-β titanium alloys are applied to various engineering fields as an important structural metal due to their high strength and toughness, excellent thermal stability, and good fatigue properties [1,2]. Most titanium alloy structural components must be plastically deformed by thermomechanical processing (TMP), including forging, hot rolling, and hot extrusion [3–6]. After TMP, the structural components exhibit good mechanical properties while obtaining the desired shape. For example, Ti-6Al-2Sn-4Zr-6Mo alloy was produced through β-processed forging, whose forging temperature is above β-transus [7]. Moreover, a titanium alloy aero-engine drum was formed by hot-deformation at a dual-phase field [8]. Thus, the TMP of such alloys may be carried out at both single-phase (β) field and dual-phase field (α + β). Moreover, an inappropriate technical process can lead to the formation of macro-defects and abnormally coarse grains, as well as harmful effects on the quality of the near-β titanium alloy structural components.

Many efforts were made to study the high-temperature deformation behavior of near β titanium alloys during TMP [9,10]. Park et al. [11] proposed the processing map and calculated the activation energy for high-temperature deformation of the near-β titanium alloy, β21S. According to the established processing map, flow instability would occur when the β21S alloy deformed at 900 °C/10 s^{-1}. Zhang et al. [12] established the constitutive relationship and processing map of Ti-6Mo-2Sn-6Al-4Zr alloy. The results suggested that deformation should occur in an area of efficient power dissipation, and the area should be within the range of temperature from 850 °C to 1000 °C, with a strain rate from 0.001 s^{-1} to 0.1 s^{-1}. Chen et al. [13] developed the Arrhenius model for hot deformation of the Ti-5.5Cr-5Mo-5V-4Al-1Nb alloy. This model exhibits an accurate prediction for flow stress with a high correlation coefficient value. Gao et al. [14] investigated the high-temperature deformed process of Ti-10Mo-3Nb-6Zr-4Sn alloy via an isothermal compression test. They quantitatively calculated a sensibility factor of strain rate and constructed the thermal processing map. These results accurately describe the high-temperature deformation behaviors of the alloy. Wang et al. [15] proposed a modified J-C constitutive model to exactly represent the flow stress behavior of Ti-22Al-23Nb-2(Mo, Zr) alloy. Zhao et al. [16] studied the high-temperature deformation behavior of ingot metallurgy Ti-5V-5Mo-3Cr-5Al alloy via a processing map. They proposed that the region of suitable deformation is in the temperature from 800 °C to 970 °C and the strain rate from $10^{-1.5}$ s^{-1} to 10^{-3} s^{-1}. Hence, there are effective methods to study the high-temperature deformation behavior of near-β titanium alloys by establishing constitutive relations and processing maps.

During high-temperature deformation, the microstructure of near-β titanium alloys can be tailored by dynamic recovery (DRV) and dynamic recrystallization (DRX) [17,18]. Meanwhile, different near-β titanium alloys exhibit distinct instability and suitable deformation regions, other flow stress behaviors, and DRV and DRX behaviors. The works reported by the literature [19–23] also performed similar research and confirm this phenomenon.

Ti-3Al-6Cr-5V-5Mo alloy is a near-β titanium alloy with good mechanical properties; it is designed by taking the Ti-5Cr-5V-8Cr-3Al alloy as the baseline alloy [24]. Its tensile strength is over 1400 MPa through TMP. The alloy shows a potential application in structural components. However, the further improvement of ductility and toughness of the alloy after hot working is often restricted by coarse grains. In addition, the available information on the flow stress behavior and processing map of the alloy is limited at present. Therefore, this work aimed to study the high-temperature deformation behavior of the Ti-3Al-6Cr-5V-5Mo alloy. Acceptable TMP conditions were traced by establishing a processing map and observing the microstructure. Furthermore, constitutive relations for deformation at the dual-phase field and single-phase field were sequentially established to predict the flow stress behavior. The purpose of this work is to provide guidance for the actual hot working of the alloy.

2. Materials and Methods

The spongy titanium, Al-Mo master alloy, Al-V master alloy, purity Al, and purity Cr were used as raw materials for smelting Ti-3Al-6Cr-5V-5Mo (wt.%) alloy. The alloy ingot was prepared by triple vacuum arc remelting. The chemical composition of the alloy ingot was analyzed by X-ray Fluorescence Spectrometer, as shown in Table 1.

Table 1. Chemical composition.

Element	Al	Cr	V	Mo	Ti
Content (wt.%)	2.8	5.9	5.2	5.2	Bal.

The cylindrical samples, which were 15 mm in height and 10 mm in diameter, were cut from the alloy ingot and then used for high-temperature compression tests. The high-temperature compression simulations were performed on a Gleeble 3800 thermomechanical simulator. A metallographic examination was adopted to measure the β-transus of the alloy. The measured β-transus is 780 °C. Five deformation temperatures and five strain rates were selected. The samples were high-temperature deformed with a height reduction of 60%, followed by water quenching. The detailed high-temperature compressing simulation parameters are shown in Table 2. After hot-temperature deformation, square sheets were cut from specimens along the compression axis to observe the deformed microstructure.

The square sheets were ground on metallographic sandpaper and electrolytically polished at ~−16 °C in a liquid of 59% methanol, 35% nbutyl alcohol, and 6% perchloric acid. The voltage of electrolytic polishing was 32 V, the current of electrolytic polishing was 1.2 A, and the electrolytic polishing time was 60 s. Then the deformed grains of the alloy were observed by electron backscattered diffraction (EBSD) installed on a ZEISS GeminiSEM300 scanning electron microscope.

Table 2. Detailed parameters of high-temperature compression simulation.

Parameters	Details
Deformation temperatures	700 °C, 730 °C, 760 °C, 790 °C, 820 °C
Strain rate	0.001 s^{-1}, 0.01 s^{-1}, 0.1 s^{-1}, 1 s^{-1}, 10 s^{-1}
Heating rate	10 °C/s
Holding time before test	300 s
Cooling method after test	water quenching
Reduction	60% of height

3. Results and Discussion

3.1. Flow Stress

The true-stress–true-strain curves during the deformation at dual-phase field and single-phase field are shown in Figure 1. According to Figure 1, the flow stress under the different deformation conditions exhibits similar characteristics. In the initial stage of deformation, a large number of dislocations are rapidly activated. The interaction between dislocations leads to the difficulty of dislocation slip, which is represented as work hardening. Thus, the flow stress rapidly rises to the peak value due to work hardening when the strain is minor [25]. After the peak of stress, the flow stress gets into a relatively stable stage of slow change or equilibrium. The competition between work-hardening and -softening mechanisms, such as DRV and DRX, causes this phenomenon. The occurrence of DRV and DRX is mainly affected by temperature and time. At a constant deformation temperature, the lower strain rate provides a condition for the event of softening. Moreover, the increase of strain rate leads to the increase in dislocation density, hindering dislocation movement [26]. Therefore, for the same deformation temperature, the flow stress increases with the increase in strain rate. In a constant time, the higher temperature provides favorable conditions for the DRV and DRX. Moreover, the higher free energy is beneficial to dislocation slip and grain-boundary migration [27]. Therefore, the flow stress decreases with the increase of deformation temperature for the same strain rate.

3.2. Constitutive Model for High-Temperature Deformation

Based on the analysis of flow stress at various deformation conditions, Arrhenius's constitutive model for high-temperature deformation at the α + β phase field and β phase field is established, sequentially.

3.2.1. Constitutive Relations

The constitutive relations between deformation conditions and stress should be obtained to determine the optimal hot-deformation parameters. An Arrhenius model [28] is adopted to establish the relationship between strain rate ($\dot{\varepsilon}$) and deformation activation energy (Q), deformation temperature (T), and stress (σ), as shown in Equation (1).

$$\dot{\varepsilon} = f(\sigma) \exp(-Q/RT) \tag{1}$$

where R is the gas constant of 8.314 J/(mol·K) [29]. Moreover, $f(\sigma)$ can be defined as a function of stress, as shown in Equation (2).

$$f(\sigma) = \begin{cases} A_1\sigma^{n_1}, & \alpha\sigma < 0.8 \\ A_2\exp(\beta\sigma), & \alpha\sigma > 1.2 \\ A[\sinh(\alpha\sigma)]^n, & for\ all\ \sigma \end{cases} \quad (2)$$

where A_1, A_2, β, α, and A are materials' constants; and n_1 and n are stress exponents.

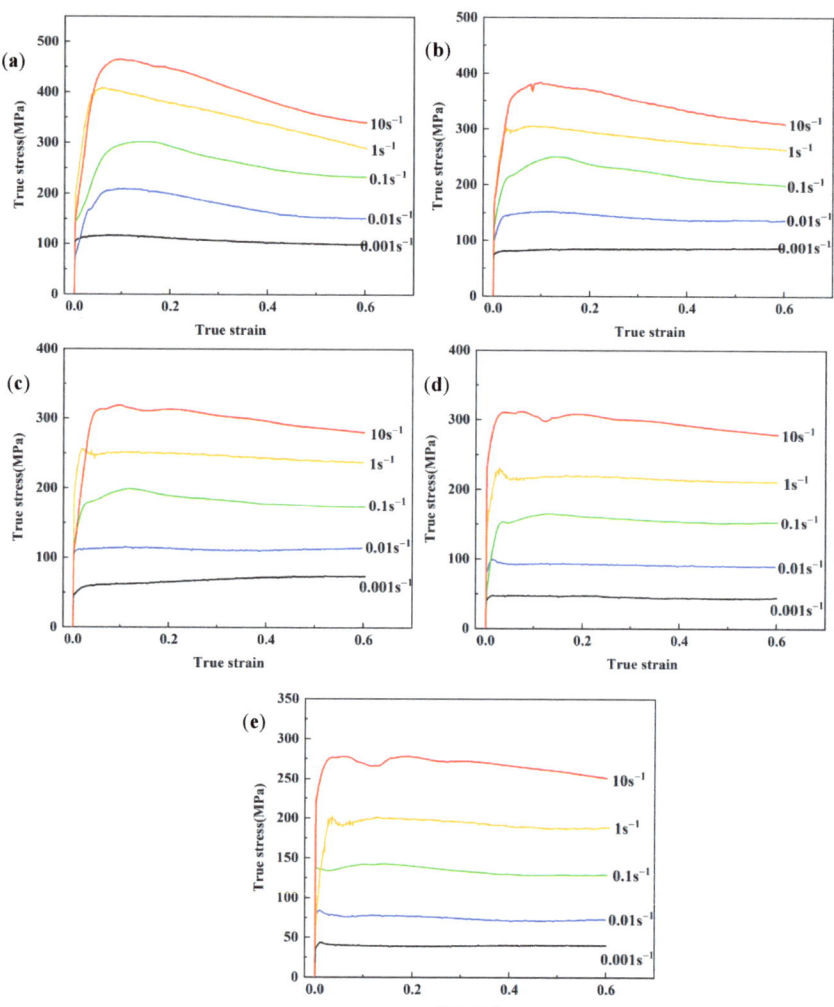

Figure 1. True-stress–true-strain curves during the deformation at different phase fields: (**a**) 700 °C (α + β), (**b**) 730 °C (α + β), (**c**) 760 °C (α + β), (**d**) 790 °C (β), and (**e**) 820 °C (β).

The peak stress is adopted to establish constitutive equations [30]. The values of peak stress are shown in Figure 2. Based on the $\dot{\varepsilon}$ values and σ values, the curves of $\ln\dot{\varepsilon} - \sigma$ and $\ln\dot{\varepsilon} - \ln\sigma$ at different phase fields are drawn in Figures 3 and 4, respectively.

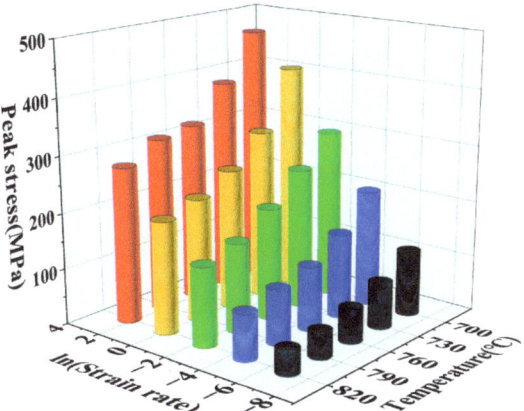

Figure 2. Peak stress adopted to establish constitutive equations.

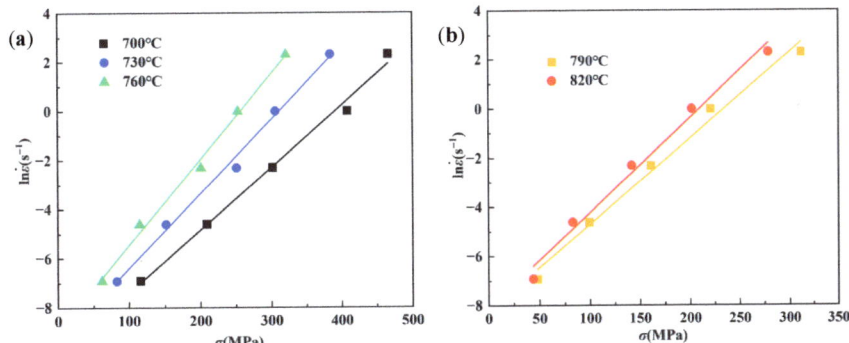

Figure 3. Curves of $\ln\dot{\varepsilon} - \sigma$ at different phase fields: (**a**) $\alpha + \beta$ and (**b**) β.

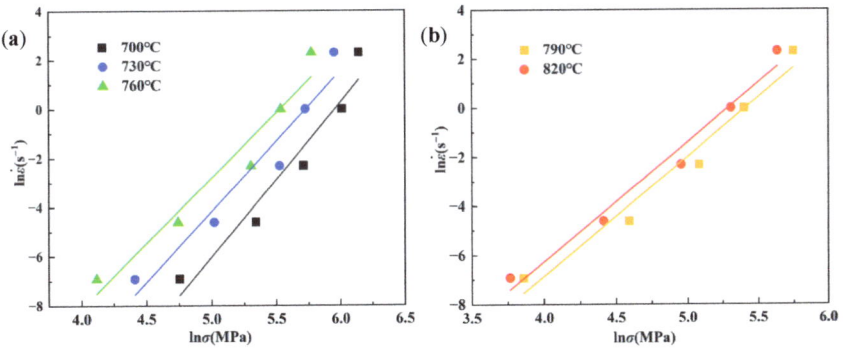

Figure 4. Curves of $\ln\dot{\varepsilon} - \ln\sigma$ at different phase fields: (**a**) $\alpha + \beta$ and (**b**) β.

The average slopes of $\ln\dot{\varepsilon} - \sigma$ curves and $\ln\dot{\varepsilon} - \ln\sigma$ curves are obtained by linear fitting. Furthermore, the logarithms of Equation (1) after bringing $f(\sigma) = A_1\sigma^{n1}$ and $f(\sigma) = A_2\exp(\beta\sigma)$ are taken, respectively, which are used to calculate the values of n_1 and β at $\alpha + \beta$ and β phase fields. Then the values of α at $\alpha + \beta$ and β phase fields can be calculated by β/n_1 [31]. The calculated values of α, β and n_1 are shown in Table 3.

Table 3. Parameters for constitutive relations.

Phase Field	α	β	n_1	n	A	Q (kJ/mol)
α + β	0.005225	0.0302	5.78	4.26352	$e^{49.13873}$	442.25
β	0.00758	0.0368	4.85	3.61766	$e^{19.79592}$	206.86

Bring $f(\sigma) = A[\sinh(\alpha\sigma)]^n$ into Equation (1), and by taking the logarithm, Equation (3) can be obtained as follows:

$$\ln \dot{\varepsilon} = \ln A + n \ln[\sinh(\alpha\sigma)] - \frac{Q}{RT} \quad (3)$$

By taking the partial differential of Equation (3), the Q is determined by Equation (4):

$$Q = R \left[\frac{\partial \ln[\sinh(\alpha\sigma)]}{\partial 1/T} \right] \left[\frac{\partial \ln \dot{\varepsilon}}{\partial \ln[\sinh(\alpha\sigma)]} \right] \quad (4)$$

According to Equation (4), the values of Q can be obtained by calculating the partial differential between $\ln[\sinh(\alpha\sigma)]/1/T$ and $\ln\dot{\varepsilon}/\ln[\sinh(\alpha\sigma)]$. Thus, the curves of $\ln[\sinh(\alpha\sigma)] - 1/T$ and $\ln\dot{\varepsilon} - \ln[\sinh(\alpha\sigma)]$ are drawn in Figures 5 and 6, respectively.

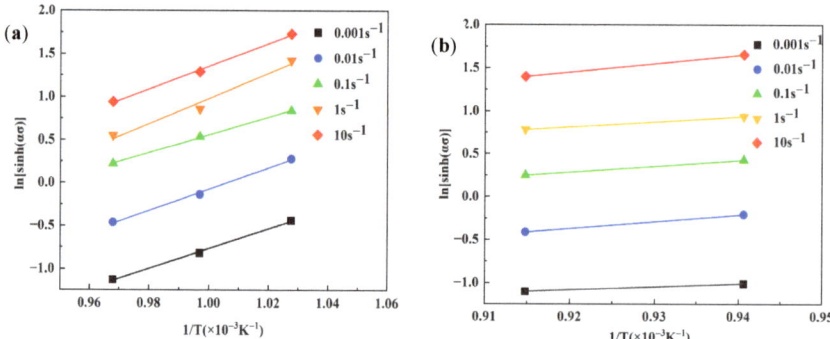

Figure 5. Curves of $\ln[\sinh(\alpha\sigma)] - 1/T$ at different phase fields: (**a**) α + β and (**b**) β.

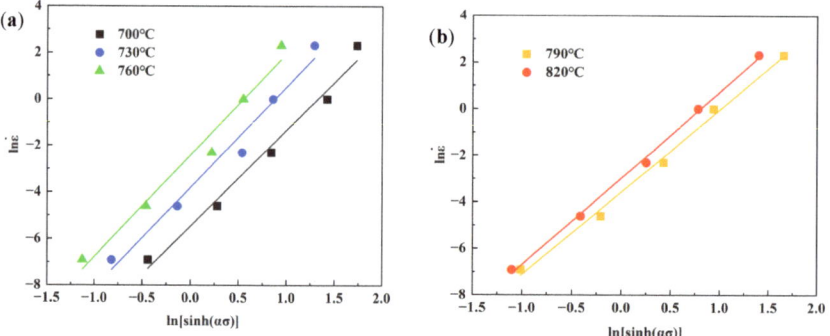

Figure 6. Curves of $\ln\dot{\varepsilon} - \ln[\sinh(\alpha\sigma)]$ at different phase fields: (**a**) α + β and (**b**) β.

According to Figures 5 and 6, the average slopes of $\ln[\sinh(\alpha\sigma)] - 1/T$ curves at α + β and β phase fields were calculated as 12.45591 and 6.870484, and those of the $\ln\dot{\varepsilon} - \ln[\sinh(\alpha\sigma)]$ curves were 4.27053 and 3.62148, which were obtained by linear fitting. The values of Q at α + β and β phase fields were calculated by introducing the abovementioned slopes into Equation (4), and they are shown in Table 3.

A Zener–Hollomon parameter Z on the relationship between strain rate, $\dot{\varepsilon}$, and deformation temperature, T, is introduced [32], as shown in Equation (5):

$$Z = \dot{\varepsilon}\exp(Q/RT) = A[\sinh(\alpha\sigma)]^n \quad (5)$$

The logarithm of Equation (5) can be described as follows:

$$\ln Z = \ln\dot{\varepsilon} + Q/RT = \ln A + n\ln[\sinh(\alpha\sigma)] \quad (6)$$

Moreover, lnZ can be calculated by taking the Q, T, and $\dot{\varepsilon}$ into Equation (6). Then the curves of $\ln[\sinh(\alpha\sigma)] - \ln Z$ at $\alpha + \beta$ and β phase fields can be obtained as shown in Figure 7. The values of n are obtained from the slopes of the curves. The values of A can be obtained from the lnA of 49.13873 ($\alpha + \beta$ phase field) and 19.79592 (β phase field), which can be determined by the intercepts of the curves. The n and A are shown in Table 3.

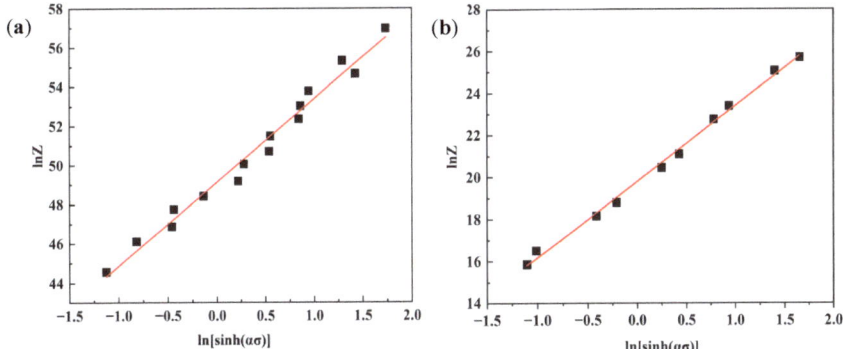

Figure 7. Curves of $\ln[\sinh(\alpha\sigma)] - \ln Z$ at different phase fields: (**a**) $\alpha + \beta$ and (**b**) β.

Moreover, the correlation coefficient of $\ln[\sinh(\alpha\sigma)] - \ln Z$ at $\alpha + \beta$ and β phase fields can be determined as 0.98 and 0.9958, respectively, which show the hyperbolic sinusoidal function coincides with the experimental data. The constitutive equations of the alloy are expressed as shown in Equation (7):

$$\dot{\varepsilon} = \begin{cases} e^{49.13873}[\sinh(0.005225\sigma)]^{4.26352}\exp(-442.25/RT), & T \text{ is at } \alpha + \beta \text{ phase field} \\ e^{19.79592}[\sinh(0.00758\sigma)]^{3.61766}\exp(-206.86/RT), & T \text{ is at } \beta \text{ phase field} \end{cases} \quad (7)$$

Generally, because high-temperature compression is a thermal-activation process, the softening mechanism can be inferred by comparing with deformation activation energy and self-diffusion activation energy of β titanium alloy (161 kJ/mol) [33]. As seen from Table 3, the self-diffusion activation energy of 161 kJ/mol is smaller than the Q values of 442.25 kJ/mol and 206.86 kJ/mol. It can be inferred that the DRX may be the main softening mechanism when the alloy deforms at a high temperature.

3.2.2. Verification of Constitutive Equations

The method for establishing the constitutive equations under the peak strain described in Section 3.2.1 can be used to obtain the constitutive equation for other strains of 0.1–0.6. The relationship between model parameters and true strain was built to accurately predict the flow stress in the strain range of 0.1–0.6, as shown in Figure 8. According to Figure 8e,f, the values of deformation activation energy, Q, under different deformed conditions are shown in Table 4. The values of Q under all strains are always larger than the value of self-diffusion activation energy in both dual-phase and single-phase fields.

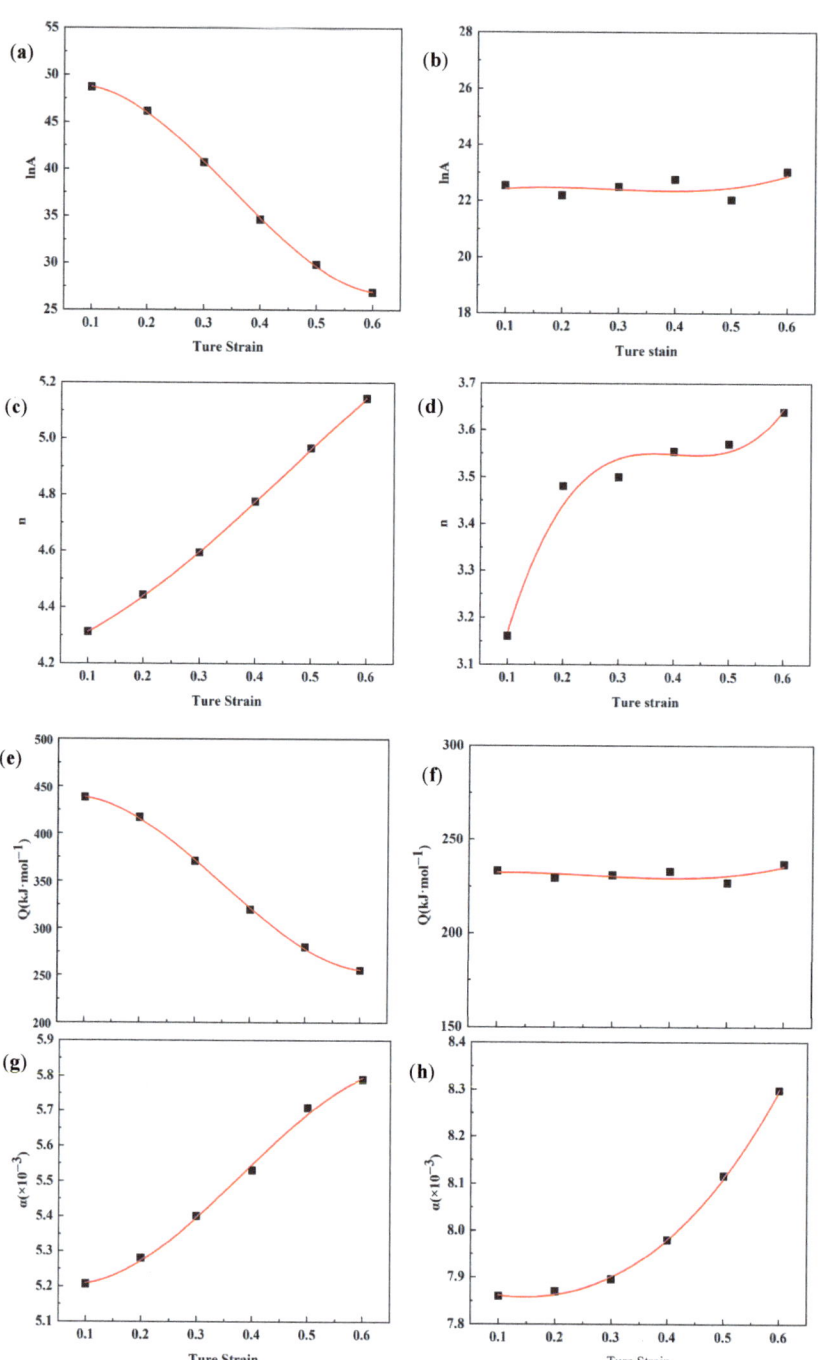

Figure 8. Relationship between model parameters and true strain: (**a**) ln$A$ at $\alpha + \beta$ phase field, (**b**) ln$A$ at β phase field, (**c**) n at $\alpha + \beta$ phase field, (**d**) n at β phase field, (**e**) Q at $\alpha + \beta$ phase field, (**f**) Q at β phase field, (**g**) α at $\alpha + \beta$ phase field, and (**h**) α at β phase field.

Table 4. Values of Q under different deformed conditions (kJ/mol).

Strain	0.1	0.2	0.3	0.4	0.5	0.6
α + β phase field	438.10	417.21	370.53	319.48	280.08	255.08
β phase field	233.18	229.48	230.90	232.88	226.98	236.68

The constitutive equations at strains from 0.1 to 0.6 can be obtained by replacing the parameters in Equation (7) with the above A, n, Q, and α. Then the predicted stress values can be calculated and compared with experimental data, as shown in Figure 9. According to Figure 9, the constitutive equations can well predict the flow stress of Ti-3Al-6Cr-5V-5Mo alloy under the TMP conditions.

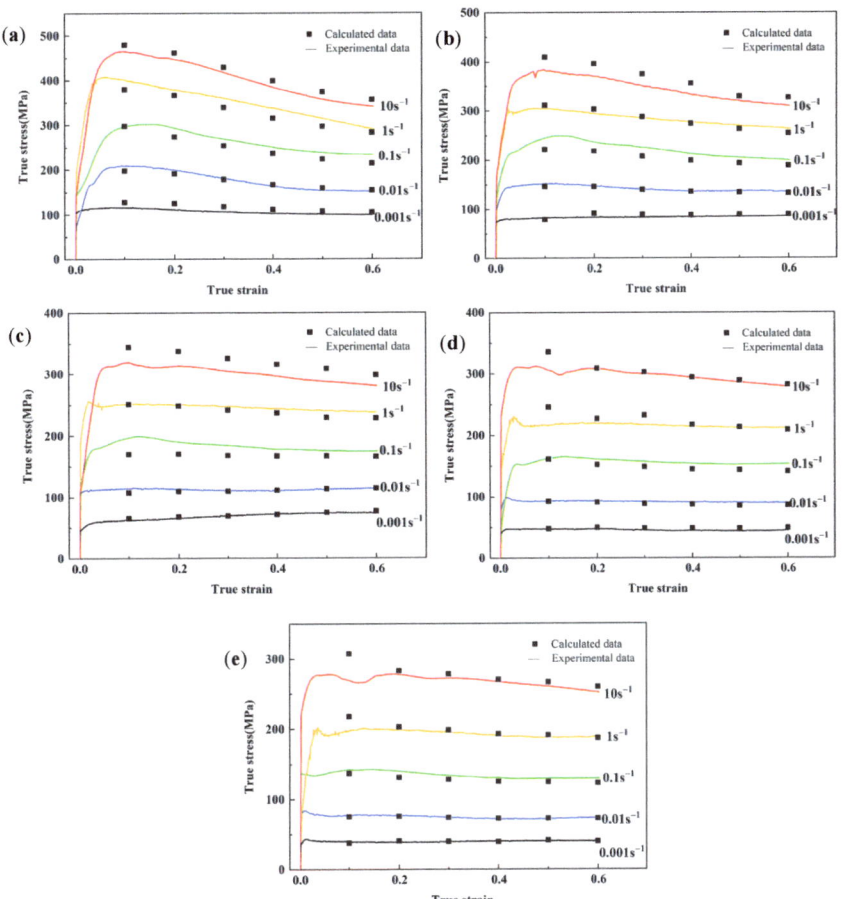

Figure 9. Predicted stress values and experimental stress values: (**a**) α + β phase field, 700 °C; (**b**) α + β phase field, 730 °C; (**c**) α + β phase field, 760 °C; (**d**) β phase field, 790 °C; and (**e**) β phase field, 820 °C.

In order to further verify the accuracy of the constitutive equations at the dual-phase field and single-phase field, respectively, a correlated coefficient (R) and a mean relative error (E) are introduced to evaluate the accuracy quantitatively. The R and E can be calculated as Equations (8) and (9), respectively [34].

$$R = \frac{\sum_{i=1}^{N}(\sigma_A - \overline{\sigma}_A)(\sigma_C - \overline{\sigma}_C)}{\sqrt{\sum_{i=1}^{N}(\sigma_A - \overline{\sigma}_A)^2}\sqrt{\sum_{i=1}^{N}(\sigma_C - \overline{\sigma}_C)^2}} \tag{8}$$

$$E = \frac{1}{N}\sum_{i=1}^{N}\left|\frac{\sigma_A - \sigma_C}{\sigma_A}\right| \times 100\% \tag{9}$$

where σ_A is the actual stress value obtained by experiment, $\overline{\sigma}_A$ is the average value of σ_A, σ_C is the predicted stress value of constitutive model, $\overline{\sigma}_C$ is the average value of σ_C, and N is the number of data used for comparison.

The correlation analysis between the actual stress values and predicted stress values at different phase fields is shown in Figure 10. At the α + β phase field, the correlated coefficient (R) is 0.986, and the mean relative error (E) is 2.7%. At the β phase field, the correlated coefficient (R) is 0.983, and the mean relative error (E) is 4.1%. Other previous research works have also used the Arrhenius model to construct constitutive equations for flow-stress predictions for titanium alloys. The literature reports the constitutive model of a near-β titanium alloy Ti-6Cr-5Mo-5V-4Al at a high temperature above 800 °C. Its correlated coefficient and mean relative error are 0.982 and 6.23%, respectively [35]. An Arrhenius constitutive equation of TA32 titanium alloy has been established, in which the correlated coefficient and mean relative error are 0.978 and 7.3%, respectively [36]. In this work, the predictions of the constitutive equation at both the α + β and β phase fields exhibit slightly larger R values and slightly smaller E values compared with those of the abovementioned Arrhenius constitutive equations for other titanium alloys. Thus, it can be further proved that the constitutive equation has good prediction accuracy for the flow stress of Ti-3Al-6Cr-5V-5Mo alloy at both dual-phase and single-phase fields.

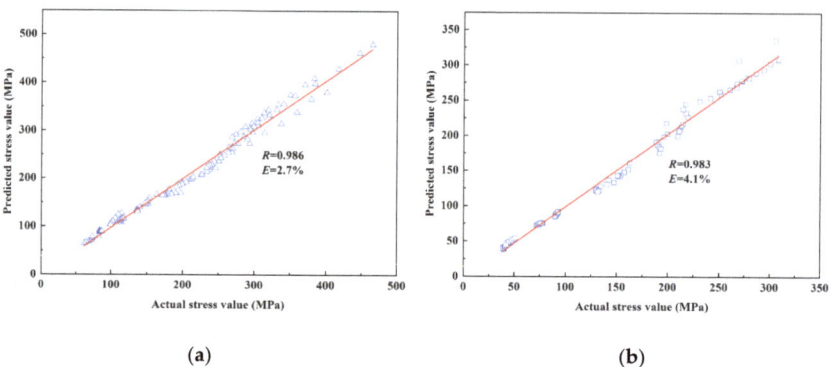

Figure 10. Correlation analysis between the actual stress values and predicted stress values at different phase fields: (**a**) α + β and (**b**) β.

3.3. Processing Map

For metals, the required energy during deformation is mainly the power dissipation, which occurs for two reasons: the plastic deformation leads to power dissipation, and the microstructural change leads to power dissipation [37]. The relationship between these parameters is shown in Equation (10) [38]:

$$P = \sigma\dot{\varepsilon} = G + J = \int_0^{\dot{\varepsilon}} \sigma d\dot{\varepsilon} + \int_0^{\dot{\varepsilon}} \dot{\varepsilon} d\sigma \qquad (10)$$

where P is the required energy during the materials' deformation, G is the power dissipation led by plastic deformation, and J is the power dissipation led by microstructural change.

For a certain strain and deformation temperature, the partial differential between J and G can be obtained by the strain-rate sensitivity index, m, as shown in Equation (11) [39]. The power dissipation map and the rheological instability map are two parts of the processing map. When the materials are in ideal dissipation, the annihilation and formation of dislocations are balanced. In this case, the value of the strain-rate sensitivity index (m) can be regarded as 1. Then $J_{max} = \sigma\dot{\varepsilon}/2$. A power-dissipation coefficient, η, generated by standardization with the ideal linear dissipation factor, is shown in Equation (12).

$$m = \frac{dJ}{dG} = \left[\frac{\partial(\ln\sigma)}{\partial(\ln\dot{\varepsilon})}\right]_{\dot{\varepsilon},T} \qquad (11)$$

$$\eta = \frac{J}{J_{max}} = \frac{2m}{m+1} \qquad (12)$$

The inequality between power dissipation and strain rate proposed by Prasad et al. [40] can be used to find the instability regions in the process of deformation. To obtain the instability conditions of TMP for the Ti-3Al-6Cr-5V-5Mo alloy, an expression including the strain rate sensitivity index and the dimensionless parameter, $\xi(\dot{\varepsilon})$, is expressed as Equation (13). According to Equation (13), when the value of $\xi(\dot{\varepsilon})$ is less than zero, flow instability would occur during deformation in these regions. Such regions should be avoided during the thermomechanical processing of the alloy.

$$\xi(\dot{\varepsilon}) = \frac{\partial \ln\left(\frac{m}{m+1}\right)}{\partial \ln \dot{\varepsilon}} + m < 0 \qquad (13)$$

By superimposing the power dissipation map and rheological instability map, a processing map can be drawn. For the alloy at a strain of 0.6, the processing map is shown in Figure 11. At $\alpha + \beta$ phase field, the power-dissipation coefficient, η, enlarges from 0.16 to 0.41 with the increase of the deformation temperature. At the β phase field, η shows the same trend and increases to 0.44. Generally, a high value of η may indicate a region suitable for deformation because of the large power dissipation [23]. For the thermomechanical processing that occurred at the $\alpha + \beta$ phase field, the range of temperature from 780 °C to 790 °C and range of strain rate from 0.001 s^{-1} to 0.016 s^{-1} lead to a high η value. A similar region for thermomechanical processing occurred at the β phase field and had a range of temperature from 790 °C to 800 °C and a range of strain rate from 0.001 s^{-1} to 0.007 s^{-1}. Compared with the two regions with a high power-dissipation-coefficient value, the η value of the β phase field is higher than that of the $\alpha + \beta$ phase field. This phenomenon indicates that the region with high power-dissipation-coefficient value of the β phase field may be more suitable for TMP. In addition, an instability region in the range of temperature from 700 °C to 780 °C, as well as the range of strain rate from 0.08 s^{-1} to 10 s^{-1}, could be found. Under such TMP conditions, it may not be suitable for the high-temperature processing of the alloy.

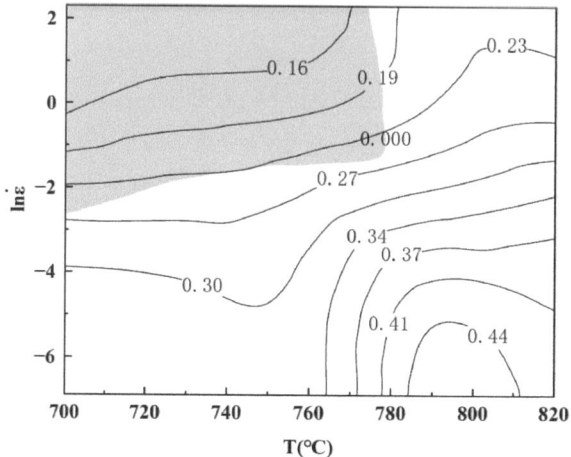

Figure 11. Processing map for 0.6 strain.

3.4. Microstructural Verification of TMP Region

An EBSD analysis was performed to further investigate the deformed grains of the alloy at the dual-phase field and single-phase field. Figure 12 shows the EBSD images of the deformed grains at the α + β phase field, as well as strain rates of 0.01 s^{-1}. According to Figure 12, a small number of DRX grains formed after deformation occurred at the α + β phase field. However, the number of DRX grains did not change significantly with the increasing deformation temperature. Figure 13 shows the EBSD images of the deformed grains at the β phase field. When deformation occurs at the β phase field, the grains exhibit completely different characteristics.

The average size of the deformed grains shown in Figures 12 and 13 was measured, as shown in Table 5. According to Table 5, the average grain size of the alloys deformed at the α + β phase field is larger than that of the alloys deformed at the β phase field. The deformed grains at the β phase field refine significantly due to DRX, compared with that of deformed grains at the α + β phase field. Deformation temperature is one of the most important factors affecting DRX [41]. The stored deformation energy increases with the increase of deformation temperature, resulting in a larger driving force for DRX [42]. It seems that the deformation at the β phase field can provide more sufficient conditions for DRX in the Ti-3Al-6Cr-5V-5Mo alloy.

When the deformation occurs at the β phase field, the average grain size increases with the increasing temperature. The boundaries of grain and subgrain are easy to migrate at higher temperatures [9]. While the temperature is higher than β-transus, the grains of near-β titanium alloys are prone to coarsening. A similar phenomenon has been reported in several works of literature [43,44]. Therefore, it can be measured from the microstructure that the average grain size of 790 °C is smaller than that of 820 °C. When the alloy is deformed under the slowest strain rate of 0.001 s^{-1} at 790 °C, the most homogeneous grains with the smallest average size of 17 μm are obtained. It is well-known that time is another main factor for recrystallization [45]. During the deformation that occurs under a lower strain rate, there is more time for dislocation rearrangement, as well as the nucleation and growth of DRX grains. The DRX integral decreases with the increasing strain rate due to the shortening of the time for grain-boundary migration. Thus, the uniform and fine grains are formed at 790 °C/0.001 s^{-1}.

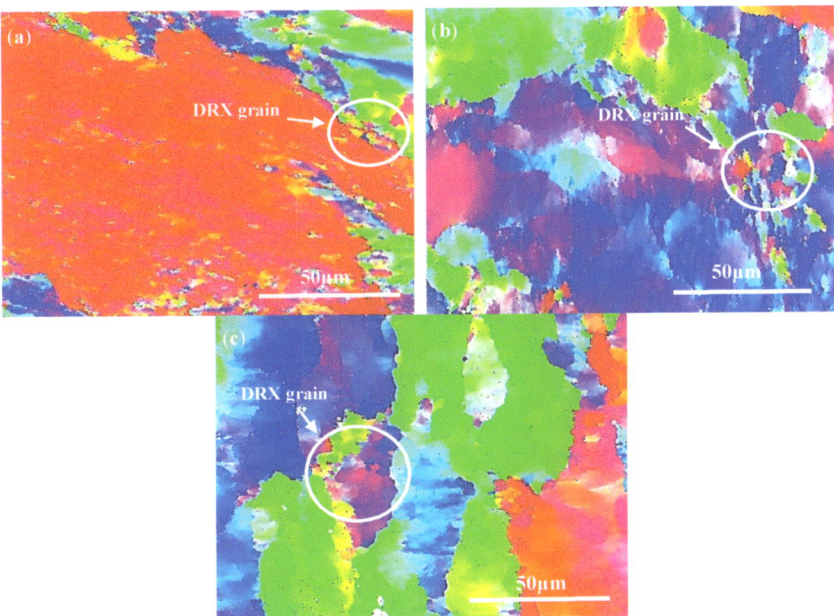

Figure 12. EBSD images of the deformed grains at α + β phase field, as well as strain rate of 0.01 s^{-1}: (**a**) 700 °C, (**b**) 730 °C, and (**c**) 760 °C.

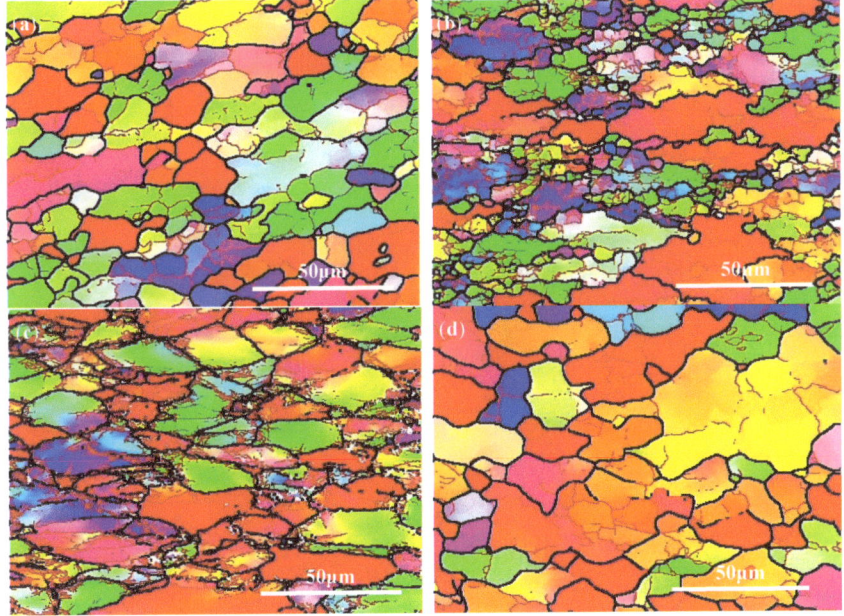

Figure 13. EBSD images of the deformed grains at β phase field: (**a**) 790 °C/0.001 s^{-1}, (**b**) 790 °C/0.1 s^{-1}, (**c**) 790 °C/10 s^{-1}, and (**d**) 820 °C/0.001 s^{-1}.

Table 5. Average grain size.

Conditions	700 °C/0.01 s^{-1}	730 °C/0.01 s^{-1}	760 °C/0.01 s^{-1}	790 °C/0.001 s^{-1}	790 °C/0.1 s^{-1}	790 °C/10 s^{-1}	820 °C/0.001 s^{-1}
Average grain size (μm)	71	56	51	17	22	26	38

The abovementioned microstructure characteristics are consistent with the high η value in the region predicted by the processing map. The reliability of the processing map for the Ti-3Al-6Cr-5V-5Mo alloy is proved.

To further investigate the deformed grains under the 0.001 s^{-1} strain rate, an EBSD analysis for DRX was performed and is shown in Figure 14. The different marked colors indicate the different microstructure of the tested alloy; that is, the DRX grains, sub-grains, and primary grains are marked by blue, yellow, and red colors, respectively. For the temperature at the $\alpha + \beta$ phase field, as well as the strain rate of 0.001 s^{-1}, the marked blue area enlarges with the increase of deformation temperature, suggesting that the number of DRX grains increases (Figure 14a–c). The deformation temperature of 790 °C and the strain rate of 0.001 s^{-1} lead to the maximum amount of dynamic recrystallization (Figure 14d). When the deformation temperature increases at the β phase field, the number of DRX grains decreases (Figure 14d,e). Such a phenomenon is consistent with the changes trend of the power dissipation coefficient η predicted by the processing map. The reliability of the processing map for the Ti-3Al-6Cr-5V-5Mo alloy is further confirmed.

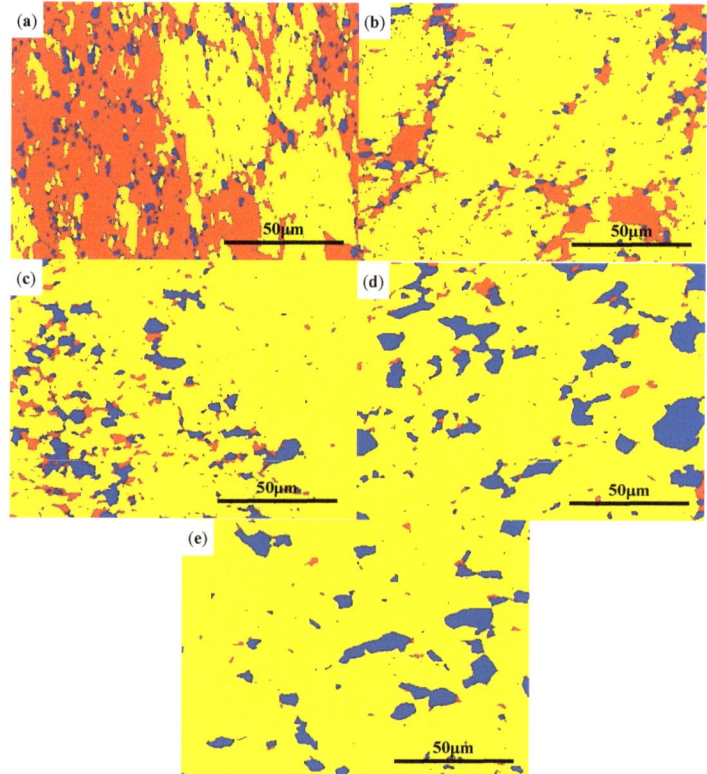

Figure 14. EBSD analysis of DRX at strain rates of 0.001 s^{-1}: (**a**) 700 °C, (**b**) 730 °C, (**c**) 760 °C, (**d**) 790 °C, and (**e**) 820 °C.

4. Conclusions

To provide guidance for the actual hot working of a near-β titanium alloy with potential application, Ti-3Al-6Cr-5V-5Mo, the constitutive relations, processing map, and deformed microstructure of the alloy were investigated. The conclusions are as follows:

(1) The constitutive models for the high-temperature deformation that occurred at the dual-phase field and single-phase field of the Ti-3Al-6Cr-5V-5Mo alloy are established, respectively.

For the α + β phase field, we have the following:

$$\dot{\varepsilon} = e^{49.138/3}[\sinh(0.005225\sigma)]^{4.26352}\exp(-442.25/RT)$$

For β phase field, we have the following:

$$\dot{\varepsilon} = e^{19.79592}[\sinh(0.00758\sigma)]^{3.61766}\exp(-206.86/RT)$$

(2) For the constitutive models established for the α + β phase field and β phase field, the correlated coefficients between actual stress and predicted stress are 0.986 and 0.983, and the mean relative errors of prediction are 2.7% and 4.1%, respectively. The accuracy of the model is slightly higher than some Arrhenius's constitutive equations for other titanium alloys. The constitutive equation has good prediction accuracy for the flow stress of Ti-3Al-6Cr-5V-5Mo alloy at both dual-phase and single-phase fields.

(3) An instability region at the range of temperature from 700 °C to 780 °C and the range of strain rates from 0.08 s^{-1} to 10 s^{-1} should be averted during thermomechanical processing. It is suggested that the alloy should be processed at a range of temperature from 790 °C to 800 °C and a range of strain rate from 0.001 s^{-1} to 0.007 s^{-1}.

(4) When the strain rate is 0.001 s^{-1}, the number of DRX grains increases with the increase of temperature during deformation occurring at α + β phase field. The most homogeneous grains with the minimum average grain size of 17 μm and maximum amount of DRX were obtained during deformation, which occurred at 790 °C/0.001 s^{-1}. It is consistent with the power-dissipation-coefficient region predicted by the processing map.

Author Contributions: Conceptualization, H.Z.; methodology, S.Z. (Shuo Zhang) and X.L.; validation, S.Z. (Shuai Zhang) and X.W.; investigation, H.Z., S.Z. (Siqian Zhang) and G.Z.; writing—original draft preparation, H.Z. All authors have read and agreed to the published version of the manuscript.

Funding: This research was funded by the "National Natural Science Foundation of China, grant number 52104379".

Data Availability Statement: Not applicable.

Conflicts of Interest: The authors declare no conflict of interest.

References

1. Wang, Y.; Hao, M.; Li, D.; Li, P.; Liang, Q.; Wang, D.; Zheng, Y.; Sun, Q.; Wang, Y. Enhanced mechanical properties of Ti-5Al-5Mo-5V-3Cr-1Zr by bimodal lamellar precipitate microstructures via two-step aging. *Mater. Sci. Eng. A* **2021**, *829*, 142117. [CrossRef]
2. Zhang, H.; Wang, C.; Zhou, G.; Zhang, S.; Chen, L. Dependence of strength and ductility on secondary α phase in a novel metastable-β titanium alloy. *J. Mater. Res. Technol.* **2022**, *18*, 5257–5266. [CrossRef]
3. Sadeghpour, S.; Abbasi, S.; Morakabati, M.; Kisko, A.; Karjalainen, L.; Porter, D. A new multi-element beta titanium alloy with a high yield strength exhibiting transformation and twinning induced plasticity effects. *Scr. Mater.* **2018**, *145*, 104–108. [CrossRef]
4. Zhang, S.H.; Deng, L.; Che, L.Z. An integrated model of rolling force for extra-thick plate by combining theoretical model and neural network model. *J. Manuf. Process.* **2022**, *75*, 100–109. [CrossRef]
5. Damodaran, D.; Shivpuri, R. Prediction and control of part distortion during the hot extrusion of titanium alloys. *J. Mater. Process. Technol.* **2004**, *150*, 70–75. [CrossRef]
6. Cai, J.; Guo, M.; Peng, P.; Han, P.; Yang, X.; Ding, B.; Qiao, K.; Wang, K.; Wang, W. Research on Hot Deformation Behavior of As-Forged TC17 Titanium Alloy. *J. Mater. Eng. Perform.* **2021**, *30*, 7259–7274. [CrossRef]

7. Meng, L.; Kitashima, T.; Tsuchiyama, T.; Watanabe, M. Effect of α precipitation on β texture evolution during β-processed forging in a near-β titanium alloy. *Mater. Sci. Eng. A* **2019**, *771*, 138640. [CrossRef]
8. Luo, S.; Yao, J.; Zou, G.; Li, J.; Jiang, J.; Yu, F. Influence of forging velocity on temperature and phase transformation characteristics of forged Ti-6Al-4V aeroengine drum. *Int. J. Adv. Manuf. Technol.* **2020**, *110*, 3101–3111. [CrossRef]
9. Zhang, W.; Yang, Q.; Tan, Y.; Ma, M.; Xiang, S.; Zhao, F. Simulation and Experimental Study of Dynamical Recrystallization Kinetics of TB8 Titanium Alloys. *Materials* **2020**, *13*, 4429. [CrossRef]
10. Li, C.; Huang, C.; Ding, Z.; Zhou, X. Research on High-Temperature Compressive Properties of Ti–10V–1Fe–3Al Alloy. *Metals* **2022**, *12*, 526. [CrossRef]
11. Park, C.W.; Choi, M.S.; Lee, H.; Yoon, J.; Javadinejad, H.R.; Kim, J.H. High-temperature deformation behavior and microstructural evolution of as-cast and hot rolled β21S alloy during hot deformation. *J. Mater. Res. Technol.* **2020**, *9*, 13555–13569. [CrossRef]
12. Zhang, J.; Xu, X.; Xue, J.; Liu, S.; Deng, Q.; Li, F.; Ding, J.; Wang, H.; Chang, H. Hot deformation characteristics and mechanism understanding of Ti–6Al–2Sn–4Zr–6Mo titanium alloy. *J. Mater. Res. Technol.* **2022**, *20*, 2591–2610. [CrossRef]
13. Chen, H.; Qin, H.; Qin, F.; Li, B.; Yu, Y.; Li, C. Hot Deformation Behavior and Microstructure Evolution of Ti–6Cr–5Mo–5V–4Al–1Nb Alloy. *Crystals* **2023**, *13*, 182. [CrossRef]
14. Guo, H.; Du, Z.; Wang, X.; Cheng, J.; Liu, F.; Cui, X.; Liu, H. Flowing and dynamic recrystallization behavior of new biomedical metastable β titanium alloy. *Mater. Res. Express* **2019**, *6*, 0865d2. [CrossRef]
15. Wang, Y.; Zhou, D.; Zhou, Y.; Sha, A.; Cheng, H.; Yan, Y. A Constitutive Relation Based on the Johnson–Cook Model for Ti-22Al-23Nb-2(Mo, Zr) Alloy at Elevated Temperature. *Crystals* **2021**, *11*, 754. [CrossRef]
16. Zhao, Q.; Yang, F.; Torrens, R.; Bolzoni, L. Comparison of hot deformation behaviour and microstructural evolution for Ti-5Al-5V-5Mo-3Cr alloys prepared by powder metallurgy and ingot metallurgy approaches. *Mater. Des.* **2019**, *169*, 107682. [CrossRef]
17. Sajadifar, S.V.; Maier, H.J.; Niendorf, T.; Yapici, G.G. Elevated Temperature Mechanical Characteristics and Fracture Behavior of a Novel Beta Titanium Alloy. *Crystals* **2023**, *13*, 269. [CrossRef]
18. Fu, M.; Pan, S.; Liu, H.; Chen, Y. Initial Microstructure Effects on Hot Tensile Deformation and Fracture Mechanisms of Ti-5Al-5Mo-5V-1Cr-1Fe Alloy Using In Situ Observation. *Crystals* **2022**, *12*, 934. [CrossRef]
19. Santosh, S.; Sampath, V.; Mouliswar, R. Hot deformation characteristics of NiTiV shape memory alloy and modeling using constitutive equations and artificial neural networks. *J. Alloy. Compd.* **2022**, *901*, 163451. [CrossRef]
20. Yang, Q.; Ma, M.; Tan, Y.; Xiang, S.; Zhao, F.; Liang, Y. Initial β Grain Size Effect on High-Temperature Flow Behavior of Tb8 Titanium Alloys in Single β Phase Field. *Metals* **2019**, *9*, 891. [CrossRef]
21. Lei, J.; Zhu, W.; Chen, L.; Sun, Q.; Xiao, L.; Sun, J. Deformation behaviour and microstructural evolution during the hot compression of Ti-5Al4Zr8Mo7V alloy. *Mater. Today Commun.* **2019**, *23*, 100873. [CrossRef]
22. Wang, J.; Wang, K.; Lu, S.; Li, X.; OuYang, D.; Qiu, Q. Softening mechanism and process parameters optimization of Ti-4.2Al-0.005B titanium alloy during hot deformation. *J. Mater. Res. Technol.* **2022**, *17*, 1842–1851. [CrossRef]
23. Han, L.; Zhang, H.; Cheng, J.; Zhou, G.; Wang, C.; Chen, L. Thermal Deformation Behavior of Ti-6Mo-5V-3Al-2Fe Alloy. *Crystals* **2021**, *11*, 1245. [CrossRef]
24. Pang, X.; Xiong, Z.; Liu, S.; Sun, J.; Misra, R.; Kokawa, H.; Li, Z. Grain refinement effect of ZrB2 in laser additive manufactured metastable β-titanium alloy with enhanced mechanical properties. *Mater. Sci. Eng. A* **2022**, *857*, 144104. [CrossRef]
25. Gu, B.; Chekhonin, P.; Xin, S.; Liu, G.; Ma, C.; Zhou, L.; Skrotzki, W. Effect of temperature and strain rate on the deformation behavior of Ti5321 during hot-compression. *J. Alloy. Compd.* **2021**, *876*, 159938. [CrossRef]
26. Liu, S.F.; Li, M.Q.; Luo, J.; Yang, Z. Deformation behavior in the isothermal compression of Ti-5Al-5Mo-5V-1Cr-1Fe alloy. *Mater. Sci. Eng. A* **2014**, *589*, 15–22. [CrossRef]
27. Wang, X.; Zhang, Y.; Ma, X. High temperature deformation and dynamic recrystallization behavior of AlCrCuFeNi high entropy alloy. *Mater. Sci. Eng. A* **2020**, *778*, 139077. [CrossRef]
28. Wang, Y.; Shao, W.; Zhen, L.; Yang, L.; Zhang, X. Flow behavior and microstructures of superalloy 718 during high temperature deformation. *Mater. Sci. Eng. A* **2008**, *497*, 479–486. [CrossRef]
29. Yong, Z.; Fukang, W.; Duo, Q.; Yongquan, N.; Min, W. High temperature deformation behavior of Ti-4.5Al-6.5Mo-2Cr-2.6Nb-2Zr-1Sn titanium alloy. *Rare Met. Mater. Eng.* **2020**, *49*, 944–949.
30. Mirzadeh, H.; Cabrera, J.M.; Najafizadeh, A. Constitutive relationships for hot deformation of austenite. *Acta Mater.* **2011**, *59*, 6441–6448. [CrossRef]
31. McQueen, H.; Yue, S.; Ryan, N.; Fry, E. Hot working characteristics of steels in austenitic state. *J. Mater. Process. Technol.* **1995**, *53*, 293–310. [CrossRef]
32. Zener, C.; Hollomon, J.H. Effect of Strain Rate Upon Plastic Flow of Steel. *J. Appl. Phys.* **1944**, *15*, 22–32. [CrossRef]
33. Shi, C.; Mao, W.; Chen, X.-G. Evolution of activation energy during hot deformation of AA7150 aluminum alloy. *Mater. Sci. Eng. A* **2013**, *571*, 83–91. [CrossRef]
34. Wang, Y.; Li, J.; Xin, Y.; Li, C.; Cheng, Y.; Chen, X.; Rashad, M.; Liu, B.; Liu, Y. Effect of Zener–Hollomon parameter on hot deformation behavior of CoCrFeMnNiC0.5 high entropy alloy. *Mater. Sci. Eng. A* **2019**, *768*, 138483. [CrossRef]
35. Li, C.; Huang, L.; Zhao, M.; Guo, S.; Li, J. Hot deformation behavior and mechanism of a new metastable β titanium alloy Ti–6Cr–5Mo–5V–4Al in single phase region. *Mater. Sci. Eng. A* **2021**, *814*, 141231. [CrossRef]

36. Feng, R.; Bao, Y.; Ding, Y.; Chen, M.; Ge, Y.; Xie, L. Three different mathematical models to predict the hot deformation behavior of TA32 titanium alloy. *J. Mater. Res.* **2022**, *37*, 1309–1322. [CrossRef]
37. Zhao, Q.; Yu, L.; Ma, Z.; Li, H.; Wang, Z.; Liu, Y. Hot Deformation Behavior and Microstructure Evolution of 14Cr ODS Steel. *Materials* **2018**, *11*, 1044. [CrossRef]
38. Poliak, E.; Jonas, J. A one-parameter approach to determining the critical conditions for the initiation of dynamic recrystallization. *Acta Mater.* **1996**, *44*, 127–136. [CrossRef]
39. Jin, Z.Y.; Li, N.N.; Yan, K.; Wang, J.; Bai, J.; Dong, H.B. Deformation mechanism and hot workability of extruded magnesium alloy AZ31. *Acta Metall. Sin. (Engl. Lett.)* **2018**, *31*, 71–81. [CrossRef]
40. Balasubrahmanyam, V.; Prasad, Y. Deformation behaviour of beta titanium alloy Ti–10V–4.5Fe–1.5Al in hot upset forging. *Mater. Sci. Eng. A* **2002**, *336*, 150–158. [CrossRef]
41. Shi, X.H.; Cao, Z.H.; Fan, Z.Y.; Li, L.; Guo, R.P.; Qiao, J.W. Isothermal Compression and Concomitant Dynamic Recrystallization Behavior of Ti-6.5Al-3.5Mo-1.5Zr-0.3Si Alloy with Initial Martensitic Microstructure. *J. Mater. Eng. Perform.* **2020**, *29*, 3361–3372. [CrossRef]
42. Chuan, W.; Liang, H. Hot deformation and dynamic recrystallization of a near-beta titanium alloy in the β single phase region. *Vacuum* **2018**, *156*, 384–401. [CrossRef]
43. Shekhar, S.; Sarkar, R.; Kar, S.K.; Bhattacharjee, A. Effect of solution treatment and aging on microstructure and tensile properties of high strength β titanium alloy, Ti–5Al–5V–5Mo–3Cr. *Mater. Des.* **2014**, *66*, 596–610. [CrossRef]
44. Li, C.-L.; Mi, X.-J.; Ye, W.-J.; Hui, S.-X.; Yu, Y.; Wang, W.-Q. Effect of solution temperature on microstructures and tensile properties of high strength Ti–6Cr–5Mo–5V–4Al alloy. *Mater. Sci. Eng. A* **2013**, *578*, 103–109. [CrossRef]
45. Fan, X.; Yang, H.; Gao, P.; Zuo, R.; Lei, P. The role of dynamic and post dynamic recrystallization on microstructure refinement in primary working of a coarse grained two-phase titanium alloy. *J. Mater. Process. Technol.* **2016**, *234*, 290–299. [CrossRef]

Disclaimer/Publisher's Note: The statements, opinions and data contained in all publications are solely those of the individual author(s) and contributor(s) and not of MDPI and/or the editor(s). MDPI and/or the editor(s) disclaim responsibility for any injury to people or property resulting from any ideas, methods, instructions or products referred to in the content.

Article

Vacuum Electrodeposition of Cu(In, Ga)Se$_2$ Thin Films and Controlling the Ga Incorporation Route

Kanwen Hou [1,3,4], Guohao Liu [1,3,4], Jia Yang [1,2,3,4,*], Wei Wang [1,3,4], Lixin Xia [1,3,4], Jun Zhang [1,3,4], Baoqiang Xu [1,2,3,4] and Bin Yang [1,2,3,4]

1. Key Laboratory for Nonferrous Vacuum Metallurgy of Yunnan Province, Kunming University of Science and Technology, Kunming 650093, China
2. State Key Laboratory of Complex Non-Ferrous Metal Resources Clean Utilization, Kunming University of Science and Technology, Kunming 650093, China
3. National Engineering Research Center of Vacuum Metallurgy, Kunming University of Science and Technology, Kunming 650093, China
4. Faculty of Metallurgical and Energy Engineering, Kunming University of Science and Technology, Kunming 650093, China
* Correspondence: yangjia0603@163.com

Abstract: The traditional electrochemical deposition process used to prepare Cu(In, Ga)Se$_2$ (CIGS) thin films has inherent flaws, such as the tendency to produce low-conductivity Ga$_2$O$_3$ phase and internal defects. In this article, CIGS thin films were prepared under vacuum (3 kPa), and the mechanism of vacuum electrodeposition CIGS was illustrated. The route of Ga incorporation into the thin films could be controlled in a vacuum environment via inhibiting pH changes at the cathode region. Through the incorporation of a low-conductivity secondary phase, Ga$_2$O$_3$ was inhibited at 3 kPa, as shown by Raman and X-ray photoelectron spectroscopy. The preparation process used a higher current density and a lower diffusion impedance and charge transfer impedance. The films that were produced had larger particle sizes.

Keywords: vacuum electrodeposition; Cu(In; Ga)Se$_2$ thin films; electrodeposition mechanism; Ga incorporation; Gallium Oxide

Citation: Hou, K.; Liu, G.; Yang, J.; Wang, W.; Xia, L.; Zhang, J.; Xu, B.; Yang, B. Vacuum Electrodeposition of Cu(In, Ga)Se$_2$ Thin Films and Controlling the Ga Incorporation Route. *Crystals* **2023**, *13*, 319. https://doi.org/10.3390/cryst13020319

Academic Editors: Stefano Carli and Giuseppe Prestopino

Received: 9 January 2023
Revised: 7 February 2023
Accepted: 10 February 2023
Published: 15 February 2023

Copyright: © 2023 by the authors. Licensee MDPI, Basel, Switzerland. This article is an open access article distributed under the terms and conditions of the Creative Commons Attribution (CC BY) license (https:// creativecommons.org/licenses/by/ 4.0/).

1. Introduction

CIGS thin films are one of the most promising photovoltaic materials for second-generation solar cells [1]. They are composite semiconductor materials with chalcopyrite structures [2] with band gaps of 1.0 eV to 1.7 eV [3]. Co-evaporation [4], magnetron sputtering [5], spray pyrolysis [6], solution gel method [7], and electrodeposition [8] are the main methods to prepare CIGS thin films.

Electrodeposition has the advantages of a low preparation cost, high efficiency, large area, and continuous preparation and is likely to be used in future industrialized production methods [9]. However, it has some inherent flaws when used to prepare metal materials. For example, hydrogen is easily generated on the cathode surface and is difficult to remove and the deposition rate can be slow [10,11].

Generally, a vacuum environment induces rapid off-gassing [12] and lower oxidation and contamination [13]. Based on these properties, researchers have conducted studies on the electrodeposition of metal thin films in a vacuum environment. In the 1940s, RCA Company [14] electrodeposited Fe and Mn in a vacuum environment and found that hydrogen quickly escaped from the films, and the oxidation of the film was avoided. This resulted in the preparation of smooth and bright metal films. In 1984, E Muttilainen et al. [15] electrodeposited Cr in a vacuum environment and found that a low-pressure environment increased the current efficiency and reduced the porosity and roughness of the coating. They concluded that low pressure was one of the most important factors affecting the quality of the coating. In a following study, S. E. Nam [16,17],

R. Su [18], and P. Ming et al. [19,20] prepared metal thin films with excellent properties using vacuum electrodeposition.

In the application of electrodeposition to prepare CIGS thin films, the Ga_2O_3 phase is also easily generated, along with the generation of hydrogen on the cathode surface [21,22]. The entry of Ga_2O_3 into the films can decrease the conductivity [23,24]. Therefore, in this article, electrodeposition was applied to prepare CIGS thin films in a vacuum environment. The effects of the vacuum environment on the deposition process and the composition, morphology, and phases in the CIGS thin films were studied.

2. Materials and Methods

The electrolyte solution contained 200 mL deionized water with 4 mM $CuCl_2$, 10 mM $InCl_3$, 10 mM $GaCl_3$, 8 mM H_2SeO_3, and a supporting electrolyte (100 mM LiCl, 100 mM NH_4Cl, and 60 mM NH_2SO_3H). The pH of the solution was adjusted to 1.8 by adding concentrated hydrochloric acid dropwise.

Electrochemical experiments, including linear scanning voltammetry (LSV), potentiostatic polarization, electrochemical impedance spectroscopy (EIS), and electrodeposition were carried out on CIGS thin films in a three-electrode system with a SnO_2/glass (1 × 1.5 cm) as the working electrode, a Pt electrode as the counter electrode, and a saturated calomel electrode (SCE) as the reference electrode.

A CorrTest CH310H electrochemical workstation was used for the LSV, potentiostatic polarization, EIS, and electrodeposition CIGS thin film studies. The LSV curves were measured at potentials between 0.1 V to −1.0 V (vs. SCE) with a scanning rate of 10 mV/s. The potentiostatic polarization curves were measured at a potential of −0.60 V (vs. SCE) and a polarization time of 30 min. EIS was performed at a potential of −0.60 V (vs. SCE) with a test frequency range of 0.01 Hz to 100 kHz. On electrodeposited CIGS thin films, the potential range was −0.1 V to −0.9 V (vs. SCE), and the polarization time was 30 min. When the experiment was carried out at 3 kPa, a vacuum pump was used to extract the air from the experimental system for 30 min. Figure 1 illustrates the electrodeposition process of CIGS thin films.

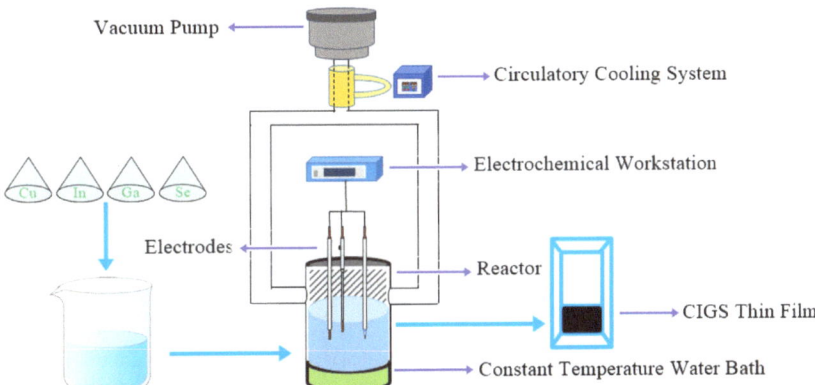

Figure 1. A schematic of the electrodeposition CIGS thin films system.

Scanning electron microscopy (SEM, TM3030Plus, Hitachi, Tokyo, Japan) was used to measure the morphology of CIGS thin films, and energy-dispersive spectroscopy (EDS, TM3030Plus, Hitachi, Tokyo, Japan) was utilized to characterize the elemental composition of the CIGS thin films. X-ray diffraction (XRD, Rigaku Ultima IV, Tokyo, Japan), Raman spectroscopy (Raman, DXRxi, Thermo Scientific, MA, USA), and X-ray photoelectron spectroscopy (XPS, Escalab Xi+, Thermo Scientific, MA, USA) were used to characterize the phases in the CIGS thin films.

3. Results and Discussion

Figure 2 illustrates a schematic of how the vacuum environment affects the electrochemical behavior of Ga^{3+}. In the cathode region, the vacuum environment inhibited the Volmer reaction of H^+ to a hydrogen atom (H_{ads}), leading to lower H^+ consumption on the cathode surface. In the anode region, the oxygen evolution reaction (OER) of H_2O to O_2 was promoted, which produced additional O_2. At the same time, the rapid release of O_2 bubbles agitated the electrolyte solution, which led to a uniform H^+ distribution in the electrolyte solution. Thus, the inhibition of the Volmer reaction and the promotion of the OER together stabilized the pH of the cathode region. This finally interrupted the route of Ga^{3+} entering the films in the form of Ga_2O_3 by the hydrolytic reaction of Ga^{3+}.

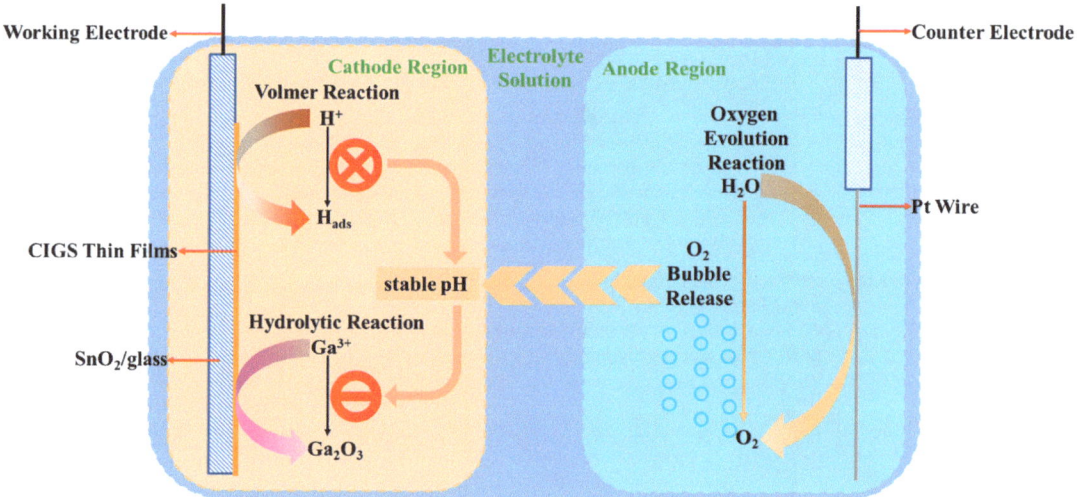

Figure 2. Schematic illustration for the inhibition of Ga^{3+} hydrolysis in 3 kPa.

3.1. LSV Analyses

Figure 3a illustrates the effects of the vacuum environment on the hydrogen evolution potentials. Evidence of an inhibited reaction of H^+ to H_{ads} was found in a vacuum environment. Peak C_1 corresponding to the reaction of H^+ to H_{ads} (Equation (1)) did not appear at 3 kPa, possibly due to a decrease in the H_2 partial pressure. Because of the reduced partial pressure of H_2, the overpotential of H^+/H_{ads} on the electrode surface was increased. This phenomenon has also been observed on Au and Pt electrodes [25,26].

$$H^+ + e^- \rightarrow H_{ads} \qquad (1)$$

Figure 3b illustrates the LSV curve of the Cu unary solution at 3 kPa. Peaks C_1, C_2, and C_3 corresponded to the reaction of Cu^{2+} to Cu^+ (Equation (2)), Cu^+ to Cu (Equation (3)), and H^+ to H_{ads} (Equation (1)), respectively [27]. Peak C_1 was indistinguishable at 80 kPa and 3 kPa, indicating that pressure did not affect the reaction of Cu^{2+} to Cu^+. Peak C_2 was reduced at 3 kPa, indicating that the formation of Cu was promoted in a low-pressure environment. Peak C_3 did not appear at 3 kPa, indicating that the reaction of H^+ to H_{ads} was inhibited in a vacuum environment.

$$Cu^{2+} + e^- \rightarrow Cu^+ \qquad (2)$$

$$Cu^+ + e^- \rightarrow Cu \qquad (3)$$

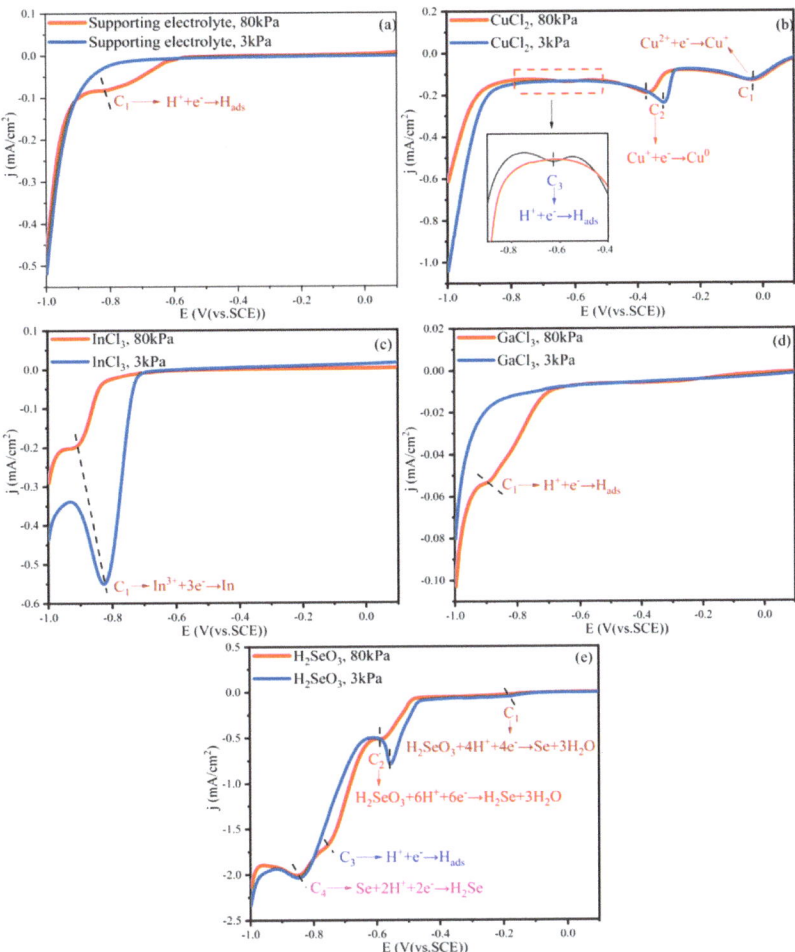

Figure 3. The LSV curves in the (**a**) supporting electrolyte, (**b**) 4 mM CuCl$_2$, (**c**) 10 mM InCl$_3$, (**d**) 10 mM GaCl$_3$, and (**e**) 8 mM H$_2$SeO$_3$ unary solutions at 3 kPa.

Figure 3c illustrates the LSV curve of the In unary solution at 3 kPa. Peak C$_1$ corresponded to the reaction of In^{3+} to In (Equation (4)) [28]. This peak was reduced at 3 kPa, showing that the formation of In was promoted in a vacuum environment.

$$\text{In}^{3+} + 3e^- \rightarrow \text{In} \tag{4}$$

Figure 3d illustrates the LSV curve of the Ga unary solution at 3 kPa. Peak C$_1$ corresponds to the reaction of H$^+$ to H$_{ads}$ [27] (Equation (1)). This peak did not appear at 3 kPa, indicating that the reaction of H$^+$ to H$_{ads}$ was inhibited in a vacuum environment. The reaction of Ga^{3+} to Ga did not occur within the range of 0.1 V to −1.0 V because the reaction required a higher potential [24].

Figure 3e illustrates the LSV curve of the Se unary solution at 3 kPa. Peaks C$_1$, C$_2$, C$_3$, and C$_4$ corresponded to the reactions of H$_2$SeO$_3$ to Se (Equation (5)), H$_2$SeO$_3$ to H$_2$Se (Equation (6)), H$^+$ to H$_{ads}$ (Equation (1)), and Se to H$_2$Se (Equation (7)), respectively [27]. The current densities of peaks C$_1$, C$_2$, and C$_4$ were greater at 3 kPa, indicating that the formation of H$_2$Se and Se was promoted. Peak C$_3$ did not appear at 3 kPa, indicating that the reaction of H$^+$ to H$_{ads}$ was inhibited in a vacuum environment.

$$H_2SeO_3 + 4H^+ + 4e^- \rightarrow Se + 3H_2O \tag{5}$$

$$H_2SeO_3 + 6H^+ + 6e^- \rightarrow H_2Se + 3H_2O \tag{6}$$

$$Se + 2H^+ + 2e^- \rightarrow H_2Se \tag{7}$$

Figure 4a illustrates the LSV curves of Cu-Se binary solutions at 3 kPa. Peaks C$_1$ and C$_2$ corresponded to the formation of Cu$^+$ and Se, respectively (Equations (2) and (5)). Peaks C$_3$ and C$_4$ corresponded to the reaction of H$_2$SeO$_3$ to H$_2$Se and Se to H$_2$Se, respectively (Equations (6) and (7)). Since the solution contained both Cu$^+$ and Cu^{2+}, the two ions were induced by H$_2$SeO$_3$ or H$_2$Se to form Cu$_2$Se or CuSe, respectively [26] (denoted as Cu$_3$Se$_2$ phase) (Equations (8)–(11)). The current densities of peaks C$_3$ and C$_4$ were greater at 3 kPa, indicating that the formation of Cu$_3$Se$_2$ was promoted in a vacuum environment.

$$2Cu^+ + H_2SeO_3 + 4H^+ + 6e \rightarrow Cu_2Se + 3H_2O \tag{8}$$

$$Cu^{2+} + H_2SeO_3 + 4H^+ + 6e \rightarrow CuSe + 3H_2O \tag{9}$$

$$2Cu^+ + H_2Se \rightarrow Cu_2Se + 2H^+ \tag{10}$$

$$Cu^{2+} + H_2Se \rightarrow CuSe + 2H^+ \tag{11}$$

Figure 4b illustrates the LSV curves of the In-Se binary solutions at 3 kPa. Peaks C$_1$ and C$_2$ corresponded to the reaction of H$_2$SeO$_3$ to H$_2$Se and Se to H$_2$Se, respectively (Equations (6) and (7)). H$_2$SeO$_3$ reacted with H$_2$Se to form Se (Equation (12)). Because In$_2$Se$_3$ has a high standard Gibbs energy of formation (-386 kJ/mol [29]), H$_2$Se induced In^{3+} to form In$_2$Se$_3$ [30] (Equation (13)). The current densities of peaks C$_1$ and C$_2$ were greater at 3 kPa, indicating that the formation of H$_2$Se, In$_2$Se$_3$, or In was promoted in a vacuum environment.

$$2H_2Se + H_2SeO_3 \rightarrow 3Se + 3H_2O \tag{12}$$

$$3H_2Se + 2In^{3+} + 6e^- \rightarrow In_2Se_3 \tag{13}$$

Figure 4c illustrates the LSV curves of Ga-Se binary solutions at 3 kPa. The peaks C$_1$, C$_2$, and C$_3$ correspond to the reaction of H$_2$SeO$_3$ to Se, H$_2$SeO$_3$ to H$_2$Se, and Se to H$_2$Se, respectively (Equations (5)–(7)). Because Ga$_2$Se$_3$ has a high standard Gibbs energy of formation (-418 kJ/mol [31]), Ga^{3+} was induced by H$_2$Se to form Ga$_2$Se$_3$ [32] (Equation (14)). The current densities of the peaks C$_2$ and C$_3$ are greater at 3 kPa, indicating that the formation of H$_2$Se or Ga$_2$Se$_3$ was promoted in a vacuum environment.

$$3H_2Se + 2Ga^{3+} \rightarrow Ga_2Se_3 + 6H^+ \tag{14}$$

Figure 4d illustrates the LSV curves of Cu-In-Se ternary solutions at 3 kPa. Peaks C$_1$, C$_2$, C$_3$, C$_4$, and C$_6$ correspond to the formation of Cu$^+$, Se, Cu$_3$Se$_2$, H$_2$Se, and In$_2$Se$_3$ or In, respectively (Equations (2), (5), (8)–(11), (6), (13) or (4)). In^{3+} was induced by Cu$_3$Se$_2$ + H$_2$SeO$_3$ or Cu$_3$Se$_2$ + H$_2$Se to produce more stable CuInSe$_2$ [32] (Equations (15) and (16)) at peak C$_5$. As the polarization potential increased, the current density became greater at 3 kPa, indicating that the formation of CIS was promoted in a vacuum environment.

$$3In^{3+} + Cu_3Se_2 + 4H_2SeO_3 + 16H^+ + 25e \rightarrow 3CuInSe_2 + 12H_2O \tag{15}$$

$$3In^{3+} + Cu_3Se_2 + 4H_2Se + e \rightarrow 3CuInSe_2 + 8H^+ \tag{16}$$

Figure 4e illustrates the LSV curves of Cu-Ga-Se ternary solutions at 3 kPa. Peaks C$_1$, C$_2$, C$_3$, and C$_4$ corresponded to the formation of Cu$^+$, Se, Cu$_3$Se$_2$, and Ga$_2$Se$_3$, respectively (Equations (2), (5), (8)–(11) and (14)). Ga^{3+} was induced by Cu$_3$Se$_2$ + H$_2$SeO$_3$

or $Cu_3Se_2 + H_2Se$ to produce $CuGaSe_2$ [31] (Equations (17) and (18)) at peak C_4. As the polarization potential increased, the current density became greater at 3 kPa, indicating that the formation of CGS was promoted in a vacuum environment.

$$3Ga^{3+} + Cu_3Se_2 + 4H_2SeO_3 + 16H^+ + 25e^- \rightarrow 3CuGaSe_2 + 12H_2O \quad (17)$$

$$3Ga^{3+} + Cu_3Se_2 + 4H_2Se + e^- \rightarrow 3CuGaSe_2 + 8H^+ \quad (18)$$

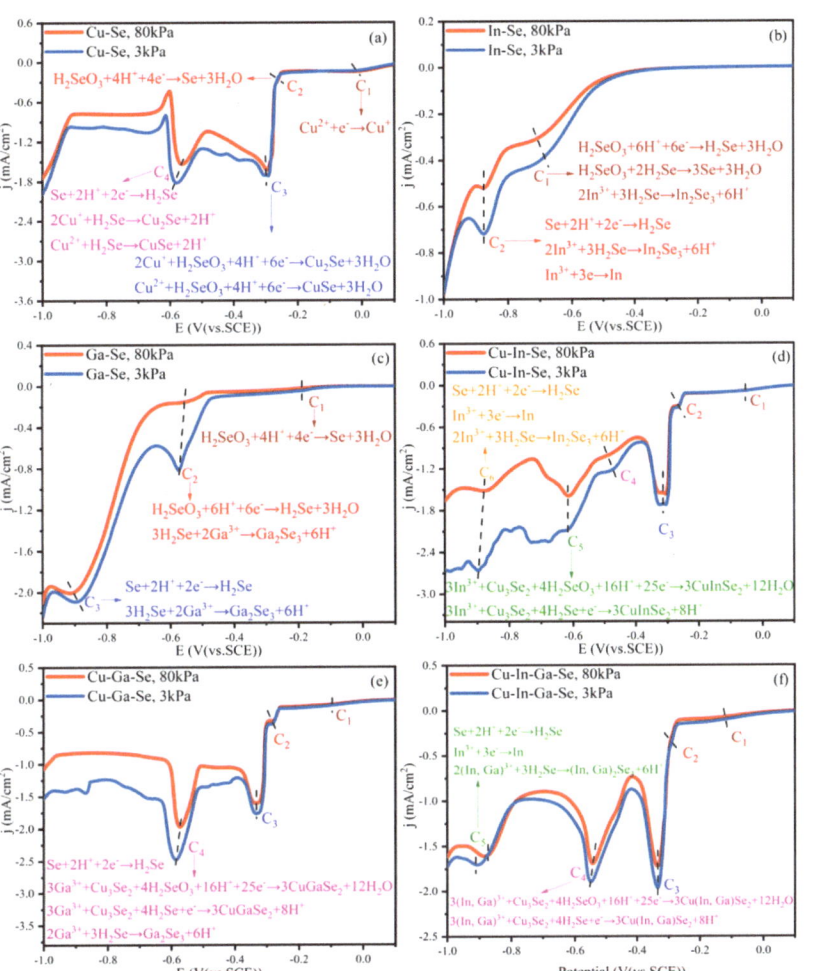

Figure 4. The LSV curves in the (**a**) Cu−Se, (**b**) In−Se (**c**) Ga−Se binary, (**d**) Cu−In−Se, (**e**) Cu−Ga−Se ternary, and (**f**) Cu−In−Ga−Se quaternary solutions at 3 kPa.

Figure 4f illustrates the LSV curves of Cu-In-Ga-Se ternary solutions at 3 kPa. Peaks C_1, C_2, and C_3 corresponded to the formation of Cu^+, Se, and Cu_3Se_2, respectively (Equations (2), (5), (8)–(11)). Ga and In are homotopic elements, and their reduction processes at the cathode are similar (noted as $(In, Ga)^{3+}$ phase). Meanwhile, both In^{3+} and Ga^{3+} were induced by $Cu_3Se_2 + H_2SeO_3$ or $Cu_3Se_2 + H_2Se$ into the films (Equations (15)–(18)). Therefore, In^{3+} and Ga^{3+} competed for reduction at the cathode. According to previous conclusions, the reduction potential of In and Ga in the CIGS thin films was −0.6 V. Therefore, peak C_4 corresponded to the formation of $CuInSe_2$, $CuGaSe_2$, or $Cu(In, Ga)Se_2$

(Equations (19) and (20)). Peak C_5 corresponded to the formation of In_2Se_3, Ga_2Se_3, or In (Equations (13), (14), or (4)). The current density of peak C_4 was greater at 3 kPa, indicating that the formation of CIGS was promoted in a vacuum environment.

$$3(In, Ga)^{3+} + Cu_3Se_2 + 4H_2SeO_3 + 16H^+ + 25e \rightarrow 3Cu(In, Ga)Se_2 + 12H_2O \quad (19)$$

$$3(In, Ga)^{3+} + Cu_3Se_2 + 4H_2Se + e \rightarrow 3Cu(In, Ga)Se_2 + 8H^+ \quad (20)$$

3.2. Potentiostatic Polarization and EIS Analyses

According to the previous conclusion, the main reduction potential of In^{3+} and Ga^{3+} in the CIGS thin films was −0.6 V. Therefore, the potentiostatic polarization potential was set to −0.6 V. Figure 5a illustrates the potentiostatic polarization curve of the CIGS thin films at 3 kPa. The current density was always greater at 3 kPa, indicating that the resistance during the preparation of CIGS thin films was lower in a vacuum environment.

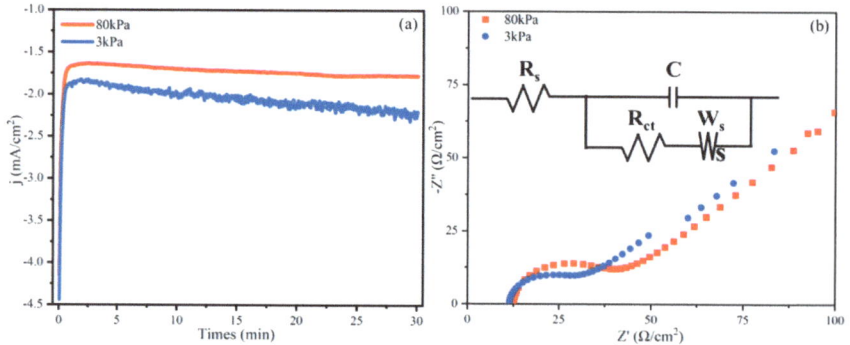

Figure 5. The (**a**) current density−time curve and (**b**) impedance Nyquist plot of electrodeposition CIGS thin films at 3 kPa.

Changes in the diffusion impedance and charge-transfer impedance during the preparation of CIGS thin films were investigated by EIS at a low pressure. Figure 5b illustrates the impedance Nyquist plot of the CIGS thin films deposition process at 3 kPa. The semicircle diameter of the curve was smaller, and the slope of the straight line was larger at 3 kPa, indicating that the electroreduction and diffusion impedances were lower, respectively. In addition, the ohmic impedance was smaller. Thus, the resistance was lower when applying vacuum electrodeposition. This corresponds to the conclusion of the current density-time curves. Both curves in Figure 5b were semicircular with straight lines, indicating that electroreduction and diffusion controlled the CIGS thin films deposition [33].

In this study, the impedance test data were combined in order to fit the equivalent circuit of the CIGS thin films deposition process. The fitted EIS data of the CIGS thin films deposition process at different pressures are shown in Table 1. W_s is the Warburg diffusion impedance, R_{ct} is the charge-transfer impedance between the interface of cathode and electrolyte solutions, C is the electrical double-layer capacitance on the electrode surface, and R_s is the electrolyte solution impedance.

Table 1. Fitting data of CIGS thin films deposition impedance at 3 kPa (Ω).

Pressure	W_s	R_{ct}	C	R_s
80 kPa	417.6	42.39	2.292×10^{-4}	18.77
3 kPa	129.1	29.83	3.546×10^{-4}	19.23

When the system pressure was reduced from 80 kPa to 3 kPa, W_s fell by 69.1%, R_{ct} fell by 29.6%, C rose by 35.4%, and R_s remained constant. Figure 6 illustrates the OER on the anode surface during the preparation of the CIGS thin films. At 3 kPa, the number of bubbles at the Pt electrode was higher, and the volume was larger. Therefore, the formation of O_2 was promoted under a vacuum. The promoted formation of O_2 indicates that the electrode reaction proceeded more smoothly. The smooth occurrence of the electrode reaction was also one of the reasons for the higher current density at 3 kPa. In addition, according to Boyle's law, the bubble volume should be larger in a vacuum environment. This is consistent with the phenomenon in Figure 6. The rapid release of O_2 bubbles enhanced the diffusion of the electrolyte solution. Therefore, the W_s value fell. Since the reaction of H^+ to H_{ads} was inhibited, and the rapid dehydrogenation occurred in the electrolyte at 3 kPa, the contact area increased between the interface of the cathode and electrolyte solution, which decreased the R_{ct} value. Because the reaction of H^+ to H_{ads} was inhibited, the deposition of major elements increased. Thus, the thickness of the CIGS thin films increased, which increased the C value at 3 kPa. Because it was determined by the electrolyte solution, the R_s value remained constant.

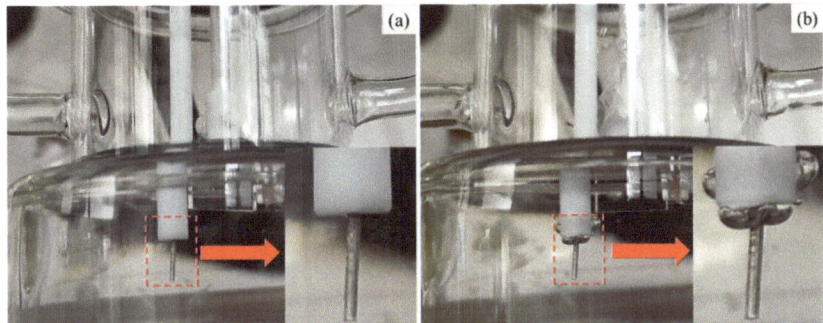

Figure 6. The picture of the OER on the anode surface at pressures of (**a**) 80 kPa and (**b**) 3 kPa.

3.3. EDS Analysis

Figure 7 illustrates the EDS composition curves of Cu, In, Ga, and Se in CIGS thin films at polarization potentials from −0.1 V to −0.9 V. Overall, changing the pressure had little effect on the element deposition pattern but had a significant impact on the elemental content. Combined with Figure 3, the first reduction potentials of Cu and Se were 0 V and −0.2 V, respectively. Therefore, Cu ions were more easily reduced when the reduction potential was −0.1 V. As shown in Figure 7a, the content of Cu in the films was highest at this time. The formation of Cu and Cu_3Se_2 was promoted in a vacuum environment, which in turn led to a high content of Cu in the films at 3 kPa. When the reduction potential was −0.2 V, the reduction of Se became easier. As shown in Figure 7d, the content of Se in the films increased. As the potential increases, the Cu and Se content tended to be relatively stable. As a whole, the content of Cu and Se in the CIGS thin films was slightly higher at 3 kPa, indicating that the vacuum environment was beneficial for the formation of Cu and Se.

As shown in Figure 7b, the potential of In in the CIGS thin films was −0.2 V at 3 kPa. However, this potential was −0.4 V at 80 kPa. The reason may be that the formation of Cu_3Se_2 was promoted in a vacuum environment and In was induced by the excess Cu_3Se_2 to produce a more stable $CuInSe_2$ phase, which in turn could enter the films at a lower potential. Meanwhile, the content of In in the films was higher at 3 kPa. As a result, the formation of In was promoted in a vacuum environment.

As shown in Figure 7c, when the potential was lower, the formation of Ga was largely unaffected due to the lower In content in the films. Because the In content in the films increased as the potential increased, and the formation of In and Ga had a competitive relationship, the formation of Ga was inhibited. This was the first reason for the low concentration of Ga in the CIGS thin films at 3 kPa. Upon increasing the potential, the production of H_2 at the cathode surface increased at 80 kPa. In turn, the consumption of H^+ in the cathode region increased. Along with the consumption of H^+, the pH at the cathode region increased, which led to the hydrolysis of Ga^{3+} to Ga_2O_3 [24]. Ga^{3+} entered the films in the form of Ga_2O_3 [23]. Because the reaction of H^+ to H_{ads} was inhibited and the OER was promoted at 3 kPa, the pH remained relatively stable at the cathode region, which prevented Ga^{3+} from entering the films in the form of Ga_2O_3. This was the second reason for the low Ga concentration in the CIGS thin films at 3 kPa. In summary, the formation of Ga was inhibited in a vacuum environment.

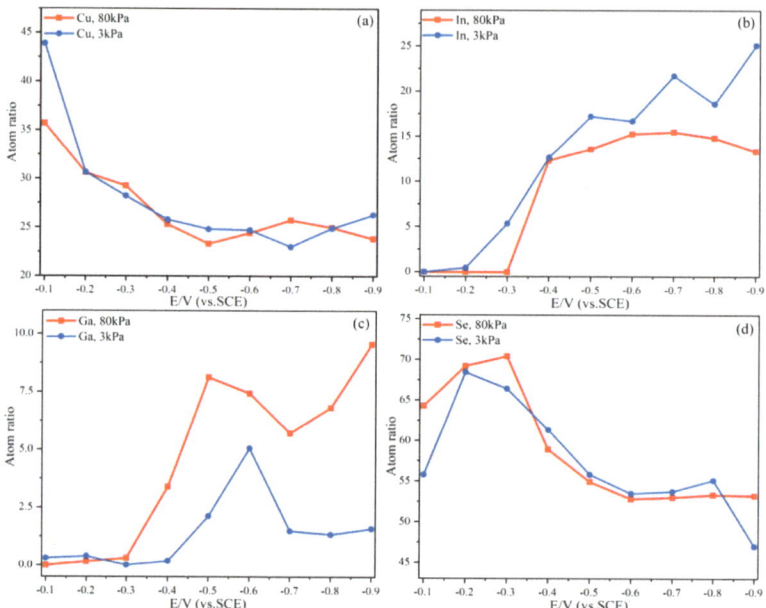

Figure 7. EDS composition curves of (**a**) Cu, (**b**) In, (**c**) Ga, and (**d**) Se in CIGS thin films.

3.4. Morphologic Analyses

SEM images showing the morphologies of the CIGS thin films prepared at various deposition potentials and pressures are displayed in Figure 8. Overall, the particles [29] in the films were larger at 3 kPa. The first reason is that the reaction of H^+ to H_{ads} was inhibited, which in turn inhibited H_2 production. Another reason is that the vacuum environment quickly removed H_2 from the surface of the CIGS thin films. The amount of H_2 adsorbed on the cathode surface decreased, and the space available for the deposition of major elements increased.

3.5. Phase Analyses

As shown in Figure 9, in the CIGS thin films, there are three main diffraction peaks due to (101), (112), and (220) planes corresponding to $CuIn_{0.7}Ga_{0.3}Se_2$. The positions of these diffraction peaks are quite similar. In addition, the difference in pressure changes the preferred crystallographic orientation. When the pressure is 80 kPa, the $CuIn_{0.7}Ga_{0.3}Se_2$ phase possesses a (101) preferred crystallographic orientation at approximately 17.27°. However, under 3 kPa, the preferred crystallographic orientation is a (112) plane located at approximately 26.90°. For the $CuIn_{0.7}Ga_{0.3}Se_2$ phase, a preference for the (112) plane is more favorable. Therefore, the main $CuIn_{0.7}Ga_{0.3}Se_2$ phase can be electrochemically deposited at the pressure of 80 kPa and even at 3 kPa, although with different preferred crystallographic orientations.

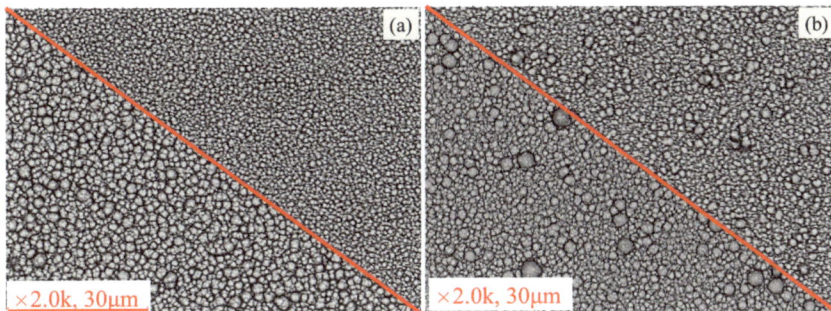

Figure 8. SEM images of the morphology of CIGS thin films prepared at the pressures of 80 kPa (above the red diagonal line) and 3 kPa (below the red diagonal line) and at the potentials of (**a**) −0.5 V and (**b**) −0.6 V.

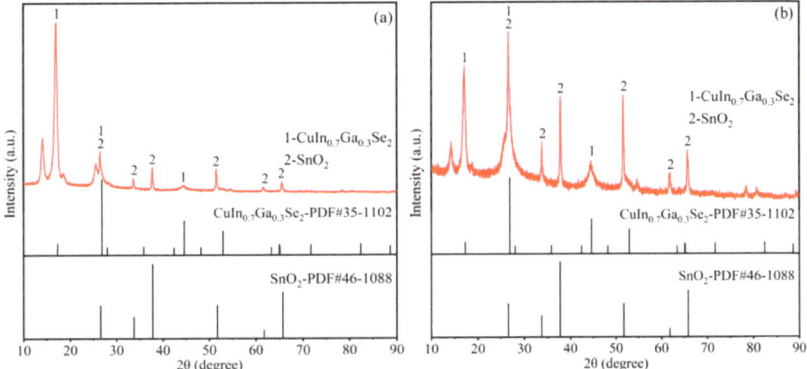

Figure 9. XRD spectra of CIGS thin film prepared at the potential of −0.6 V and pressures of: (**a**) 80 kPa and (**b**) 3 kPa.

The surface phases were characterized by Raman spectroscopy, as shown in Figures 10 and 11. According to Figure 10a, the vibrational peaks of the Ga_2O_3 phase (200 cm^{-1} [34]) appeared in the potential range from −0.1 V to −0.2 V at both 80 kPa and 3 kPa. This was because the reaction of H_2SeO_3 to Se consumed H$^+$ in the cathode region, which in turn hydrolyzed Ga to Ga_2O_3 in the films [35]. Meanwhile, little O_2 formed on the anode surface due to the low potential and could not increase the mass transfer rate of the solution at 3 kPa. This was why the Ga_2O_3 phase appeared at a low potential at 3 kPa.

When the potential was −0.3 V, the main phase of the films changed from the Cu-Se phase (260 cm^{-1} [36]) to the CIGS A1 phase (175 cm^{-1} [37]) at 3 kPa. The formation of the CIGS A1 phase led to a decrease in the consumption of H$^+$. Increasing the potential led to a gradual increase in the O_2 production, which in turn led to a gradual increase in the mass transfer rate, resulting in a stable pH in the cathode region. Ga^{3+} hydrolysis was inhibited, and the Ga_2O_3 phase disappeared. However, at 80 kPa, when the potential was −0.4 V, only the main phase of the films began to transform. In correspondence with the higher potential, the hydrogen evolution reaction began to occur. Therefore, the vibrational peak of the Ga_2O_3 phase was always present when the potential was in the range of −0.3 V to −0.5 V.

As shown in Figure 11, Ga entered the films as Ga-Se phase (214 cm^{-1} [38]) at 3 kPa. Combined with the above conclusions, the formation of Ga_2O_3 was inhibited at 3 kPa.

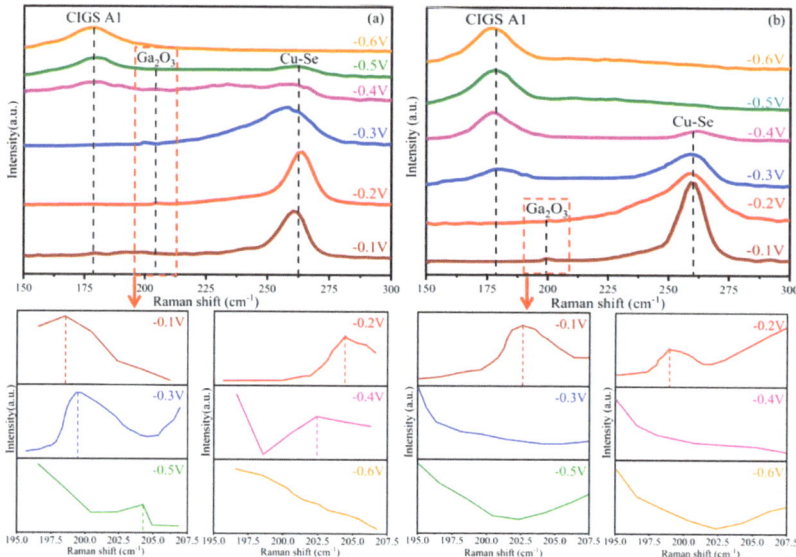

Figure 10. Raman spectra of CIGS thin film prepared at potentials of −0.1 V to −0.6 V and pressures of: (**a**) 80 kPa and (**b**) 3 kPa.

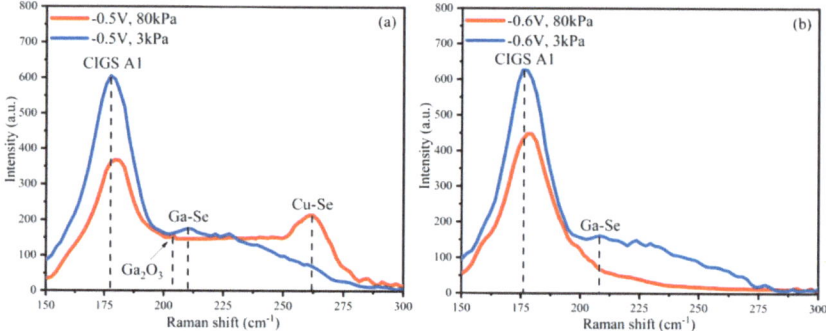

Figure 11. Raman spectra of CIGS thin films prepared at pressures of 80 kPa and 3 kPa and at potentials of (**a**) −0.5 V and (**b**) −0.6 V.

Since poorly-crystallized Ga_2O_3 did not produce a Raman response, this article determined the phase of Ga from the Ga 3d binding energy. The XPS spectra of the CIGS thin films are presented in Figure 12. According to Figure 11, at 80 kPa, the binding energy of 19.73 eV and 20.51 eV are related to Ga of CIGS thin films, and at 3 kPa, 19.07 eV and 19.87 eV are related to Ga of CIGS thin films. According to the literature, 19.07 eV corresponds to GaSe (19.00 eV, Ga 3d) [39], and 19.73 eV and 19.87 eV correspond to Ga_2Se_3 (19.70 eV or 19.90 eV, Ga 3d5/2) [40]. In addition, 20.51 eV is close to Ga_2O_3 (20.50 eV, Ga 3d5/2) [41]. In summary, the phases of the films were Ga_2Se_3 and Ga_2O_3 at 80 kPa, while GaSe and Ga_2Se_3 were the phases at 3 kPa. That is to say, the formation of Ga_2O_3 must be inhibited in a vacuum environment.

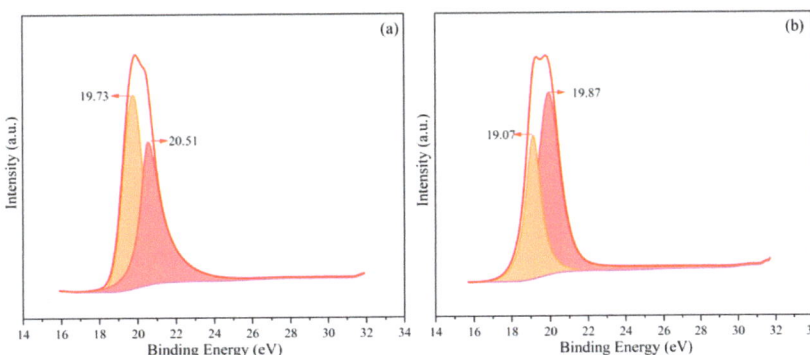

Figure 12. Binding energy of Ga 3d in CIGS thin films prepared at the potential of −0.6 V and pressures of (**a**) 80 kPa and (**b**) 3 kPa.

4. Conclusions

In this article, CIGS films without a low-conductivity Ga_2O_3 phase were prepared using vacuum electrodeposition, as supported by the conclusions of Raman and XPS. There were two main reasons for the inhibited Ga_2O_3 formation. The first reason is the inhibition of the H^+ to H_{ads} reaction, which decreased the consumption of H^+ on the cathode surface, as confirmed by the disappearance of the H_{ads} generation peak in the LSV curve. The second reason is that the OER reaction was promoted by a smoother electrode reaction and a decrease in the electrochemical impedance during deposition. The larger current density in the current density-time curves and the smaller semicircle diameter in the impedance Nyquist plots confirmed this. The inhibition of the H^+ to H_{ads} reaction and the promotion of the OER together stabilized the pH in the cathode region, which in turn inhibited Ga^{3+} hydrolysis. The path of Ga^{3+} into the films in the form of Ga_2O_3 was blocked.

In a vacuum environment, the content of Cu, In, and Se in the films increased, the current density of the preparation process increased, and the resistance decreased. The increase in the deposition space of the main elements produced larger particles in the films.

Author Contributions: In this joint work, each author was in charge of their expertise and capability: Conceptualization, G.L.; methodology, K.H.; validation, J.Z.; formal analysis, K.H.; investigation, K.H.; resources, L.X.; data curation, W.W.; writing—original draft preparation, G.L.; writing—review and editing, J.Y.; supervision, B.Y.; project administration, B.X.; funding acquisition, J.Y. All authors have read and agreed to the published version of the manuscript.

Funding: This research was funded by the National Natural Science Foundation of China grant number 52104350 and 52264038, the Natural Science Foundation of Yunnan Province grant number 202001AT070045, National Key Research and Development Program of China grant number 2022YFC2904204.

Data Availability Statement: The data presented in this study are available on request from the corresponding author. The data are not publicly available due to unfinished related ongoing further studies.

Conflicts of Interest: The authors declare no conflict of interest. The funders had no role in the design of the study, in the collection, analyses, or interpretation of data, in the writing of the manuscript, or in the decision to publish the results.

References

1. Kang, C.; Lee, G.; Lee, W.; Cho, D.H.; Maeng, I.; Chung, Y.D.; Kee, C.S. Terahertz Emission and Ultrafast Carrier Dynamics of Ar-Ion Implanted Cu(In, Ga)Se$_2$ Thin Films. *Crystals* **2021**, *11*, 411. [CrossRef]
2. Jiang, J.; Giridharagopal, R.; Jedlicka, E.; Sun, K.; Yu, S.; Wu, S.; Gong, Y.; Yan, W.; Ginger, D.S.; Green, M.A.; et al. Highly efficient copper-rich chalcopyrite solar cells from DMF molecular solution. *Nano Energy* **2020**, *69*, 104438. [CrossRef]

3. Hedayati, M.; Olyaee, S. High-Efficiency p-n Homojunction Perovskite and CIGS Tandem Solar Cell. *Crystals* **2022**, *12*, 703. [CrossRef]
4. Huang, C.H.; Chuang, W.J.; Lin, C.P.; Jan, Y.L.; Shih, Y.C. Deposition Technologies of High-Efficiency CIGS Solar Cells: Development of Two-Step and Co-Evaporation Processes. *Crystals* **2018**, *8*, 296. [CrossRef]
5. Ma, Q.; Zhang, W.; Jiang, Z.; Ma, D.; Zhang, Y.; Lu, C.; Fan, Z. The Formation Mechanism of Cu(In$_{0.7}$Ga$_{0.3}$)Se$_2$ Nanoparticles and the Densification Trajectory of the Se-Rich Quaternary Target by Hot Pressing. *Crystals* **2018**, *8*, 135. [CrossRef]
6. Altaf, C.T.; Sahsuvar, N.S.; Abdullayeva, N.; Coskun, O.; Kumtepe, A.; Karagoz, E.; Sankir, M.; Sankir, N.D. Inverted Configuration of Cu(In, Ga)S$_2$/In$_2$S$_3$ on 3D-ZnO/ZnSnO$_3$ Bilayer System for Highly Efficient Photoelectrochemical Water Splitting. *ACS. Sustain. Chem. Eng.* **2020**, *8*, 15209–15222. [CrossRef]
7. Matur, U.C.; Baydogan, N. Changes in gamma attenuation behaviour of sol-gel derived CIGS thin film irradiated using Co-60 radioisotope. *J. Alloys Compd.* **2017**, *695*, 1405–1413. [CrossRef]
8. Ao, J.; Fu, R.; Jeng, M.J.; Bi, J.; Yao, L.; Gao, S.; Sun, G.; He, Q.; Zhou, Z.; Sun, Y.; et al. Formation of Cl-Doped ZnO Thin Films by a Cathodic Electrodeposition for Use as a Window Layer in CIGS Solar Cells. *Materials* **2018**, *11*, 953. [CrossRef]
9. Oliveri, R.L.; Patella, B.; Pisa, F.D.; Mangione, A.; Aiello, G.; Inguanta, R. Fabrication of CZTSe/CIGS Nanowire Arrays by One-Step Electrodeposition for Solar-Cell Application. *Materials* **2021**, *14*, 2778. [CrossRef]
10. Péter, L.; Fekete, É.; Kapoor, G.; Gubicza, J. Influence of the preparation conditions on the microstructure of electrodeposited nanocrystalline Ni–Mo alloys. *Electrochim. Acta* **2021**, *382*, 138352. [CrossRef]
11. Beltowska-Lehman, E.; Bigos, A.; Indyka, P.; Kot, M. Electrodeposition and characterisation of nanocrystalline Ni–Mo coatings. *Surf. Coat. Technol.* **2012**, *211*, 67–71. [CrossRef]
12. Ren, Y.; Ma, W.; Wei, K.; Yu, W.; Dai, Y.; Morita, K. Degassing of aluminum alloys via the electromagnetic directional solidification. *Vacuum* **2014**, *109*, 82–85. [CrossRef]
13. Fromm, E. Maximum rate of sorption and degassing processes in vacuum metallurgical treatments. *Vacuum* **1971**, *21*, 585–586. [CrossRef]
14. Pessel, L. Apparatus for electroplating metal. U.S. Patent 2465747, 30 April 1945.
15. Muttilainen, E.; Tunturi, P.J. Hard chromium plating under reduced pressure improves corrosion resistance. *Anti-Corros. Method. M.* **1984**, *31*, 13–15. [CrossRef]
16. Nam, S.E.; Lee, S.H.; Lee, K.H. Preparation of a palladium alloy composite membrane supported in a porous stainless steel by vacuum electrodeposition. *J. Membrane. Sci.* **1999**, *153*, 163–173. [CrossRef]
17. Nam, S.E.; Seong, Y.K.; Lee, J.W.; Lee, K.H. Preparation of highly stable palladium alloy composite membranes for hydrogen separation. *Desalination* **2007**, *236*, 51–55. [CrossRef]
18. Su, R.; Lü, Z.; Chen, K.; Ai, N.; Li, S.; Wei, B.; Su, W. Novel in situ method (vacuum assisted electroless plating) modified porous cathode for solid oxide fuel cells. *Electrochem. Commun.* **2008**, *10*, 844–847. [CrossRef]
19. Ming, P.; Zhu, D.; Hu, Y.; Zeng, Y. Micro-electroforming under periodic vacuum-degassing and temperature-gradient conditions. *Vacuum* **2009**, *83*, 1191–1199. [CrossRef]
20. Ming, P.; Li, Y.; Wang, S.; Li, S.; Li, X. Microstructure and properties of nickel prepared by electrolyte vacuum boiling electrodeposition. *Surf. Coat. Technol.* **2012**, *213*, 299–306. [CrossRef]
21. Hibberd, C.J.; Chassaing, E.; Liu, W.; Mitzi, D.B.; Lincot, D.; Tiwari, A.N. Non-vacuum methods for formation of Cu(In, Ga)(Se, S)$_2$ thin film photovoltaic absorbers. *Prog. Photovoltaics.* **2010**, *18*, 434–452. [CrossRef]
22. Estela Calixto, M.; Dobson, K.D.; McCandless, B.E.; Birkmire, R.W. Controlling Growth Chemistry and Morphology of Single-Bath Electrodeposited Cu(In, Ga)Se$_2$ Thin Films for Photovoltaic Application. *J. Electrochem. Soc.* **2006**, *153*, G521–G528. [CrossRef]
23. Yang, J.; Huang, C.; Jiang, L.; Liu, F.; Lai, Y.; Li, J.; Liu, Y. Effects of hydrogen peroxide on electrodeposition of Cu(In, Ga)Se$_2$ Thin films and band gap controlling. *Electrochim. Acta* **2014**, *142*, 208–214. [CrossRef]
24. Flamini, D.O.; Saidman, S.B.; Bessone, J.B. Electrodeposition of gallium onto vitreous carbon. *J. Appl. Electrochem.* **2007**, *37*, 467–471. [CrossRef]
25. Murray, R.; Sigmund, S. The H$^+$/H$_2$ equilibrium potential dependence on H$_2$ partial pressure on gold electrodes. *Electrochim. Acta* **1973**, *18*, 687–690.
26. Warner, T.B.; Schuldiner, S. Potential of a Platinum Electrode at Low Partial Pressures of Hydrogen or Oxygen. *J. Electrochem. Soc.* **1965**, *112*, 853–856. [CrossRef]
27. Liu, J.; Liu, F.; Lai, Y.; Zhang, Z.; Li, J.; Liu, Y. Effects of sodium sulfamate on electrodeposition of Cu(In, Ga)Se$_2$ thin film. *Electroanal. Chem.* **2011**, *651*, 191–196. [CrossRef]
28. Lai, Y.; Liu, F.; Zhang, Z.; Liu, J.; Li, Y.; Kuang, S.; Li, J.; Liu, Y. Cyclic voltammetry study of electrodeposition of Cu(In, Ga)Se$_2$ thin films. *Electrochim. Acta* **2008**, *54*, 3004–3010. [CrossRef]
29. Lincot, D.; Guillemoles, J.F.; Taunier, S.; Guimard, D.; Sicx-Kurdi, J.; Chaumont, A.; Roussel, O.; Ramdani, O.; Hubert, C.; Fauvarque, J.P.; et al. Chalcopyrite thin film solar cells by electrodeposition. *Sol. Energy.* **2004**, *77*, 725–737. [CrossRef]
30. Lai, Y.; Liu, J.; Yang, J.; Wang, B.; Liu, F.; Zhang, Z.; Li, J.; Liu, Y. Incorporation Mechanism of Indium and Gallium during Electrodeposition of Cu(In, Ga)Se$_2$ Thin Film. *J. Electrochem. Soc.* **2011**, *158*, D704–D709. [CrossRef]
31. Lai, Y.; Liu, F.; Li, J.; Zhang, Z.; Liu, Y. Nucleation and growth of selenium electrodeposition onto tin oxide electrode. *Electroanal. Chem.* **2010**, *639*, 187–192. [CrossRef]

32. Kemell, M.; Ritala, M.; Saloniemi, H.; Leskelä, M.; Sajavaara, T.; Rauhala, E. One-Step Electrodeposition of $Cu_{2-x}Se$ and $CuInSe_2$ Thin Films by the Induced Co-deposition Mechanism. *J. Electrochem. Soc.* **2000**, *147*, 1080–1087. [CrossRef]
33. You, R.; Lew, K.K.; Fu, Y.P. Effect of indium concentration on electrochemical properties of electrode-electrolyte interface of $CuIn_{1-x}Ga_xSe_2$ prepared by electrodeposition. *Mater. Res. Bull.* **2017**, *96*, 183–187. [CrossRef]
34. Gonzalo, A.; Nogales, E.; Lorenz, K.; Víllora, E.G.; Shimamura, K.; Piqueras, J.; Méndez, B. Raman and cathodoluminescence analysis of transition metal ion implanted Ga_2O_3 nanowires. *J. Lumin.* **2017**, *191*, 56–60. [CrossRef]
35. Liu, F.; Yang, J.; Zhou, J.; Lai, Y.; Jia, M.; Li, J.; Liu, Y. One-step electrodeposition of $CuGaSe_2$ thin films. *Thin Solid Films* **2012**, *520*, 2781–2784. [CrossRef]
36. Ren, T.; Yu, R.; Zhong, M.; Shi, J.; Li, C. Microstructure evolution of $CuInSe_2$ thin films prepared by single-bath electrodeposition. *Sol. Energ. Mat. Sol. C* **2010**, *95*, 510–520. [CrossRef]
37. Insignares-Cuello, C.; Izquierdo-Roca, V.; López-García, J.; Calvo-Barrio, L.; Saucedo, E.; Kretzschmar, S.; Unold, T.; Broussillou, C.; Goislard de Monsabert, T.; Bermudez, V.; et al. Combined Raman scattering/photoluminescence analysis of $Cu(In, Ga)Se_2$ electrodeposited layers. *Solar Energy* **2014**, *103*, 89–95. [CrossRef]
38. Bergeron, A.; Ibrahim, J.; Leonelli, R.; Francoeur, S. Oxidation dynamics of ultrathin GaSe probed through Raman spectroscopy. *Appl. Phys. Lett.* **2017**, *110*, 241901. [CrossRef]
39. Lang, O.; Tomm, Y.; Schlaf, R.; Pettenkofer, C.; Jaegermann, W. Single crystalline $GaSe/WSe_2$ heterointerfaces grown by van der Waals epitaxy. II. Junction characterization. *J. Appl. Phys.* **1994**, *75*, 7814–7820. [CrossRef]
40. Iwakuro, H.; Tatsuyama, C.; Ichimura, S. XPS and AES Studies on the Oxidation of Layered Semiconductor GaSe. *Jpn. J. Appl. Phys.* **1982**, *21*, 94–99. [CrossRef]
41. Carli, R.; Bianchi, C.L. XPS analysis of gallium oxides. *Appl. Surf. Sci.* **1994**, *74*, 99–102. [CrossRef]

Disclaimer/Publisher's Note: The statements, opinions and data contained in all publications are solely those of the individual author(s) and contributor(s) and not of MDPI and/or the editor(s). MDPI and/or the editor(s) disclaim responsibility for any injury to people or property resulting from any ideas, methods, instructions or products referred to in the content.

Article

Elastic Constitutive Relationship of Metallic Materials Containing Grain Shape

Zhiwen Lan [1], Hanjie Shao [1], Lei Zhang [2], Hong Yan [2], Mojia Huang [1] and Tengfei Zhao [3,*]

[1] Institute of Engineering Mechanics, Nanchang University, 999 Xuefu Avenue, Nanchang 330031, China
[2] College of Architectural Engineering and Planning, Jiujiang University, 551 Qianjin East Road, Jiujiang 332005, China
[3] College of City Construction, Jiangxi Normal University, 99 Ziyang Avenue, Nanchang 330022, China
* Correspondence: 005539@jxnu.edu.cn

Abstract: The grain shape and orientation distribution of metal sheets at mesoscales are usually irregular, which has an impact on the elastic properties of metal materials. A grain shape function (GSF) is constructed to represent the shape of grains. The expansion coefficient of GSF on the basis of the Wigner D function is called the shape coefficient. In this paper, we study the influence of average grain shape on the elastic constitutive relation of orthogonal polycrystalline materials, and obtain a new expression of the elastic constitutive relation of polycrystalline materials containing grain shape effects. The seven string method is proposed to fit the shape of irregular grains. Experiments show that the GSF can better describe the shape of irregular grains. Using the microscopic images of the grains, we carried out the experimental measurement of micro and macrostrain at grain scale. The experimental results show that the grain shape parameter (slenderness ratio) is consistent with the theoretical results of the material macroscopic mechanical properties.

Keywords: grain shape function; elastic constitutive; seven string method; microscopic images; experimental measurement

Citation: Lan, Z.; Shao, H.; Zhang, L.; Yan, H.; Huang, M.; Zhao, T. Elastic Constitutive Relationship of Metallic Materials Containing Grain Shape. *Crystals* **2022**, *12*, 1768. https://doi.org/10.3390/cryst12121768

Academic Editor: Pavel Lukáč

Received: 7 November 2022
Accepted: 1 December 2022
Published: 5 December 2022

Copyright: © 2022 by the authors. Licensee MDPI, Basel, Switzerland. This article is an open access article distributed under the terms and conditions of the Creative Commons Attribution (CC BY) license (https://creativecommons.org/licenses/by/4.0/).

1. Introduction

The constitutive relation of metal materials reflects the law of the change of stress with strain under certain deformation conditions. The grains of polycrystalline metal materials contain such microstructure information as grain orientation, grain shape, grain size, and grain boundary distribution [1–5]. The microstructure information reflects the micro structure characteristics of polycrystals, and affect the macro elastic and plastic mechanical properties of polycrystalline materials.

General constitutive theoretical models, such as the Miller [6] model, Walker [7] model, Bonder Partom [8] model, Chaboche [9] model, and Sadovskii model [10], have the following common features [11] despite their various forms: the theoretical basis of each equation is the basic law of thermodynamics. The strain of any point in the material can be regarded as the sum of elastic strain and inelastic strain. The elastic deformation conforms to Hooke's law, and the inelastic deformation conforms to the flow equation. The mechanical properties of materials are determined by two completely independent basic internal variables, i.e., isotropic hardening of the internal variable and kinematic hardening of the internal variable. The above macro phenomenological metal constitutive relation theory is based on macro phenomena and simulates macro mechanical behavior to determine parameters. The equations obtained are often semi theoretical and semi empirical, and the morphology and changes of material damage cannot be understood from the fine and micro structure levels. Therefore, these studies are difficult to go deep into the nature of metal material deformation.

Many studies [12–16] believe that among the micro structural characteristics of metallic materials, the macro elastic constitutive relationship of polycrystalline metallic materials is mainly affected by the grain orientation distribution. Guenoun et al. [17] followed the grain orientation during in situ crystallization experiments with a fine time resolution. Tang et al. [18] investigated the anisotropic hardness, elastic modulus, and dislocation behavior of AlN grains by Berkovich nanoindentation. Şahin et al. [19] studied the limiting role of grain domain orientation on the modulus and strength of aramid fibers. Tang et al. [20] described the constitutive relation of individual grains in the micro-scale reconstructed models with the single-crystal-scale plasticity model. Gu et al. [21] proposed a method considering grain size and shape effects based on the classical crystal plasticity finite element method. Trusov et al. [22] analyzed the kinematic relations and constitutive laws in crystal plasticity in the case of elastic deformation. Lakshmanan et al. [23] presented a computational framework to include the effects of grain size and morphology in the crystal plasticity. Agius et al. [24] incorporated a length scale dependence into classical crystal plasticity simulations, and the effectiveness of the method was proved by experiments. The grain orientation distribution is expressed by the ODF function [25,26], and the stress distribution of single grains presented obvious orientation dependence during deformation [27]. The relationship between material texture coefficient and the elastic constitutive relationship can reflect the influence of grain orientation distribution on the elastic constitutive relationship. Equations (1) and (2) give the single-crystal-scale plasticity model and polycrystal-scale plasticity model [18,28–38].

$$\left.\begin{array}{l}\hat{\sigma}^e = C : D^e \\ \hat{\sigma} = \hat{\sigma}^e - (W^p \cdot \sigma - \sigma \cdot W^p) \\ \hat{\sigma} = C : D - \sum_{\alpha=1}^{n}\left(C : P^{(\alpha)} + B^{(\alpha)}\right)\gamma^{(\alpha)}\end{array}\right\} \quad (1)$$

$$\begin{aligned}\hat{\sigma} &= \overline{C} : \left(D - \sum_{k=1}^{N_2}\sum_{\alpha=1}^{N_1}\overline{C} : P^{(\alpha)}\right)\gamma^{(\alpha)}f_k + \sum_{k=1}^{N_2}\sum_{\alpha=1}^{N_1}B^{(\alpha)}\gamma^{(\alpha)}f_k \\ &= \overline{C} : D - \sum_{k=1}^{N_2}\sum_{\alpha=1}^{N_1}\left(\overline{C} : P^{(\alpha)} + B^{(\alpha)}\right)\gamma^{(\alpha)}f_k\end{aligned} \quad (2)$$

However, up to now, people have not given the expression of the polycrystalline elastic constitutive relation containing the grain shape effect. The work of this paper is to study the influence of the grain shape parameters of polycrystalline materials on the elastic constitutive relation, and to carry out the parametric study of the grain shape of grain materials.

2. Elastic Constitutive Relationship of Metallic Materials Containing Grain Shape Effect

2.1. Grain Shape Function and Shape Coefficient Expression

We define polycrystalline $\Omega \subset \mathbf{R}^3$ as a collection of many small grains Ω_p ($p = 1, 2, \ldots, N$)

$$\Omega = \text{int}\left(\cup_{p=1}^{N}\overline{\Omega}_p\right), \Omega_p \cap \Omega_q = \varnothing, \forall p \neq q \quad (3)$$

where p is taken from all grains of polycrystals, the grains p occupy the domain Ω_p (open set), $\overline{\Omega}_p$ is the closed set of Ω_p and is the inner product, and N is the total number of polycrystal grains.

We assume

$$r_p(\mathbf{n}) = |\mathbf{x} - \mathbf{c}_p|, \mathbf{x} \in \partial\Omega_p \quad (4)$$

Equation (4) represents the distance from the center of the crystal grain Ω_p to the point $\mathbf{x} \in \partial\Omega_p$, where $d\mathbf{x} \in dx_1 dx_2 dx_3$ and $\mathbf{n} = (\mathbf{x} - \mathbf{c}_p)/|\mathbf{x} - \mathbf{c}_p|, \mathbf{x} \in \partial\Omega_p$, the plane calculation model is shown in Figure 1. The direction $\mathbf{n}$ in (4) can be represented by Euler angles α and β.

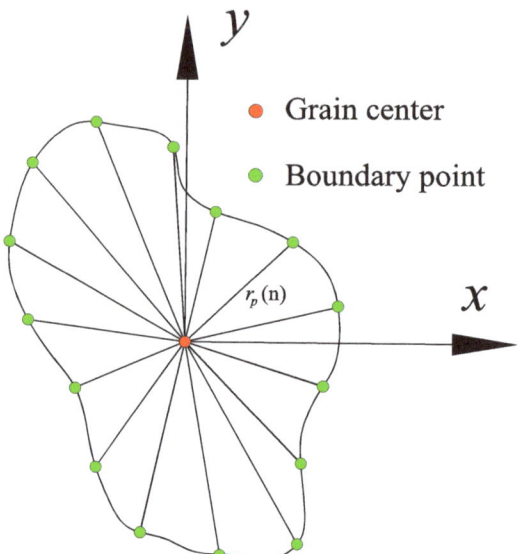

Figure 1. The plane calculation model of the grain.

$$\mathbf{n}(\alpha, \beta) = [\sin\beta\cos\alpha, \sin\beta\sin\alpha, \cos\beta]^T = \mathbf{R}(\alpha, \beta, 0)\mathbf{e}_3 \tag{5}$$

where $\mathbf{R}(\alpha, \beta, 0)$ is the rotation tensor, and $\mathbf{e}_3 = [0, 0, 1]^T$.

The average size $r(\mathbf{n})$ of crystal grain $r_p(\mathbf{n})$ in polycrystalline Ω is

$$r(\mathbf{n}) = \frac{1}{N} \sum_{p=1}^{N} r_p(\mathbf{n}) \tag{6}$$

The average shape and size of crystal grains are defined as

$$(\Omega_{cr})_{mean} = \left\{ x \in \mathbf{R}^3 \middle| \begin{array}{l} \mathbf{x} = l(\mathbf{n}) \times [\sin\beta\cos\alpha, \sin\beta\sin\alpha, \cos\beta]^T \\ 0 \leq \alpha \triangleleft 2\pi, 0 \leq \beta \leq \pi, 0 \leq l(\mathbf{n}) \triangleleft r(\mathbf{n}) \end{array} \right\} \tag{7}$$

The elastic constitutive relation is independent of grain size and tiny physical unit, hence

$$\Re(\mathbf{n}(\alpha, \beta)) = \frac{r(\mathbf{n})}{\bar{r}} \tag{8}$$

where $\bar{r}$ is the average radius of polycrystalline grains, and the GSF can describe the average shape of polycrystalline grains.

$$\bar{r} = \frac{1}{4\pi} \int_0^{2\pi} \int_0^{\pi} r(\mathbf{n}(\alpha, \beta)) \sin\beta d\beta d\alpha \tag{9}$$

In Equation (8), the GSF $\Re(\mathbf{n}(\alpha, \beta))$ can be extended to infinite series $Y_m^l(\alpha, \beta)$ or Wigner D function

$$\begin{aligned} \Re(\mathbf{n}(\alpha, \beta)) &= \Re_{\text{sphere}} + \sum_{l=1}^{\infty} \sum_{m=-l}^{l} \sqrt{\frac{4\pi}{2l+1}} s_{m0}^l Y_m^l(\alpha, \beta) \\ &= \Re_{\text{sphere}} + \sum_{l=1}^{\infty} \sum_{m=-l}^{l} s_{m0}^l D_{m0}^l(\mathbf{R}(\alpha, \beta, 0)) \end{aligned} \tag{10}$$

where $\Re_{sphere} = 1$, $s_{m0}^l = (-1)^m (s_{\bar{m}0}^l)^*$, and $s_{m0}^l \in \mathbf{C}(l \geq 1)$ is the shape factor.

The average grain shape is not anisotropic and $s_{m0}^l = 0$, the relationship between the ball function and Wigner D function is

$$Y_m^l(\alpha, \beta) = \sqrt{\frac{2l+1}{4\pi}} D_{m0}^l(\mathbf{R}(\alpha, \beta, 0)) \tag{11}$$

According to the orthogonal averaging condition

$$\int_0^{2\pi} \int_0^\pi Y_m^l(\alpha, \beta)(Y_{m'}^{l'}(\alpha, \beta))^* \sin\beta \, d\beta \, d\alpha = \delta_{ll'}\delta_{mm'} \tag{12}$$

If $\left\|\Re(\mathbf{n}(\alpha, \beta, 0)) - \Re_{sphere}\right\|$ is a small amount, then the average grain shape is weakly anisotropic. The polycrystalline undergoes a rotation Q, and the average shape of grains can be described by a new GSF $\Re(\mathbf{n}(\alpha, \beta))$.

$$\begin{aligned}
\Re(\mathbf{n}(\alpha, \beta)) &= \Re(\mathbf{Q}^{-1}\mathbf{n}(\alpha, \beta)) = \Re_{sphere} + \sum_{l=1}^{\infty} \sum_{m=-l}^{l} s_{p0}^l D_{p0}^l(\mathbf{Q}^{-1}\mathbf{R}(\alpha, \beta, 0)) \\
&= \Re_{sphere} + \sum_{l=1}^{\infty} \sum_{m=-l}^{l} s_{p0}^l \times \sum_{m=-l}^{l} D_{pm}^l(\mathbf{Q}^{-1}) D_{m0}^l(\mathbf{R}(\alpha, \beta, 0)) \\
&= \Re_{sphere} + \sum_{l=1}^{\infty} \sum_{m=-l}^{l} \tilde{s}_{m0}^l \times D_{m0}^l(\mathbf{R}(\alpha, \beta, 0))
\end{aligned} \tag{13}$$

where $\tilde{s}_{m0}^l = \sum_{p=-l}^{l} s_{p0}^l D_{pm}^l(\mathbf{Q}^{-1})$.

We assume that the polycrystal is an aggregate of orthorhombic grains, and the shape coefficient of the polycrystal satisfies

$$s_{m0}^l = \sum_{p=-l}^{l} s_{p0}^l D_{pm}^l(\mathbf{Q}^{-1}), \forall \mathbf{Q}^{-1} \in \{\mathbf{I}, \mathbf{R}(0, \pi, \pi), \mathbf{R}(0, \pi, 0), \mathbf{R}(0, 0, \pi)\} \tag{14}$$

We obtain

$$\begin{cases} s_{m0}^l = (-1)^l s_{\overline{m}0}^l, & m \in even \\ s_{m0}^l = 0, & m \in odd \end{cases} \tag{15}$$

Equation (13) holds for all grain aggregates of the orthorhombic system, because

$$\begin{aligned}
s_{m0}^l &= \sum_{p=-l}^{l} s_{p0}^l D_{pm}^l(\mathbf{R}(0, \pi, 0)) \\
&= \sum_{p=-l}^{l} s_{p0}^l d_{pm}^l(\pi) \\
&= \sum_{p=-l}^{l} s_{p0}^l (-1)^{l+p} d_{p\overline{m}}^l(0) \\
&= \sum_{p=-l}^{l} s_{p0}^l (-1)^{l+p} \delta_{p\overline{m}} = s_{\overline{m}0}^l (-1)^{l-m}
\end{aligned} \tag{16}$$

$$\begin{aligned}
s_{m0}^l &= \sum_{p=-l}^{l} s_{p0}^l D_{pm}^l(\mathbf{R}(0, 0, \pi)) \\
&= \sum_{p=-l}^{l} s_{p0}^l d_{pm}^l(0) e^{-im\pi} \\
&= s_{m0}^l \cos(m\pi)
\end{aligned} \tag{17}$$

By (13)
$$s_{m0}^l = (-1)^m (s_{\overline{m}0}^l)^* = (-1)^l (s_{m0}^l)^* \tag{18}$$

when the polycrystals have orthogonal symmetry, the dual number l of the shape coefficient is a real number.

2.2. Elastic Constitutive Relation Considering Average Grain Shape

We assume that the polycrystalline constitutive relation $\mathbf{C}^{eff}$ depends on ODF and GSF (e.g., $\mathbf{C}^{eff} = \mathbf{C}^{eff}(w, \Re)$), and the objectivity of the material limits the constitutive relation

$$\mathbf{C}^{eff}(w(\mathbf{Q}^{-1}\mathbf{R}), \Re(\mathbf{Q}^{-1}\mathbf{n})) = \mathbf{Q}^{\otimes 4}\mathbf{C}^{eff}(w(\mathbf{R}), \Re(\mathbf{n})), \forall \mathbf{Q} \in SO(3) \tag{19}$$

$$(\mathbf{Q}^{\otimes 4}\mathbf{A})_{ijkl} = Q_{im}Q_{jn}Q_{kp}Q_{lq}A_{mnpq} \tag{20}$$

where, the polycrystalline constitutive relation $\mathbf{C}^{eff}$ has primary and secondary symmetry, and $\mathbf{C}^{eff}(w, \Re) = \mathbf{C}^{eff}(c_{mn}^l, s_{m0}^l)$ is expanded into a series with c_{mn}^l and s_{m0}^l.

$$\begin{aligned}\mathbf{C}^{eff}(c_{mn}^l, s_{m0}^l) &= \mathbf{C}^{eff}(0,0) + \sum_{l=1}^{\infty}\sum_{n}\sum_{m=-l}^{l} \frac{\partial \mathbf{C}^{eff}(0,0)}{\partial c_{mn}^l} c_{mn}^l \\ &+ \sum_{l=1}^{\infty}\sum_{m=-l}^{l} \frac{\partial \mathbf{C}^{eff}(0,0)}{\partial s_{m0}^l} s_{m0}^l + o(|c_{mn}^l|) + o(|s_{m0}^l|)\end{aligned} \tag{21}$$

If the polycrystal is weak texture (e.g., $\|w - w_{iso}\|$ is small), and the average grain shape is weak anisotropy (e.g., $\|\Re - \Re_{sphere}\|$ is small), then $o(|c_{mn}^l|)$ and $o(|s_{m0}^l|)$ in (21) can be removed, we obtain

$$\mathbf{C}^{eff}(w(\mathbf{R}), \Re(\mathbf{n})) = \mathbf{C}^{eff}(c_{mn}^l, s_{m0}^l) = \mathbf{C}^{(0)} + \mathbf{C}^{(1)}(w(\mathbf{R})) + \mathbf{C}^{(2)}(\Re(\mathbf{n})) \tag{22}$$

where $\mathbf{C}^{(0)} = \mathbf{C}^{eff}(0,0)$.

$$\mathbf{C}^{(1)}(w(\mathbf{R})) = \sum_{l=1}^{\infty}\sum_{n}\sum_{m=-l}^{l} \frac{\partial \mathbf{C}^{eff}(0,0)}{\partial c_{mn}^l} c_{mn}^l \tag{23}$$

$$\mathbf{C}^{(2)}(\Re(\mathbf{n})) = \sum_{l=1}^{\infty}\sum_{m=-l}^{l} F_{m0}^l s_{m0}^l, \quad F_{m0}^l = \frac{\partial \mathbf{C}^{eff}(0,0)}{\partial s_{m0}^l} \tag{24}$$

By (19) and (22), we obtain

$$\begin{aligned}\mathbf{C}^{(0)} &= \mathbf{Q}^{\otimes 4}\mathbf{C}^{(0)} \\ \mathbf{C}^{(1)}(w(\mathbf{Q}^{-1}\mathbf{R})) &= \mathbf{Q}^{\otimes 4}\mathbf{C}^{(1)}(w(\mathbf{R})) \\ \mathbf{C}^{(2)}(\Re(\mathbf{Q}^{-1}\mathbf{n})) &= \mathbf{Q}^{\otimes 4}\mathbf{C}^{(2)}(\Re(\mathbf{n}))\end{aligned} \tag{25}$$

For cubic grain orthogonal system, by (25), we obtain the tensor form of $\mathbf{C}^{(0)}$ and $\mathbf{C}^{(1)}$ under the change of Voigt symbol.

The isotropic part can be expressed as

$$\mathbf{C}^{(0)} = \lambda \mathbf{B}^{(1)} + 2\mu \mathbf{B}^{(2)}$$

$$= \begin{bmatrix} \lambda + 2\mu & \lambda & \lambda & 0 & 0 & 0 \\ \lambda & \lambda + 2\mu & \lambda & 0 & 0 & 0 \\ \lambda & \lambda & \lambda + 2\mu & 0 & 0 & 0 \\ 0 & 0 & 0 & \mu & 0 & 0 \\ 0 & 0 & 0 & 0 & \mu & 0 \\ 0 & 0 & 0 & 0 & 0 & \mu \end{bmatrix} \tag{26}$$

where λ and μ are the undetermined material constants, $\mathbf{B}^{(1)}_{ijkl} = \delta_{ij}\delta_{kl}$, $\mathbf{B}^{(2)}_{ijkl} = \frac{1}{2}(\delta_{ik}\delta_{jl} + \delta_{il}\delta_{jk})$. The anisotropic part can be expressed as

$$\mathbf{C}^{(1)} = c\mathbf{\Phi}(w)$$

$$= c \begin{bmatrix} -a_2 - a_3 & a_3 & a_2 & 0 & 0 & 0 \\ a_3 & -a_3 - a_1 & a_1 & 0 & 0 & 0 \\ a_2 & a_1 & -a_1 - a_2 & 0 & 0 & 0 \\ 0 & 0 & 0 & a_1 & 0 & 0 \\ 0 & 0 & 0 & 0 & a_2 & 0 \\ 0 & 0 & 0 & 0 & 0 & a_3 \end{bmatrix} \quad (27)$$

where $a_1 = -\frac{32\pi^2}{105}(c^4_{00} + \sqrt{\frac{5}{2}}c^4_{20})$, $a_2 = -\frac{32\pi^2}{105}(c^4_{00} - \sqrt{\frac{5}{2}}c^4_{20})$, $a_3 = \frac{8\pi^2}{105}(c^4_{00} + \sqrt{70}c^4_{40})$, since the polycrystals are cubic crystal orthogonal systems, the texture coefficients c^4_{00}, c^4_{20} and c^4_{40} are both real numbers [39,40].

In (24), the tensor $\mathbf{F}^l_{m0}$ satisfies

$$\mathbf{F}^l_{\overline{m}0} = (-1)^m (\mathbf{F}^l_{m0})^* \quad (28)$$

Hence, the fourth-order tensor $\mathbf{C}^{(2)}$ is a real number, and by the relationshape $s^l_{\overline{m}0} = (-1)^m (s^l_{m0})^*$ in (10), we obtain

$$\mathbf{F}^l_{m0}s^l_{m0} + \mathbf{F}^l_{\overline{m}0}s^l_{\overline{m}0} = \mathbf{F}^l_{m0}s^l_{m0} + (\mathbf{F}^l_{m0}s^l_{m0})^* \quad (29)$$

Combining (24) and (25), we obtain

$$\sum_{m=-l}^{l} \tilde{s}^l_{m0} \mathbf{F}^l_{m0} = \sum_{m=-l}^{l} \sum_{p=-l}^{l} s^l_{p0} D^l_{pm}(\mathbf{Q}^{-1}) \mathbf{F}^l_{m0}$$

$$= \sum_{m=-l}^{l} s^l_{m0} \mathbf{Q}^{\otimes 4} \mathbf{F}^l_{m0}, \forall \mathbf{Q} \in SO(3) \quad (30)$$

Equation (30) is applicable to any shape factor $s^l_{m0} \in \mathbf{C}$, let $s^l_{00} \neq 0$ and $s^l_{m0} = 0$, we obtain

$$\mathbf{Q}^{\otimes 4} \mathbf{F}^l_{00} = \sum_{m=-l}^{l} D^l_{0m}(\mathbf{Q}^{-1}) \mathbf{F}^l_{m0} \quad (31)$$

In order to meet the requirement that $\mathbf{F}^l_{m0}$ holds true for $\mathbf{Q} \in SO(3)$, we assume $\mathbf{F}^{(l)}_{00} = \mathbf{F}^l_{00}$, where $\mathbf{F}^{(l)}$ is a fourth-order tensor satisfying primary and secondary symmetry. We multiply $(D^l_{0k}(\mathbf{Q}^{-1}))^*$ on both sides of (31) and integrate in SO(3) space, then

$$\int_{SO(3)} \mathbf{Q}^{\otimes 4} \mathbf{F}^{(l)} (D^l_{0m}(\mathbf{Q}^{-1}))^* dg = \sum_{m=-l}^{l} \int_{SO(3)} D^l_{0m}(\mathbf{Q}^{-1})(D^l_{0k}(\mathbf{Q}^{-1}))^* dg \mathbf{F}^l_{m0} \quad (32)$$

$\mathbf{F}^l_{m0}$ can be expressed as

$$\mathbf{F}^l_{k0} = \frac{2l+1}{8\pi^2} \int_{SO(3)} \mathbf{Q}^{\otimes 4} \mathbf{F}^{(l)} D^l_{k0}(\mathbf{Q}) dg, \; k \in [0, \pm 1, \pm 2, \ldots, \pm l] \quad (33)$$

The component is represented as

$$(\mathbf{F}^l_{k0})_{mnpq} = \frac{2l+1}{8\pi^2} \int_0^{2\pi} \int_0^{\pi} \int_0^{2\pi} Q_{mu} Q_{nv} \times Q_{ps} Q_{qt} \mathbf{F}^{(l)}_{uvst} D^l_{k0}(\mathbf{Q}) \sin\theta d\psi d\theta d\varphi \quad (34)$$

The fourth-order tensor $\mathbf{F}^{(l)}$ satisfies the primary and secondary symmetry, and has 21 independent material constants, in (33), the fourth-order tensor $\mathbf{F}^l_{m0}$ satisfies

$$\mathbf{Q}^{\otimes 4}\mathbf{F}_{m0}^l = \sum_{k=-l}^{l}(D_{km}^l(\mathbf{Q}))^*\mathbf{F}_{k0}^l \tag{35}$$

We obtain

$$\begin{aligned}
\sum_{k=-l}^{l}(D_{km}^l(\mathbf{Q}))^*\mathbf{F}_{k0}^l &= \sum_{k=-l}^{l}(D_{mk}^l(\mathbf{Q}^{-1}))\frac{2l+1}{8\pi^2}\times\int_{SO(3)}\mathbf{Q}_1^{\otimes 4}\mathbf{F}^{(l)}D_{k0}^l(\mathbf{Q}_1)dg(\mathbf{Q}_1) \\
&= \frac{2l+1}{8\pi^2}\int_{SO(3)}\mathbf{Q}_1^{\otimes 4}\mathbf{F}^{(l)}\sum_{k=-l}^{l}D_{mk}^l(\mathbf{Q}^{-1})D_{k0}^l(\mathbf{Q}_1)dg(\mathbf{Q}_1) \\
&= \frac{2l+1}{8\pi^2}\mathbf{Q}^{\otimes 4}\int_{SO(3)}(\mathbf{Q}^{-1})^{\otimes 4}\mathbf{Q}_1^{\otimes 4}\mathbf{F}^{(l)}D_{m0}^l(\mathbf{Q}^{-1}\mathbf{Q}_1)dg(\mathbf{Q}_1) \\
&= \mathbf{Q}^{\otimes 4}(\frac{2l+1}{8\pi^2}\int_{SO(3)}\mathbf{Q}_2^{\otimes 4}\mathbf{F}^{(l)}D_{m0}^l(\mathbf{Q}_2)dg(\mathbf{Q}_2)) \\
&= \mathbf{Q}^{\otimes 4}\mathbf{F}_{m0}^l
\end{aligned} \tag{36}$$

By $\mathbf{F}_{k0}^l$ in (33), Equation (30) holds for all shape coefficients $s_{m0}^l \notin \mathbf{C}$, hence

$$\begin{aligned}
\sum_{k=-l}^{l}s_{m0}^l\mathbf{Q}^{\otimes 4}\mathbf{F}_{m0}^l &= \sum_{k=-l}^{l}s_{m0}^l[\sum_{p=-l}^{l}(D_{mk}^l(\mathbf{Q}))^*\mathbf{F}_{p0}^l] \\
&= \sum_{p=-l}^{l}s_{p0}^l\sum_{m=-l}^{l}(D_{mp}^l(\mathbf{Q}))^*\mathbf{F}_{m0}^l \\
&= \sum_{m=-l}^{l}(\sum_{p=-l}^{l}s_{p0}^lD_{pm}^l(\mathbf{Q}^{-1}))\mathbf{F}_{m0}^l
\end{aligned} \tag{37}$$

In the rectangular coordinate system, the rotation tensor Q can be represented by a linear combination of Wigner D function $D_{mk}^l(\mathbf{Q})$ ($l=1$) [41–43], rewriting each component of tensor $\int_{SO(3)}\mathbf{Q}^{\otimes 4}\mathbf{F}^{(l)}D_{k0}^l(\mathbf{Q})dg$ [39] as a linear combination of integral $\int_{SO(3)}D_{qr}^p(\mathbf{Q})D_{k0}^l(\mathbf{Q})dg$ ($0 \le p \le 4, l \ge 1$). Using the orthogonality of Wigner D function, we obtain

When $l \ge 5$

$$\mathbf{F}_{k0}^l = \frac{2l+1}{8\pi^2}\int_{SO(3)}\mathbf{Q}^{\otimes 4}\mathbf{F}^{(l)}D_{k0}^l(\mathbf{Q})dg = 0 \tag{38}$$

By (38), we can write $\mathbf{C}^{(2)}(\Re(\mathbf{n}))$ in (24) as

$$\mathbf{C}^{(2)}(\Re(\mathbf{n})) = \sum_{l=1}^{4}\sum_{m=-1}^{l}\mathbf{F}_{m0}^ls_{m0}^l \tag{39}$$

By integrating, we obtain

$$\begin{aligned}
\sum_{m=-1}^{1}\mathbf{F}_{m0}^1 s_{m0}^1 &= 0 \\
\sum_{m=-2}^{2}\mathbf{F}_{m0}^2 s_{m0}^2 &= s_1\Theta_1(\Re) + s_2\Theta_2(\Re) \\
\sum_{m=-3}^{3}\mathbf{F}_{m0}^3 s_{m0}^3 &= 0 \\
\sum_{m=-4}^{4}\mathbf{F}_{m0}^4 s_{m0}^4 &= s_3\Theta_3(\Re)
\end{aligned} \tag{40}$$

where

$$s_1 = \frac{1}{42}(\mathbf{F}^{(2)}_{1111} + \mathbf{F}^{(2)}_{2222} - 2\mathbf{F}^{(2)}_{3333} + 2\mathbf{F}^{(2)}_{2233} + 2\mathbf{F}^{(2)}_{1133} - 4\mathbf{F}^{(2)}_{1122} - 3\mathbf{F}^{(2)}_{2323} - 3\mathbf{F}^{(2)}_{3131} + 6\mathbf{F}^{(2)}_{1212})$$

$$s_2 = \frac{1}{42}(\mathbf{F}^{(2)}_{1111} + \mathbf{F}^{(2)}_{2222} - 2\mathbf{F}^{(2)}_{3333} - 5\mathbf{F}^{(2)}_{2233} - 5\mathbf{F}^{(2)}_{1133} + 10\mathbf{F}^{(2)}_{1122} + 4\mathbf{F}^{(2)}_{2323} + 4\mathbf{F}^{(2)}_{3131} - 8\mathbf{F}^{(2)}_{1212}) \quad (41)$$

$$s_3 = \frac{3}{64\pi^2}(3\mathbf{F}^{(4)}_{1111} + 3\mathbf{F}^{(4)}_{2222} + 8\mathbf{F}^{(4)}_{3333} - 8\mathbf{F}^{(4)}_{2233} - 8\mathbf{F}^{(4)}_{1133} + 2\mathbf{F}^{(4)}_{1122} - 16\mathbf{F}^{(4)}_{2323} - 16\mathbf{F}^{(4)}_{3131} + 4\mathbf{F}^{(4)}_{1212})$$

$$\Theta_1(\Re) = \begin{bmatrix} 4s_{00}^2 - 4\sqrt{6}s_{20}^2 & 0 & 0 & 0 & 0 & 0 \\ 0 & 4s_{00}^2 + 4\sqrt{6}s_{20}^2 & 0 & 0 & 0 & 0 \\ 0 & 0 & -8s_{00}^2 & 0 & 0 & 0 \\ 0 & 0 & 0 & -s_{00}^2 + \sqrt{6}s_{20}^2 & 0 & 0 \\ 0 & 0 & 0 & 0 & -s_{00}^2 - \sqrt{6}s_{20}^2 & 0 \\ 0 & 0 & 0 & 0 & 0 & 2s_{00}^2 \end{bmatrix}$$

$$\Theta_2(\Re) = \begin{bmatrix} 2s_{00}^2 - 2\sqrt{6}s_{20}^2 & 2s_{00}^2 & -s_{00}^2 - \sqrt{6}s_{20}^2 & 0 & 0 & 0 \\ 2s_{00}^2 & 2s_{00}^2 + 2\sqrt{6}s_{20}^2 & -s_{00}^2 + \sqrt{6}s_{20}^2 & 0 & 0 & 0 \\ -s_{00}^2 - \sqrt{6}s_{20}^2 & -s_{00}^2 + \sqrt{6}s_{20}^2 & -4s_{00}^2 & 0 & 0 & 0 \\ 0 & 0 & 0 & 0 & 0 & 0 \\ 0 & 0 & 0 & 0 & 0 & 0 \\ 0 & 0 & 0 & 0 & 0 & 0 \end{bmatrix} \quad (42)$$

$$\Theta_3(\Re) = \begin{bmatrix} -b_2 - b_3 & b_3 & b_2 & 0 & 0 & 0 \\ b_3 & -b_1 - b_3 & b_1 & 0 & 0 & 0 \\ b_2 & b_1 & -b_1 - b_2 & 0 & 0 & 0 \\ 0 & 0 & 0 & b_1 & 0 & 0 \\ 0 & 0 & 0 & 0 & b_2 & 0 \\ 0 & 0 & 0 & 0 & 0 & b_3 \end{bmatrix}$$

$$b_1 = -\frac{32\pi^2}{105}(s_{00}^4 + \sqrt{\frac{5}{2}}s_{20}^4), \quad b_2 = -\frac{32\pi^2}{105}(s_{00}^4 - \sqrt{\frac{5}{2}}s_{20}^4), \quad b_3 = \frac{8\pi^2}{105}(s_{00}^4 - \sqrt{70}s_{40}^4) \quad (43)$$

By (19), for orthogonal polycrystals, in (41)–(43), the shape coefficients $s_{00}^2, s_{20}^2, s_{00}^4, s_{20}^4, s_{20}^4$ are real numbers. Substitute (40) into (39), and we obtain the anisotropic part of the constitutive relation according to the anisotropy of the average grain shape

$$\mathbf{C}^{(2)}(\Re(\mathbf{n})) = s_1\Theta_1(\Re) + s_2\Theta_2(\Re) + s_3\Theta_3(\Re) \quad (44)$$

If the polycrystals are orthorhombic aggregates of weakly textured cubic grains, and the average grain shape is weakly anisotropic, according to the principle of material objectivity, considering the effects of GSF and grain orientation function ODF, by (22), (42), (43), and (44), we obtain the elastic constitutive relationship $\mathbf{C}^{eff}$ of polycrystals

$$\mathbf{C}^{eff} = \lambda\mathbf{B}^{(1)} + 2\mu\mathbf{B}^{(2)} + c\Phi(w) + s_1\Theta_1(\Re) + s_2\Theta_2(\Re) + s_3\Theta_3(\Re) \quad (45)$$

where $\lambda, \mu, c, s_1, s_2$ and s_3 are undetermined material constants, which can be determined by experiment, physical model, or numerical simulation.

3. Parameterization and Experiment of Metal Material Grain Shape

On the meso scale, when discussing the relationship between the grain shape and the mechanical properties of metal materials, most of the grain shapes are expressed as ellipses. In fact, the grain shapes of polycrystalline metal materials are very different, and there is no unified shape description method for irregular grains. This section discusses the mathematical description of the grain shape, introduces the parameters that represent the grain shape, uses the digital image analysis method to realize the parametric representation of the evolution of the micro and macrograin shape of metal materials under stress deformation, and carries out experimental research on the mechanical properties of metal materials containing the grain shape effect.

3.1. Extraction of Grain Image

Through the test of several common metal materials (e.g., the low carbon steel, pure copper, and pure aluminum), we found that using pure aluminum sheets can obtain relatively clear grain images, so the grain images analyzed in this paper are taken from a pure aluminum sheet. We process the rolled pure aluminum plate into a dumbbell-shaped sample, conduct high temperature annealing at 600 °C for 30 min, and cool it in the furnace to room temperature. Then we use mixed acid solution (15 mL HF, 15 mL HNO_3, 25 mL HCl, 25 mL H_2O) to etch the sample, remove the oxide layer on the surface of pure aluminum plate, and expose the grain structure. These grains can also be used as natural speckles on the sample surface, as shown in Figure 2.

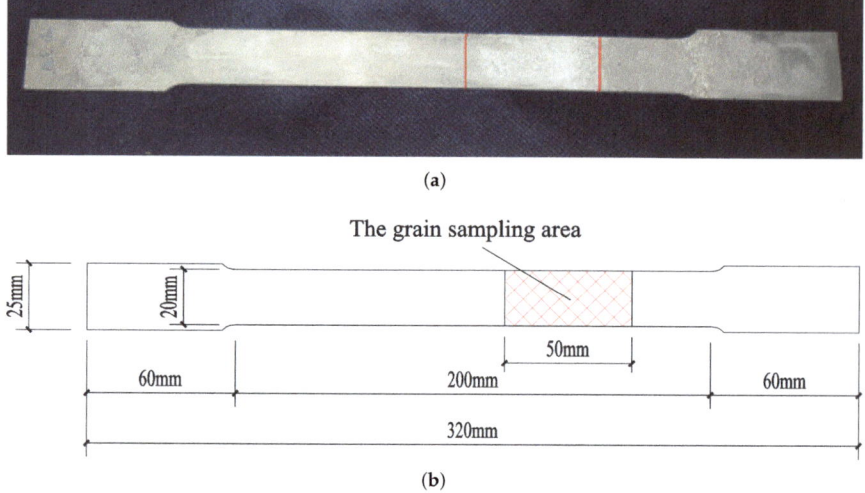

Figure 2. (a) The original grain on pure aluminum plate specimen (natural speckle); (b) the dimensions of the plate specimen.

In order to facilitate the research, the following five images (as shown in Figure 3) with clear grain boundaries and easily recognizable grain shapes during loading are selected from a large number of sample grain images, which are marked with GRAB1, GRAB5, GRAB10, GRAB14, and GRAB15, respectively. Three grains with obvious boundary contours are selected from these five grain images for shape parameterization research.

First, we use the 2D Gaussian filtering template to smooth the image, calculate the amplitude and direction of the filtered image gradient, apply non-maximum suppression to the gradient amplitude, find out the local maximum points in the image gradient, set other non-local maximum points to zero to obtain the refined edge, and use the double threshold algorithm to detect and connect the edges, use two thresholds to find each line segment and extend them in two directions to find the fracture at the grain boundary edge, and connect these fractures.

From Figure 3, taking GRAB1 as an example, three grains with clear grain boundaries were selected as the research object, and the boundary diagrams of the three grains under different loads were intercepted by using the aforementioned boundary extraction method, as shown in Figure 4.

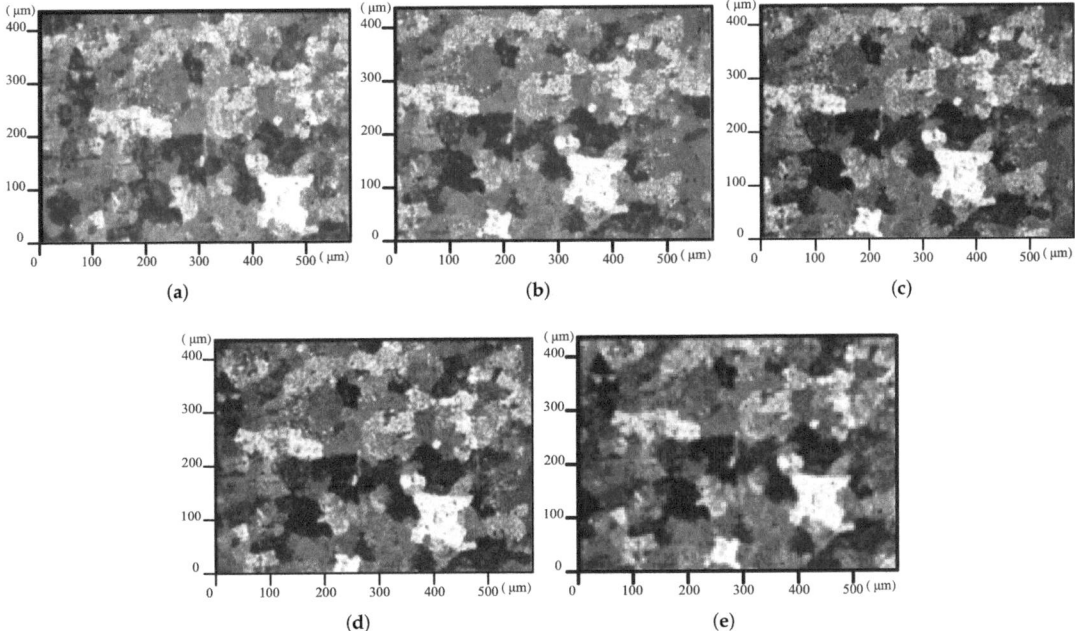

Figure 3. The original grain image of pure aluminum plate sample: (**a**) GRAB 1; (**b**) GRAB 5; (**c**) GRAB 10; (**d**) GRAB 14; (**e**) GRAB 15.

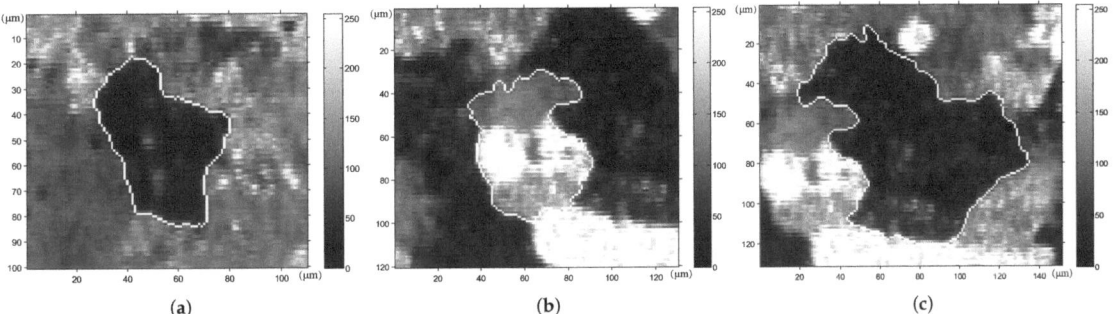

Figure 4. The boundary diagram of three grains in GRAB1: (**a**) grain 1; (**b**) grain 2; (**c**) grain 3.

3.2. Multi-String Method for Grain Shape Parameterization

The grain shape parameters can be fitted by the external rectangle, ellipse, and the multi-chord method for grain segmentation. By comparison, we find that the external rectangle and ellipse are sometimes not the closest fitting figure to the actual shape of the grain for grains with different shapes, and the multi-chord method for grain segmentation is more accurate and effective for the boundary fitting of irregular grains.

Taking the digital speckle image of grain 1, grain 2, and grain 3 provided in Figure 4 as the data source, and taking the minimum circumscribed rectangle at the edge of the grain as the object, we divide the minimum circumscribed rectangle into eight equal parts in the height and width directions respectively, then we obtain seven strings in the vertical and horizontal directions inside the grain, calculate the length of each string respectively, and take the ratio of the average length of the vertical and horizontal chords as the slenderness ratio of the grain, i.e., the grain shape parameter. Figure 5 is the schematic diagram of irregular grain boundary multi-chord method fitting.

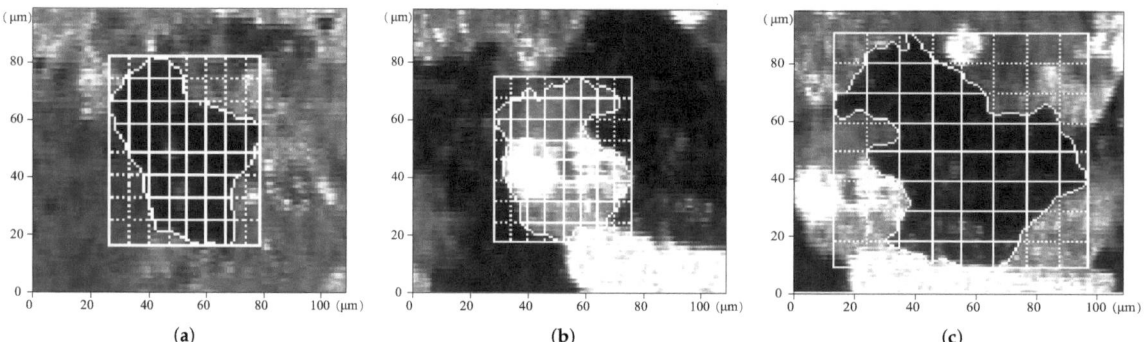

Figure 5. The multi-chord fitting of three grain boundaries: (a) grain 1; (b) grain 2; (c) grain 3.

We fit and calculate each chord (solid line part) in each grain in Figure 5. The calculation data and shape parameters are listed in Table 1 according to the number of chords in each grain.

Since the number of chords for dividing grains can be set manually, the appropriate number of chords depends on whether the shape of grains can be truly reflected and the efficiency of calculation. Therefore, this paper analyzes and compares the number of chords in the multi-chord method. Tables 1–3 list the fitting data and corresponding shape parameters of the grain image after grain segmentation with 1, 3, 5, and 7 chords in three grains.

Table 1. The multi-chord grain fitting data and shape parameters (grain 1).

Chord Number	1 Chord			3 Chord			5 Chord			7 Chord		
Grain Image Number	H.[1]	W.[2]	S.R.[3]	H.[1]	W.[2]	S.R.[3]	H.[1]	W.[2]	S.R.[3]	H.[1]	W.[2]	S.R.[3]
GRAB1	67	45	1.489	65.67	45.33	1.449	67.4	46.4	1.453	68.14	46.71	1.459
GRAB5	69	44	1.568	70	44.67	1.567	68.8	44.2	1.557	70.57	45.13	1.564
GRAB10	70	44	1.591	70.33	44	1.598	70.8	44	1.609	71.28	44.85	1.59
GRAB14	75	41	1.829	75.67	40.67	1.861	75.6	41	1.844	76.11	40.14	1.896
GRAB15	77	40	1.925	77	40.33	1.909	77	40.6	1.897	77.57	39.42	1.968

[1] This is the height of the crystal. [2] This is the width of the crystal. [3] This is the slenderness ratio of the crystal.

Table 2. The multi-chord grain fitting data and shape parameters (grain 2).

Chord Number	1 Chord			3 Chord			5 Chord			7 Chord		
Grain Image Number	H.[1]	W.[2]	S.R.[3]	H.[1]	W.[2]	S.R.[3]	H.[1]	W.[2]	S.R.[3]	H.[1]	W.[2]	S.R.[3]
GRAB1	51	44	1.159	53	39.67	1.336	53.8	40.7	1.322	53.37	40.93	1.304
GRAB5	56	45	1.244	53	40	1.325	53.6	40.6	1.321	53.73	40.87	1.314
GRAB10	50	46	1.087	51	41	1.244	53.2	40.4	1.317	53.84	40.23	1.338
GRAB14	59	46	1.283	55.33	40.67	1.361	54.4	41.1	1.324	54.33	41.54	1.307
GRAB15	55	45	1.222	55.67	40.33	1.38	55.2	41.7	1.324	55.87	41.67	1.341

[1] This is the height of the crystal. [2] This is the width of the crystal. [3] This is the slenderness ratio of the crystal.

Table 3. The multi-chord grain fitting data and shape parameters (grain 3).

Chord Number	1 Chord			3 Chord			5 Chord			7 Chord		
Grain Image Number	H.[1]	W.[2]	S.R.[3]	H.[1]	W.[2]	S.R.[3]	H.[1]	W.[2]	S.R.[3]	H.[1]	W.[2]	S.R.[3]
GRAB1	103	109	0.945	106.67	109.33	0.976	102.8	108.6	0.947	103.57	109.17	0.949
GRAB5	105	108	0.972	108	107.67	1.003	106.2	108.2	0.982	104.58	108.71	0.962
GRAB10	108	106	1.019	109.33	106	1.031	108.4	105.8	1.025	109.17	105.14	1.038
GRAB14	111	103	1.078	110.67	103.33	1.071	112.2	102.6	1.094	111.34	102.47	1.087
GRAB15	113	100	1.13	114.33	100	1.143	113	99.2	1.139	114.02	99.42	1.147

[1] This is the height of the crystal. [2] This is the width of the crystal. [3] This is the slenderness ratio of the crystal.

According to the grain shape fitting data in Tables 1–3, we find that the results of grain shape fitting for the same grain under the same load state are different with a different number of split chords. When the number of split chords is 5 and 7, the grain fitting data shows better convergence, which is significantly different from the grain fitting data when the number of split chords is 1 and 3. Theoretically, the more the number of segmenting chords of the grains, the more information reflecting the shape of the grains, and the closer the description of the shape of the irregular grains should be. However, there is a problem of computational efficiency at this time. After comparison, the seven chords can be used as the more appropriate chord number of the multi-chord method for segmenting grains.

3.3. Experimental Study on Grain Shape Coefficient

In Section 2, according to the ratio of the arbitrary radius value to the average radius value in the grain, we establish the GSF to describe the average shape of the grain. The GSF can be expanded into infinite series on the basis of Wigner D function, and the expanded coefficients are called grain shape coefficients s_{m0}^l (Equations (10)–(13)). Equation (10) degenerates to the form of a plane, then

$$r(\theta) = 1 + 2\sum_{i=1}^{4}[a_i \cos(i\theta) + b_i \sin(i\theta)] \qquad (46)$$

where $r(\theta)$ is the shape function of the grain, a_i and b_i are the coefficients of the expansion series of the GSF, i.e., the grain shape coefficient.

We selecte the first loading step G1 of grain 1, find the centroid of the grain through the image analysis technology in Section 3.1, and establish a rectangular coordinate system with the centroid as the coordinate origin (as shown in Figure 6), segment the grain with the seven chord method, and obtain 28 information points at the transverse chord end and vertical chord end on the grain boundary.

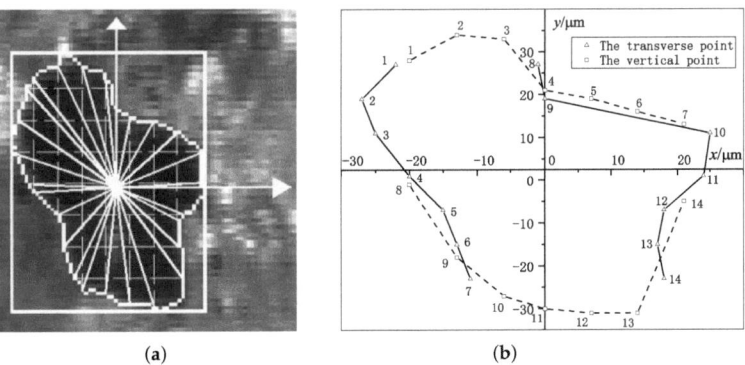

Figure 6. (a) The grain 1 (G1) in the rectangular coordinate system; (b) the transverse and vertical points.

Further, we obtain the X and Y coordinate values of the 28 information points, as shown in Tables 4 and 5.

Table 4. The coordinates of grain 1 (G1) boundary information points (transverse point).

Point Number	X	Y	Point Number	X	Y
Transverse point 1	−22	27	Transverse point 8	−1	27
Transverse point 2	−27	19	Transverse point 9	0	19
Transverse point 3	−25	11	Transverse point 10	25	11
Transverse point 4	−20	1	Transverse point 11	24	1
Transverse point 5	−15	−7	Transverse point 12	18	−7
Transverse point 6	−13	−15	Transverse point 13	17	−15
Transverse point 7	−11	−23	Transverse point 14	18	−23

Table 5. The coordinates of grain 1 (G1) boundary information points (vertical point).

Point Number	X	Y	Point Number	X	Y
Vertical point 1	−20	28	Vertical point 8	−20	−1
Vertical point 2	−13	34	Vertical point 9	−13	−18
Vertical point 3	−6	33	Vertical point 10	−6	−27
Vertical point 4	0	21	Vertical point 11	0	−30
Vertical point 5	7	19	Vertical point 12	7	−31
Vertical point 6	14	16	Vertical point 13	14	−31
Vertical point 7	21	13	Vertical point 14	21	−5

Using the same processing method, each grain has five analysis steps (G1, G5, G10, G14, G15), and we obtain 15 data sheets of three grains. According to the coordinate value of each information point, we calculate the distance R from the information point to the centroid, the average distance $\bar{r}$, and the included angle θ between the centroid line of the information point and the X axis. If $r = R/\bar{r}$, r is the shape function value of the function $r(\theta)$ corresponding to θ. In (46), we can establish

$$\Pi = \sum_{i=1}^{28} [r(\theta_i) - r_i]^2 \quad (47)$$

where $r(\theta_i)$ is the value of function $r(\theta)$ when $\theta = \theta_i$, and r_i is the experimental value corresponding to $\theta = \theta_i$.

According to $\frac{\partial \Pi}{\partial a_i} = 0$, $\frac{\partial \Pi}{\partial b_i} = 0$ ($i = 1 \cdots n$, n is the number of experimental data points), we obtain a_i, b_i.

Using the seven string method, the results of grain 1 shape fitting are obtained as shown in Figure 7. Figure 7a shows the shape function curve of grain 1, and the fitting curve is in good agreement with the experimental data points. Figure 7b reflects the function fitting curve of grain 1 under different loading steps. The five grain curves are not completely coincident, reflecting the changes of grain shape under load. Such fitting results show that the GSF can better describe the shape of irregular grains, where the grain shape coefficients a_i and b_i are coefficients of the expansion of the GSF, and different grain shape coefficients correspond to different grain shapes.

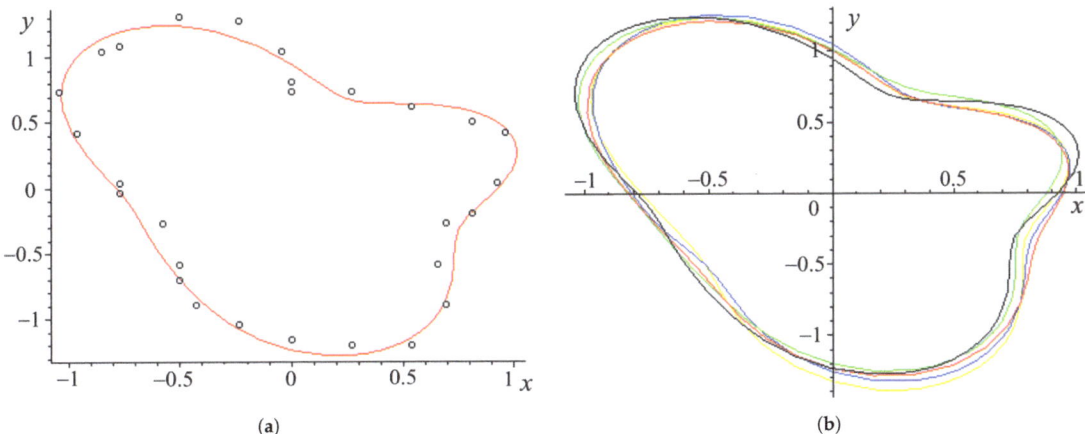

Figure 7. (a) The shape function fitting curve of grain; (b) the function fitting curve of grain 1 under different loading steps.

4. Parametric Experimental Study on Grain Shape Evolution

In Section 3, we studied the parametric shape of grains with different shapes in the micro structures of polycrystalline metal materials. In this section, taking the axial tensile deformation experiment of pure aluminum plate as an example, we discussed the micro shape evolution of grains under macro tension from the micro scale of metal grain shape parameters.

Figure 8 shows the local grain diagram in the pure aluminum plate sample. There are a large number of grains with irregular shapes within the grain image range. We selected grains with relatively regular shape and near ellipse as the analysis object of this experimental study. The target grain (red) is shown in Figure 8b.

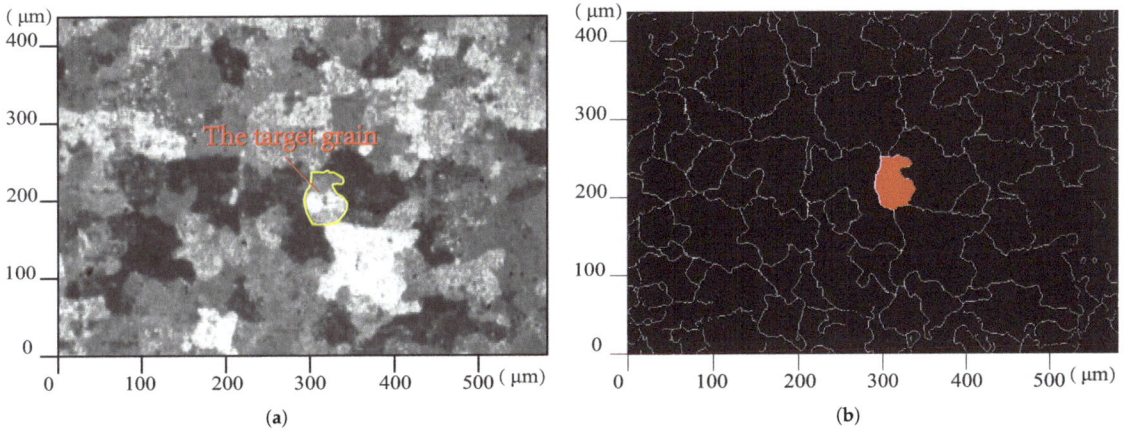

Figure 8. (a) The target grain of the grain diagram when $F = 200$ N; (b) the corresponding boundary diagram.

During the tensile loading of a pure aluminum sheet, the grain distribution on the surface of the sheet is synchronously collected. The loading range is 0 N~3800 N. Images are collected every 200 N. The images collected under various loads are processed, and the corresponding grain boundary diagram is obtained, as shown in Figure 9.

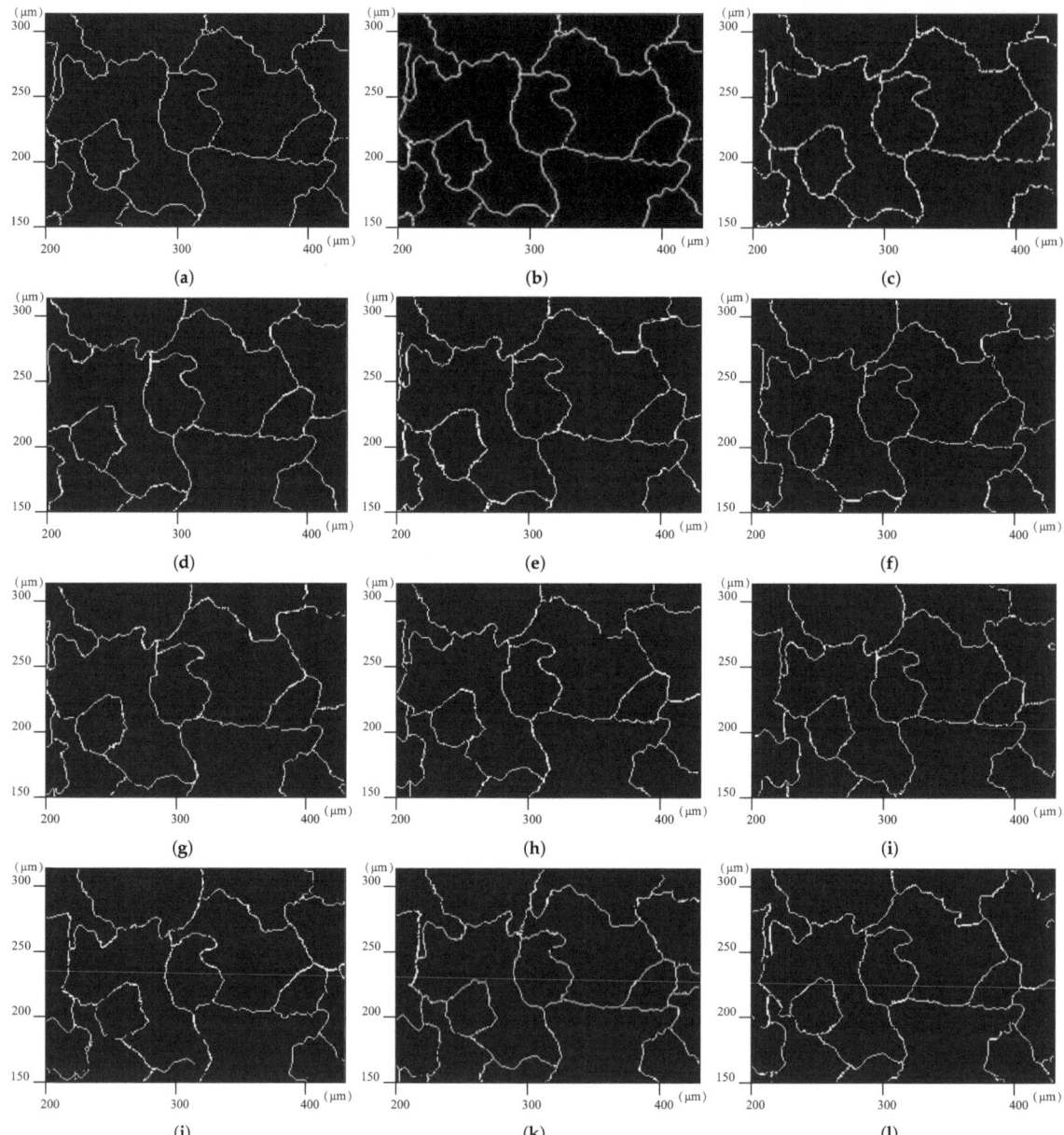

Figure 9. *Cont.*

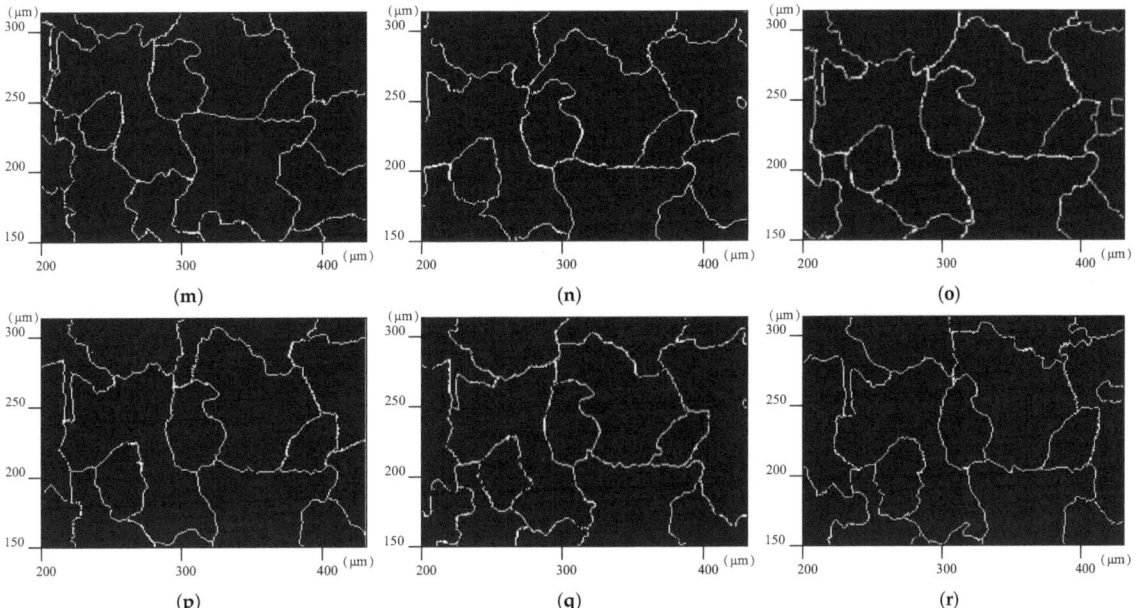

Figure 9. The grain boundary observed under monotonically increasing tensile loading F: (**a**) $F = 0$ N; (**b**) $F = 200$ N; (**c**) $F = 600$ N; (**d**) $F = 1000$ N; (**e**) $F = 1200$ N; (**f**) $F = 1400$ N; (**g**) $F = 1600$ N; (**h**) $F = 1800$ N; (**i**) $F = 2000$ N; (**j**) $F = 2200$ N; (**k**) $F = 2400$ N; (**l**) $F = 2600$ N; (**m**) $F = 2800$ N; (**n**) $F = 3000$ N; (**o**) $F = 3200$ N; (**p**) $F = 3400$ N; (**q**) $F = 3600$ N; (**r**) $F = 3800$ N.

Since the shape of the target grain is close to the ellipse, according to Section 3, we use the seven chord method to fit, and take the slenderness ratio as the shape parameter of the grain to describe the shape of the target grain. The load of the target grain and the corresponding shape parameter (slenderness ratio) are shown in Table 6. Through comparative calculation, the fitting results of the seven chord method are very close.

Table 6. The load and corresponding shape parameters of target grain (slenderness ratio).

Load (N)	Shape Parameter	Load (N)	Shape Parameter	Load (N)	Shape Parameter
0	0.717802	1600	0.713215	2800	0.695813
200	0.721352	1800	0.720267	3000	0.67768
600	0.721482	2000	0.703568	3200	0.681628
1000	0.717047	2200	0.723909	3400	0.687682
1200	0.717419	2400	0.704046	3600	0.622458
1400	0.711743	2600	0.718436	3800	0.545502

Further, we obtain the relationship curve between the load of the target grain and the corresponding shape parameter (slenderness ratio), as shown in Figure 10.

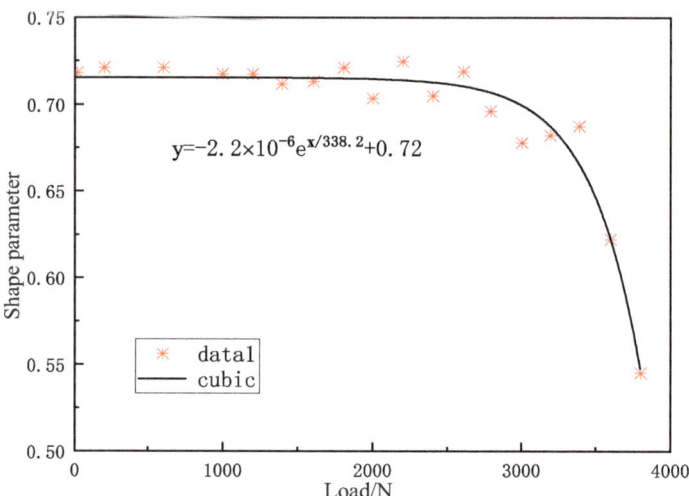

Figure 10. The relation curve between the load of target grain and corresponding shape parameter (slenderness ratio).

According to the results in Table 6 and Figure 10, the aluminum plate tensile process has gone through the elastic and plastic stages. In the elastic deformation stage, the load is less than 2500 N, and the shape parameters (slenderness ratio) of the selected target grains in the aluminum plate are between 0.70 and 0.72, basically keeping constant. When the load is greater than 2600 N, the aluminum plate enters the plastic deformation stage, and the slenderness ratio of the target grain becomes significantly smaller, and the slenderness ratio finally approaches 0.5. The experimental results of the shape evolution of a single typical grain are consistent with the theoretical results of the macroscopic mechanical properties of aluminum metal plates, e.g., the Poisson's ratio in the elastic phase is constant, and the volume of plastic deformation is constant, which also indicates that the experimental research based on grain size and digital image analysis technology in this paper reveals the relationship between the microscopic mechanical properties of grains and the macroscopic mechanical properties of metal plates.

5. Conclusions

1. Polycrystalline metallic materials are composed of small grains, and their constitutive relations must be related to the characteristics of grains, e.g., the average shape and orientation distribution of grains. According to the principle of no difference in the material frame, we choose the ratio of the arbitrary radius value in the grain to the average radius value of the grain, and establish the GSF, which can be used to describe the average shape of the grain. The GSF can be expanded into infinite series on the basis of Wigner D function, and the expanded coefficient is defined as the grain shape coefficient s_{m0}^l.
2. We discuss the shape coefficients of special grains with weak anisotropy, and obtain the expression of the shape coefficients. Considering the average grain shape effect, we study the elastic constitutive relation of metallic multi-grain materials, and derive the analytical formula of elastic constitutive relation containing the grain shape coefficient s_{m0}^l.
3. By using the power transformation, we improve the linear contrast of the digital image of the grain in polycrystalline metal materials, adopt the open and close operations in mathematical morphology to smooth the image, and then carry out histogram equalization and filtering noise removal processing to obtain a more ideal grain boundary. The approximate boundary of the image is extracted by the Canny operator, and then the boundary image is linearly expanded and refined. Then, using the internal

function of MATLAB, a single, complete grain with clear and accurate boundary is obtained, and a total of 15 grain images of three grains are extracted under five loading steps.

4. We discuss the mathematical description method of grain shape, and propose the multi-chord method to segment grains to represent the grain shape. When the grain shape is particularly irregular, the seven chord method is more reasonable. Furthermore, we carry out the experimental research on the shape function and shape coefficient of grains, fit the shape function of irregular grains, and prove that the shape function of grains can better describe the shape of irregular grains.

5. Using the digital image analysis method of grains (e.g., the grain boundary processing, grain image acquisition, and grain shape parameterization), we track the shape evolution of the target grains in the metal materials under stress, obtain the parameterized representation of grain deformation, and analyze the relationship between the metal materials' micro deformation and the materials' macro mechanical properties.

Author Contributions: Conceptualization, Z.L.; software, H.S. and M.H.; formal analysis, L.Z. and H.Y.; investigation, H.S.; resources, Z.L.; data curation, T.Z. and M.H.; writing—original draft preparation, Z.L. and T.Z.; writing—review and editing, T.Z. and Z.L.; project administration, Z.L. All authors have read and agreed to the published version of the manuscript.

Funding: This research was funded by the Jiangxi Province Education Science "14th Five-Year Plan" Project (Awards Nos. 22QN007) and the National Natural Science Foundation of China (Awards Nos. 51568046).

Data Availability Statement: The data of this study are available from the corresponding author upon request.

Conflicts of Interest: The authors declare no conflict of interest.

References

1. Adams, B.L. Orientation imaging microscopy: Emerging and future applications. *Ultramicroscopy* **1997**, *67*, 11–17. [CrossRef]
2. Thakur, A.; Gangopadhyay, S. State-of-the-art in surface integrity in machining of nickel-based super alloys. *Int. J. Mach. Tools Manuf.* **2016**, *100*, 25–54. [CrossRef]
3. Veenhuizen, K.; McAnany, S.; Nolan, D.; Aitken, B.; Dierolf, V.; Jain, H. Fabrication of graded index single crystal in glass. *Sci. Rep.* **2017**, *7*, 44327. [CrossRef]
4. Pan, Q.; Yang, D.; Dong, G.; Qiu, J.; Yang, Z. Nanocrystal-in-glass composite (NGC): A powerful pathway from nanocrystals to advanced optical materials. *Prog. Mater. Sci.* **2022**, *130*, 100998. [CrossRef]
5. Trusov, P.; Kondratev, N.; Podsedertsev, A. Description of Dynamic Recrystallization by Means of An Advanced Statistical Multilevel Model: Grain Structure Evolution Analysis. *Crystals* **2022**, *12*, 653. [CrossRef]
6. Miller, A.K. *Unified Constitutive Equations for Creep and Plasticity*; Springer: Berlin/Heidelberg, Germany, 2012.
7. Walker, K.P. *Research and Development Program for Non-Linear Structural Modelling with Advanced Time-Temperature Dependent Constitutive Relationships*; NASA CR-165533, Report; National Aeronautics and Space Administration: Washington, DC, USA, 1981.
8. Bodner, S.R. *Unified Plasticity—An Engineering Approach (Final Report)*; Faculty of Mechanical Engineering, Technion-Israel Institute of Technology: Haifa, Israel, 2000; p. 32000.
9. Mahmoudi, A.H.; Pezeshki-Najafabadi, S.M.; Badnava, H. Parameter determination of Chaboche kinematic hardening model using a multi objective Genetic Algorithm. *Comput. Mater. Sci.* **2011**, *50*, 1114–1122. [CrossRef]
10. Sadovskii, V.M.; Sadovskaya, O.V.; Petrakov, I.E. On the theory of constitutive equations for composites with different resistance in compression and tension. *Compos. Struct.* **2021**, *268*, 113921. [CrossRef]
11. Feng, M.; Lu, H.; Lin, G.; Guo, Y. A unified visco-elastic-plastic constitutive model for concrete deformation. *Eng. Mech.* **2002**, *19*, 1–6. [CrossRef]
12. Man, C.S.; Du, W. Recasting Classical Expansion of Orientation Distribution Function as Tensorial Fourier Expansion. *J. Elast.* **2022**, 1–23. [CrossRef]
13. Xu, B.; Mao, N.; Zhao, Y.; Tong, L.; Zhang, J. Polarized Raman spectroscopy for determining crystallographic orientation of low-dimensional materials. *J. Phys. Chem. Lett.* **2021**, *12*, 7442–7452. [CrossRef]
14. Li, S.; Zhang, X.; Wang, C.; Jian, X. Texture evolution during uniaxial tension of aluminum sheet. *Chin. J. Nonferr. Met.* **1999**, *9*, 45–48.
15. Liu, W.C.; Man, C.S.; Morris, J.G. Lattice rotation of the cube orientation to the β fiber during cold rolling of AA 5052 aluminum alloy. *Scr. Mater.* **2001**, *45*, 807–814. [CrossRef]

16. Ivasishin, O.M.; Shevchenko, S.V.; Vasiliev, N.L.; Semiatinb, S.L. A 3-D Monte-Carlo (Potts) model for recrystallization and grain growth in polycrystalline materials. *Mater. Sci. Eng. A* **2006**, *433*, 216–232. [CrossRef]
17. Guenoun, G.; Schmitt, N.; Roux, S.; Régnier, G. Crystalline orientation assessment in transversely isotropic semicrystalline polymer: Application to oedometric compaction of PTFE. *Polym. Eng. Sci.* **2021**, *61*, 107–114. [CrossRef]
18. Tang, P.; Feng, J.; Wan, Z.; Huang, X.; Yang, S.; Lu, L.; Zhong, X. Influence of grain orientation on hardness anisotropy and dislocation behavior of AlN ceramic in nanoindentation. *Ceram. Int.* **2021**, *47*, 20298–20309. [CrossRef]
19. Şahin, K.; Clawson, J.K.; Singletary, J.; Horner, S.; Zheng, J.; Pelegri, A.; Chasiotis, I. Limiting role of crystalline domain orientation on the modulus and strength of aramid fibers. *Polymer* **2018**, *140*, 96–106. [CrossRef]
20. Tang, H.; Huang, H.; Liu, C.; Liu, Z.; Yan, W. Multi-Scale modelling of structure-property relationship in additively manufactured metallic materials. *Int. J. Mech. Sci.* **2021**, *194*, 106185. [CrossRef]
21. Gu, S.; Zhao, J.; Ma, T.; Xu, T.; Yu, B. Modeling and analysis of grain morphology effects on deformation response based on crystal plasticity finite element method. *Mater. Und Werkst.* **2022**, *53*, 770–780. [CrossRef]
22. Trusov, P.; Shveykin, A.; Kondratev, N. Some Issues on Crystal Plasticity Models Formulation: Motion Decomposition and Constitutive Law Variants. *Crystals* **2021**, *11*, 1392. [CrossRef]
23. Lakshmanan, A.; Yaghoobi, M.; Stopka, K.S.; Sundararaghavan, V. Crystal plasticity finite element modeling of grain size and morphology effects on yield strength and extreme value fatigue response. *J. Mater. Res. Technol.* **2022**, *19*, 3337–3354. [CrossRef]
24. Agius, D.; Kareer, A.; Al Mamun, A.; Truman, C.; Collins, D.M.; Mostafavi, M.; Knowles, D. A crystal plasticity model that accounts for grain size effects and slip system interactions on the deformation of austenitic stainless steels. *Int. J. Plast.* **2022**, *152*, 103249. [CrossRef]
25. Nabergoj, M.; Urevc, J.; Halilovič, M. Function-based reconstruction of the fiber orientation distribution function of short-fiber-reinforced polymers. *J. Rheol.* **2022**, *66*, 147–160. [CrossRef]
26. Zhu, S.; Zhao, M.; Mao, J.; Liang, S.Y. Study on Hot Deformation Behavior and Texture Evolution of Aluminum Alloy 7075 Based on Visco-Plastic Self-Consistent Model. *Metals* **2022**, *12*, 1648. [CrossRef]
27. Zheng, C.; Xu, L.; Feng, X.; Huang, Q.; Li, Y.; Zhang, Z.; Yang, Y. Influence of Grain Orientation and Grain Boundary Features on Local Stress State of Cu-8Al-11Mn Alloy Investigated Using Crystal Plasticity Finite Element Method. *Materials* **2022**, *15*, 6950. [CrossRef] [PubMed]
28. Huang, Y. *A User-Material Subroutine Incroporating Single Crystal Plasticity in the ABAQUS Finite Element Program*; Harvard University: Cambridge, MA, USA, 1991.
29. Xiao, J.; Cui, H.; Zhang, H.; Wen, W.; Zhou, J. A physical-based constitutive model considering the motion of dislocation for Ni3Al-base superalloy. *Mater. Sci. Eng. A* **2020**, *772*, 138631. [CrossRef]
30. Zhang, M.; Zhang, J.; McDowell, D.L. Microstructure-based crystal plasticity modeling of cyclic deformation of Ti–6Al–4V. *Int. J. Plast.* **2007**, *23*, 1328–1348. [CrossRef]
31. Bandyopadhyay, R.; Mello, A.W.; Kapoor, K.; Reinhold, M.P.; Broderick, T.F.; Sangid, M.D. On the crack initiation and heterogeneous deformation of Ti-6Al-4V during high cycle fatigue at high R ratios. *J. Mech. Phys. Solids* **2019**, *129*, 61–82. [CrossRef]
32. Bridier, F.; McDowell, D.L.; Villechaise, P.; Mendez, J. Crystal plasticity modeling of slip activity in Ti–6Al–4V under high cycle fatigue loading. *Int. J. Plast.* **2009**, *25*, 1066–1082. [CrossRef]
33. Ma, A.; Dye, D.; Reed, R.C. A model for the creep deformation behaviour of single-crystal superalloy CMSX-4. *Acta Mater.* **2008**, *56*, 1657–1670. [CrossRef]
34. Rodas, E.A.E.; Neu, R.W. Crystal viscoplasticity model for the creep-fatigue interactions in single-crystal Ni-base superalloy CMSX-8. *Int. J. Plast.* **2018**, *100*, 14–33. [CrossRef]
35. Mori, T.; Tanaka, K. Average stress in matrix and average elastic energy of materials with misfitting inclusions. *Acta Metall.* **1973**, *21*, 571–574. [CrossRef]
36. Tandon, G.P.; Weng, G.J. The effect of aspect ratio of inclusions on the elastic properties of unidirectionally aligned composites. *Polym. Compos.* **1984**, *5*, 327–333. [CrossRef]
37. Mura, T. *Micromechanics of Defects in Solids (Mechanics of Eastic and Inelastic Solids)*; Kluwer Academic Publishers: Amsterdam, The Netherlands, 1987.
38. Benveniste, Y. A new approach to the application of Mori-Tanaka's theory in composite materials. *Mech. Mater.* **1987**, *6*, 147–157. [CrossRef]
39. Bunge, H.J. *Texture Analysis in Material Science: Mathematical Methods*; Butterworths: London, UK, 1982.
40. Huang, M.; Man, C.S. Constitutive relation of elastic polycrystal with quadratic texture dependence. *J. Elast.* **2003**, *72*, 183–212. [CrossRef]
41. Man, C.S. On the Constitutive Equations of Some Weakly-Textured Materials. *Arch. Ration. Mech. Anal.* **1998**, *143*, 77–103. [CrossRef]
42. Huang, M. Elastic constants of a polycrystal with an orthorhombic texture. *Mech. Mater.* **2004**, *36*, 623–632. [CrossRef]
43. Huang, M. Perturbation approach to elastic constitutive relations of polycrystals. *J. Mech. Phys. Solids* **2004**, *52*, 1827–1853. [CrossRef]

Electronic Structure and Optical Properties of Cu$_2$ZnSnS$_4$ under Stress Effect

Xiufan Yang *, Xinmao Qin, Wanjun Yan, Chunhong Zhang, Dianxi Zhang and Benhua Guo

College of Physics and Electronic Science, Anshun University, Anshun 561000, China
* Correspondence: xnmdyxf@163.com

Abstract: By using the pseudopotential plane-wave method of first principles based on density functional theory, the band structure, density of states and optical properties of Cu$_2$ZnSnS$_4$ under isotropic stress are calculated and analyzed. The results show that Cu$_2$ZnSnS$_4$ is a direct band gap semiconductor under isotropic stress, the lattice is tetragonal, and the band gap of Cu$_2$ZnSnS$_4$ is 0.16 eV at 0 GPa. Stretching the lattice causes the bottom of the conduction band of Cu$_2$ZnSnS$_4$ to move toward lower energies, while the top of the valence band remains unchanged and the band gap gradually narrows. Squeezing the lattice causes the bottom of the conduction band to move toward the high-energy direction, while the top of the valence band moves downward toward the low-energy direction, and the Cu$_2$ZnSnS$_4$ band gap becomes larger. The static permittivity, absorption coefficient, reflectivity, refractive index, electrical conductivity, and energy loss function all decrease when the lattice is stretched, and the above optical parameters increase when the lattice is compressed. When the lattice is stretched, the optical characteristic peaks such as the dielectric function shift to the lower-energy direction, while the optical characteristic peak position shifts to the higher-energy direction when the lattice is compressed.

Keywords: Cu$_2$ZnSnS$_4$; stress; electronic structure; optical properties; first principles

1. Introduction

With the increasing scarcity of non-renewable energy sources such as fossils, the energy crisis and environmental pollution problems are becoming more and more prominent. The development of new energy sources, such as solar energy and wind energy, is an effective means to solve the energy crisis and environmental pollution problems [1]. Among many new energy sources, solar energy has received great attention due to its advantages of greenness, environmental protection, and high efficiency [2–6]. Solar cells are the main means of utilizing solar energy. Solar cells include silicon-based solar cells, perovskite solar cells, organic solar cells, dye-sensitized cells, compound thin-film solar cells, etc. [7–11]. Compound thin-film solar cells are effective devices for utilizing solar energy due to their small size, light weight, and flexibility. Compound thin-film solar cells mainly include [12–14] CdTe cells, GaAs cells, Cu(In-Ga)(Se,S)$_2$ cells, and so on. Among them, Cu(In-Ga)(Se,S)$_2$ cells shavehigh battery conversion efficiency due to their spike-like conduction band valence and they do not damage the short-circuit current [15] and are widely used. Cu(In-Ga)(Se,S)$_2$ cells contain rare metals In and Ga, which are limited in large-scale commercial applications. Therefore, seeking environmentally friendly solar cell materials is one of the strategies to solve the difficulty of the large-scale application of Cu(In-Ga)(Se,S)$_2$ cells. The band gap of the semiconductor material Cu$_2$ZnSnS$_4$ (CZTS) is 1.4~1.6 eV, which is strongly matched with the required band gap of the solar cell absorber material [16]. The physical and chemical properties of CZTS are very close to those of Cu(In-Ga)(Se,S)$_2$. The theoretical conversion efficiency of CZTS as an absorber layer of solar cells is 32.2% [17], and CZTS is rich in various elements in the crust, as well as being non-toxic and non-polluting. Therefore, CZTS is considered as a high-efficiency, low-cost, and environmentally friendly solar cell material.

At present, the experimentally reported value of the conversion efficiency of CZTS solar cells reaches 12.6% [18], but there is still a large gap between this and the theoretical calculation value, and the conversion efficiency of the cell has large room for improvement. The main factors affecting the conversion efficiency of CZTS solar cells are the low open-circuit voltage Voc and low filling factor of the cell [19,20]. Since the ionic radii of Cu and Zn are very close, substitution defects of Cu and Zn are easily formed in the CZTS system. Anti-occupancy defects with Cu and Sn and Zn anti-occupancy defects [21] will introduce energy level defects to different degrees and inhibit the open-circuit voltage Voc of the battery. It was found [22] that the introduction of metal ions in CZTS can play a role in the passivation of defects; for example, the introduction of a small amount of Ge can suppress intrinsic deep-level Sn_{Zn} defects. We note that the doping of monovalent metals such as silver can change the defects of Cu_2ZnSnS_4, thus adjusting the band gap. Due to the high formation energy of Ag_{Zn} defects, when the Ag doping concentration is high, the defect concentration of the CZTS system is low, which is conducive to increasing the carrier concentration. Affected by the electronic configuration of elements, the atomic radii of Ag atoms and Cu atoms differ greatly, which causes the bending coefficient of the energy band to change. Low-concentration doping causes the energy band to change slowly, while high-concentration doping causes the energy band to change obviously [23].

In addition, the introduction of stress into the material can play the role of changing the lattice constant, regulating the energy band structure, and changing the optical properties. The 4-metal alloy of Cu_2ZnSnS_4 is derived from the binary alloy of ZnS [24]; due to its good band gap matching and strong light absorption, it is an ideal solar cell absorption layer material. However, Cu_{Zn} and Sn_{Zn} defect energy levels exist when Cu_2ZnSnS_4 alloy films are actually prepared, which affects the photoelectric properties of the materials. Therefore, we control the band gap and photoelectric properties of Cu_2ZnSnS_4 alloy films by applying stress. Stressing is an effective means to regulate the heterojunction carrier dynamics [25]. As a solar cell absorbing layer material, CZTS needs to form a solar cell system together with other materials, so it is necessary to consider the transport of electrons and holes at the heterojunction interface. Adjusting the stress can change the K point energy level of the material. When the K point energy level is coupled with other energy levels, it can provide an intermediate-state energy level for electron and hole transfer, making the transfer of electrons and holes between layers easier, and improving the utilization of photogenerated carriers. In addition, adjusting the stress can change the orbital occupancy of electrons, thereby changing the photoelectric characteristics of the film. At the same time, adjusting the stress can also change the phase transition temperature of the film [26]. Therefore, it can be predicted that the introduction of stress is an effective means to adjust the photoelectric properties and phase transition temperature of CZTS. Based on the above considerations, we study the effect of stress on the electronic structure and optical properties of CZTS.

There have been extensive reports on the photoelectric properties of materials by introducing stress [27–30]. However, there are few reports on the effect of stress on the electronic structure and optical properties of CZTS. In this paper, by using the pseudopotential plane-wave method of first principles based on density functional theory, the band structure, density of states, and optical properties of Cu_2ZnSnS_4 under isotropic stress are calculated and analyzed.

2. Methods and Calculations

The calculation model selected in this paper is Cu_2ZnSnS_4 with a kesterite structure, the space group is $I\bar{4}(No.82)$, the lattice constants are $a = b = 0.54628$ nm, $c = 1.0864$ nm, and the crystal plane angle is $\alpha = \beta = \gamma = 90°$ [31]. The unit cell structure of Cu_2ZnSnS_4 is shown in Figure 1.

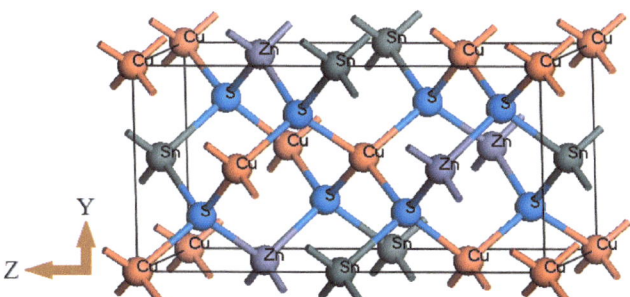

Figure 1. Cu$_2$ZnSnS$_4$ unit cell structure.

The first-principles-based pseudopotential method was used for the calculations; all calculations were performed using the Cambridge Serial Total Energy Package (CASTEP) [32] in Materials Studio software, and processed using the Perdew–Burke–Eurke–Ernzerh(PBE) [33] functional of the generalized gradient approximation (GGA) for the exchange correlation energy between electrons. An ultrasoft pseudopotential [34] was used to deal with the interaction between ionic solids and electrons. The truncation energy was set to 400 eV, the self-consistent convergence accuracy was 1.0×10^{-7} eV/atom, and the integral of the Brillouin zone was divided by $5 \times 5 \times 2$ of Monkhorst–Pack.

3. Results and Discussion

3.1. Geometry Optimization

Table 1 shows the lattice constants of Cu$_2$ZnSnS$_4$ with a kesterite structure obtained after applying isotropic tensile stress and compressive stress optimization, where "−" represents tensile stress. It can be seen from Table 1 that with the increase inisotropic tensile stress, the lattice constant of CZTS increases, the volume of the unit cell increases, and the lattice of CZTS is tetragonal. With the increase inisotropic compressive stress, the lattice constant of CZTS decreases, the unit cell volume decreases, and the lattice still maintains the tetragonal system. The results calculated by us are in good agreement with those of Liu [35], which shows that the applied stress has a significant effect on the lattice constant and cell volume of Cu$_2$ZnSnS$_4$.

For comparison, the theoretical calculations [36] and experimental values [37] of lattice constant and cell volume are added in Table 1. It can be seen from Table 1 that when no stress is applied, the difference between the a and b values and theoretical calculation is 0.0%, and the difference between them and the laboratory values is 0.7%. The difference between the c value and theoretical calculation is 0.2%, and the difference between the c value and experimental value is 0.8%. The unit cell volume is 0.2% different from the theoretical calculation and 2.4% different from the experimental value. The results in Table 1 show that our calculation results are in good agreement with the calculation and experimental values of other research groups.

Table 1. Cu$_2$ZnSnS$_4$ lattice constants of applied tensile and compressive stresses.

Stress/GPa	a/Å	b/Å	c/Å	V/Å^3
−6	5.688	5.688	11.377	368.043
−4	5.602	5.602	11.202	351.220
−2	5.528	5.528	11.061	338.034
0	5.469	5.469	10.944	327.377
Theoretical [36]	5.469	5.469	10.921	326.647
Experimental [37]	5.427	5.427	10.854	319.676
6	5.338	5.338	10.673	304.123
12	5.240	5.240	10.480	287.732
20	5.149	5.149	10.242	271.568

3.2. Electronic Structure

3.2.1. Band Structure

Figure 2 shows the Cu_2ZnSnS_4 band structure obtained by applying tensile stress. It can be seen from the figure that CZTS is a typical direct band gap semiconductor, and the minimum band gap value is obtained at the highly symmetrical G point position. The band gap of CZTS is 0.16 eV when no stress is applied (0 GPa), which is consistent with the research results of Zhao [38].

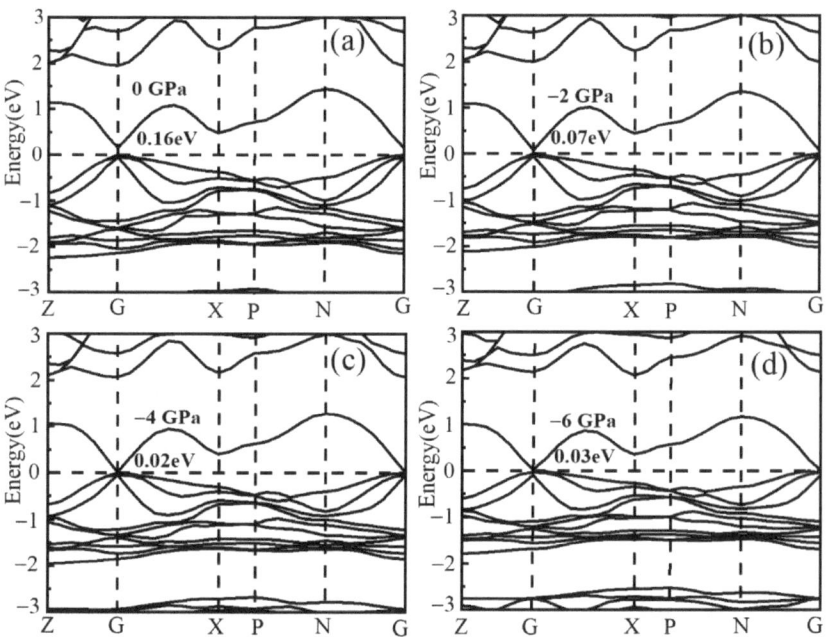

Figure 2. Band structure of Cu_2ZnSnS_4 under tensile stress for (**a**) 0 GPa band structure; (**b**) −2 GPa band structure; (**c**) −4 GPa band structure; (**d**) −6 GPa band structure.

With the increase intensile stress, the bottom of the conduction band moves to the lower-energy direction, while the top of the valence band remains unchanged; the band gap decreases with the increase intensile stress, and the band gap reaches a minimum value of 0.02 eV when the tensile stress is −4 GPa. After this, the band gap widens with the increase intensile stress. This is because the bottom of the conduction band of CZTS moves toward the high-energy direction after the tensile stress continues to increase, while the top of the valence band moves down toward the low-energy direction, and the band gap widens.

Figure 3 shows the band structure of Cu_2ZnSnS_4 obtained by applying compressive stress. It can be seen from Figure 3 that the CZTS is still suitable as a direct bandgap semiconductor; the minimum bandgap value is still obtained at the highly symmetrical G point position.

With the increase incompressive stress, the bottom of the conduction band of CZTS moves towards high energy and the top of the valence band remains unchanged. The band gap of CZTS widens with the increase incompressive stress. When the applied compressive stress is 20 GPa, the band gap of CZTS reaches 0.71 eV.

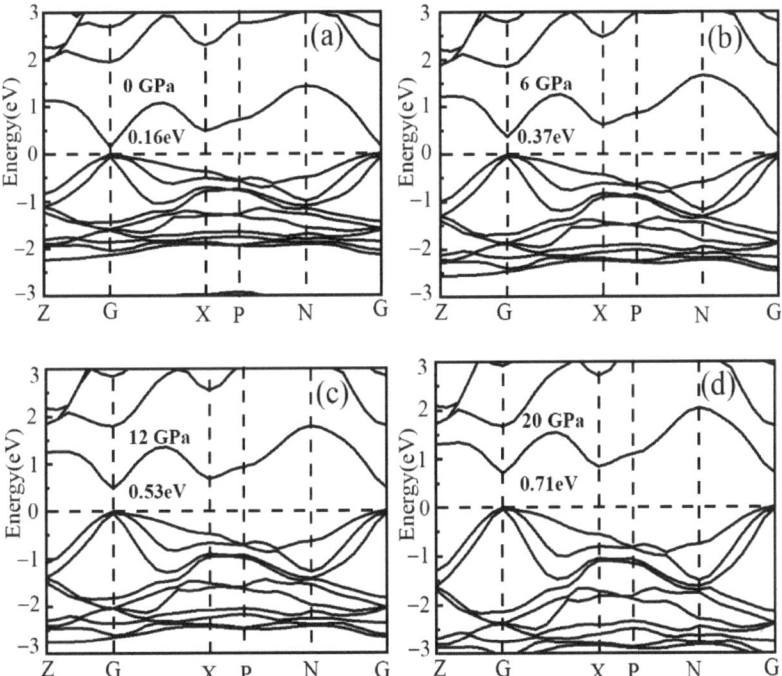

Figure 3. Band structure of Cu_2ZnSnS_4 under compressive stress for (**a**) 0 GPa band structure; (**b**) 6 GPa band structure; (**c**) 12 GPa band structure; and (**d**) 20 GPa band structure.

3.2.2. Density of Electronic States

The total density of states and the density of states of each atomic partial wave under the action of tensile stress are shown in Figure 4. The main contributions to the CZTS density of states are the Cu 3d configuration, Zn 3d configuration, Sn 5s and 5p configuration, and S 3s and 3p configuration, and the 3p configuration of Cu and Zn also has a small contribution.

It can be seen from Figure 4 that the lower valence band region of −14.5~−12.3 eV is mainly contributed by the 3s state of S and a small amount of 5p state electrons of Sn. The mid-valence band region of −8.4~−5.5 eV is mainly contributed by the 3d state of Zn, the 5s state of Sn, and a small amount of the 3p state of S and 4s state of Zn. The upper valence band region of −5.5~0 eV is mainly contributed by the 3d state of Cu, a small amount of the Sn 5p state, and the 3p state of S. The conduction band part is mainly contributed by the 5s and 5p states of Sn and a small amount of the 3p state of S.

With the increase intensile stress, the peak position of the density of states in the valence band of CZTS shifts to the higher-energy direction. The −13 eV peak is shifted to the high-energy band by 0.45 eV, the −6.88 eV peak is shifted to the high-energy band by 0.27 eV, the −3.69 eV peak is shifted to the high-energy band by 0.8 eV, and the −1.73 eV peak is shifted to the high-energy band by 0.42 eV. The peak position of the conduction band is shifted to the lower-energy direction, and the shift amount is essentially the same at around 0.13 eV.

The Cu_2ZnSnS_4 density of states and the partial wave density of states under the action of compressive stress are shown in Figure 5.

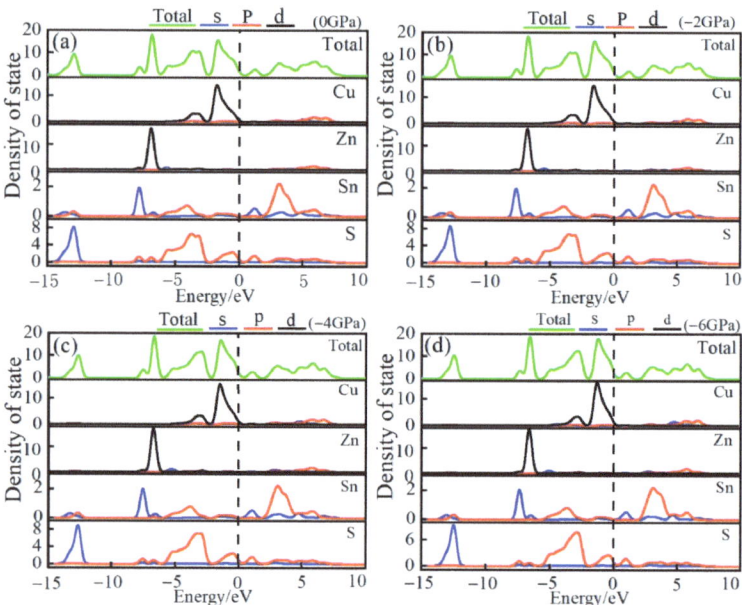

Figure 4. Cu_2ZnSnS_4 density of states under tensile stress for (**a**) 0 GPa density of states and fractional density of states; (**b**) −2 GPa density of states and fractional density of states; (**c**) −4 GPa density of states and fractional density of states; (**d**) −6 GPa density of states and partial wave density of states.

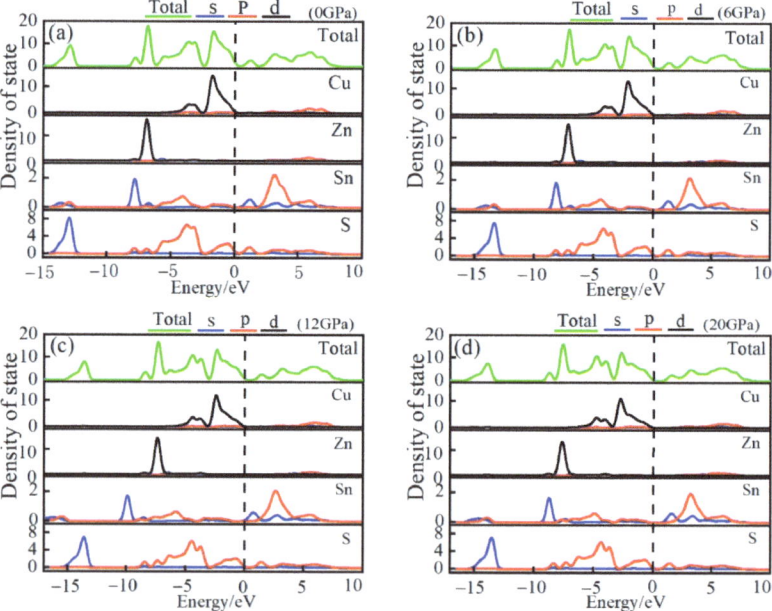

Figure 5. Cu_2ZnSnS_4 density of states under compressive stress for (**a**) 0 GPa density of states and fractional density of states; (**b**) 6 GPa density of states and fractional density of states; (**c**) 12 GPa density of states and fractional density of states; (**d**) 20 GPa density of states and partial wave density of states.

It can be seen from Figure 5 that the electronic configurations contributing to the valence and conduction bands of CZTS are essentially the same as those under tensile stress. The peak positions of the density of states in the valence band of CZTS shifted to the lower-energy direction with the increase in compressive stress, with an average offset of 0.64 eV, and the peak positions of the density of states in the conduction band shifted to the high-energy direction with an average offset of 0.17 eV.

We believe that when the lattice is subjected to tensile stress, the interatomic distance increases, the coupling degree of valence electrons decreases, the electrostatic repulsion decreases, the attractive force between valence electrons near the Fermi level and electrons in the conduction band increases, and some valence electrons break free. The tendency for nuclear confinement to occupy higher energy levels becomes more pronounced with increasing tensile stress.

Since valence electrons occupy higher energy levels, the energy required for their transition to the conduction band decreases and the band gap narrows. On the contrary, when the lattice is subjected to compressive stress, the atomic spacing decreases, the coupling degree of valence electrons increases, and the electrostatic repulsion increases, which hinders valence electrons from occupying high-energy states and pushes valence electrons to move to lower-energy orbitals. Therefore, the energy required for the transition of valence electrons to the conduction band increases, and the band gap becomes wider.

On the other hand, when the material is stressed, the Coulomb force changes due to the change in atomic spacing. These changes lead to an increase or decrease in exciton binding energy [39], which affects the dielectric shielding effect and ultimately affects the electron transport, making the band gap wider or narrower.

3.3. Optical Properties

3.3.1. Complex Dielectric Function

The dielectric function $\varepsilon(\omega) = \varepsilon_1(\omega) + i\varepsilon_2(\omega)$ is a complex number. The imaginary part ε_2 indicates that the polarization of the molecules inside the material cannot keep up with the change in the external electric field; it represents the loss. The real part ε_1 indicates the ability of the material to bind charges. The dielectric function reflects the band structure of the solid and its spectral information [40], and the dielectric peak of the dielectric function is mainly determined by the band structure and density of states. Figure 6 is a graph showing the change in the CZTS real part ε_1 and imaginary part ε_2 with photon energy under stress.

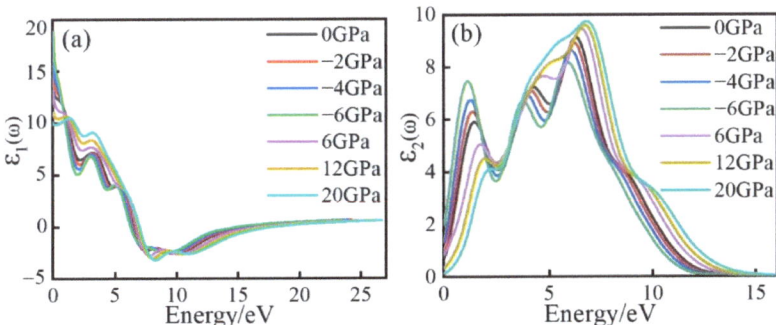

Figure 6. The complex dielectric function of Cu_2ZnSnS_4 under stress for (a) the real part of the complex dielectric function; (b) the imaginary part of the complex dielectric function.

It can be seen from Figure 6a that when the tensile stress is 0, −2, −4, and −6 GPa, the corresponding static dielectric constants are 13.74, 14.98, 16.48, and 18.88. When the compressive stress is 6, 12, and 20 GPa, the corresponding static dielectric constants are 13.09, 11.09, and 10.04. It can be seen from the figure that the static permittivity of CZTS increases when the lattice is subjected to tensile stress, while the static permittivity of CZTS decreases when it is subjected to compressive stress. At 0 GPa, the imaginary part of the complex dielectric function of CZTS has four distinct characteristic peaks at 1.44, 4.23, 6.26, and 8.58 eV. Combining the density of states in Figure 4a, it can be seen that the dielectric peak at 1.44 eV has mainly transitioned from Cu 3d orbital electrons to Sn 5s orbital electrons, and the dielectric peak at 4.23 eV has mainly transitioned from Cu 3d orbital electrons to Sn 5p orbital electrons, at 6.26 eV. The dielectric peak at 8.58 eV mainly transitions from Cu 3d orbital electrons to Sn 5p orbital electrons, and the dielectric peak at 8.58 eV mainly transitions from Zn 3d orbital electrons to Sn 5s orbital electrons. When the energy is lower than 3.78 eV, the dielectric peak shifts to the high-energy direction with the increase instress, and the peak decreases with the increase instress. When the energy is greater than 3.78 eV, the dielectric peak also shifts to the high-energy direction with the increase instress, but, at this time, the peak increases with the increase instress. As the stress increases, the CZTS band gap increases, so the dielectric peak shifts toward higher energies. The above conclusions are consistent with the results of the literature [41]; this is in agreement with the relation between the band gap energy and the dielectric response, which implies that the larger band gap energy results in a small dielectric constant.

3.3.2. Absorption and Reflection Spectra

The absorption spectrum of Cu_2ZnSnS_4 is shown in Figure 7a. It can be seen from the figure that the absorption spectrum of CZTS is mainly divided into three parts: the visible light region of 0.5~4.8 eV, the ultraviolet light absorption region of 4.8~14.2 eV, and the absorption range of more than 14.2 eV in the high-energy absorption area. In the range of 0.5~9.9 eV, the absorption coefficient gradually increased with the increase in incident light energy, and the light absorption decreased sharply when the energy was greater than 9.9 eV, while CZTS almost no longer absorbed the spectrum when the energy was greater than 16.19 eV. It can be seen from the figure that the absorption coefficient of CZTS in the visible light band is greater than 10^4 cm^{-1} [42], and the light absorption is good.

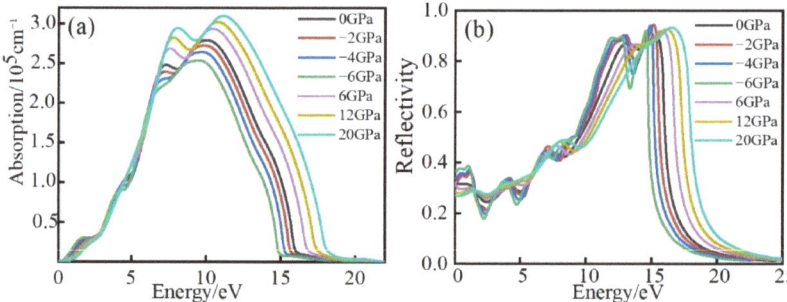

Figure 7. Absorption and reflection spectra for (**a**) the absorption spectrum and (**b**) the reflection spectrum.

With the increase incompressive stress, the absorption peak of CZTS shifted to the high-energy direction, and the peak value gradually increased. With the increase intensile stress, the absorption peak of CZTS shifted to the lower-energy direction, and the peak value gradually decreased. The results obtained in this paper are very consistent with the research results of Kahlaoui [43], and the phenomena could be attributed to the decrease in the electronic transition energy with the strain, which results in more photons being absorbed.

In this work, it is believed that when the lattice is subjected to tensile stress, the atomic spacing becomes larger, and the light wave of the same energy is "short-wave" relative to the lattice, the lattice scattering effect is enhanced, and the light absorption is weakened. In contrast, when the lattice is subjected to compressive stress, the atomic spacing decreases, and the light wave with the same energy is "long-wave" relative to the lattice, the lattice scattering effect is weakened, and the light absorption is enhanced.

Figure 7b shows the reflectance spectrum of CZTS. The reflectance in the visible light region of 0.5~4.8 eV is lower than 30%, indicating that CZTS has good light absorption in the visible light region. With the increase in incident light energy, the reflectivity gradually increased and reached a peak value of 90% near 15.30 eV, and high reflectivity appeared in the ultraviolet and high-energy regions. As the compressive stress increases, the reflection peak shifts towards higher energy. However, as the tensile stress increases, the reflection peak shifts toward the lower-energy direction.

3.3.3. Complex Refractive Index

Figure 8 shows the Cu_2ZnSnS_4 complex refractive index; (a) is the refractive index, and (b) is the extinction coefficient.

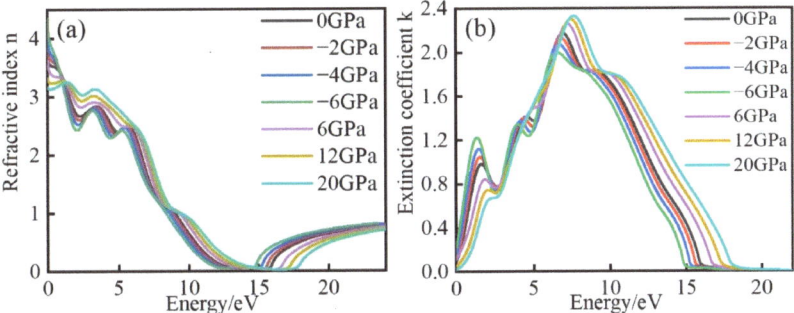

Figure 8. Complex refractive index for (**a**) the refractive index and (**b**) the extinction coefficient.

It can be seen from Figure 8 that when the energy is greater than 16 eV, the refractive index is essentially constant, and the extinction coefficient is around 0, indicating that the absorption of CZTS is weak at high frequencies, which is consistent with the conclusion of the absorption spectrum. It can also be seen from the figure that the refractive index decreases with increasing tensile stress and increases with increasing compressive stress. This is mainly because the unit cell volume becomes larger when tensile stress is applied, the density becomes worse, and the refractive index decreases. When compressive stress is applied, the volume of the unit cell decreases, the density of CZTS becomes better, and therefore the refractive index increases.

3.3.4. Complex Conductivity

It can be seen from Figure 9a that when the energy is less than 0.5 eV and greater than 12.6 eV, the real part of the complex conductivity is around 0—that is, there is almost no dissipation. It can be seen from the figure that the peak of the real part of the conductivity appears in the energy range of 1.5~8.8 eV, and the peak of the imaginary part appears in the energy range of 1.5~11.35 eV, which corresponds to the absorption spectrum. The real part of the conductivity reaches the maximum value near 6.4 eV. Combined with the analysis of the density of states map, it can be seen that these interband transition sources are related to the transition of Cu 3d orbital electrons to Sn 5p orbital electrons. With the increase intensile stress, the peak value of the real part of conductivity shifts to the direction of low energy, and with the increase incompressive stress, the peak value of the real part of conductivity shifts to the direction of high energy.

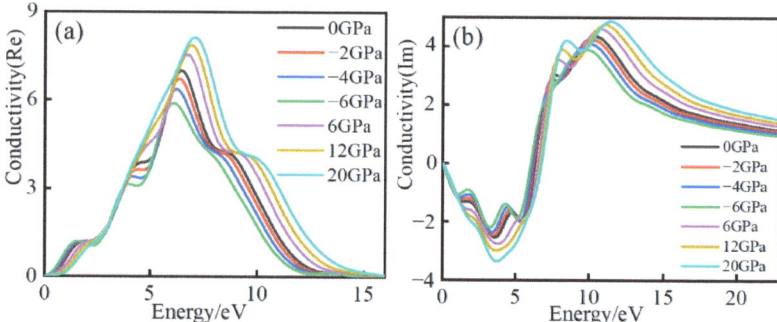

Figure 9. Complex conductivity for (**a**) real part and (**b**) imaginary part.

3.3.5. The Energy Loss Function

The energy loss function describes the energy loss of electrons when passing through a homogeneous medium. It can be seen from Figure 10 that the energy loss of Cu_2ZnSnS_4 reaches a maximum value of 57.22 near 15.89 eV at 0 GPa. When the stress is −6 GPa, the maximum energy loss is 149.23, and the minimum energy loss is 47.37 when the stress is 20 GPa.

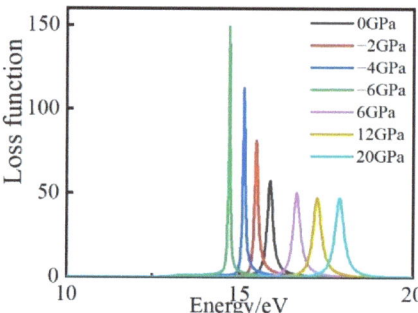

Figure 10. The energy loss function.

With the increase intensile stress, the peak position of the energy loss function shifts to the lower-energy direction, and the peak value increases significantly with the increase intensile stress. With the increase incompressive stress, the peak position of the energy loss function shifts to the high-energy direction, and the peak value decreases slowly with the increase incompressive stress. This is mainly because the CZTS band gap changes significantly when the lattice is stretched, while the CZTS band gap changes slowly when the lattice is compressed. Therefore, applying stress can tune the peak position and peak value of the CZTS energy loss function.

4. Conclusions

By using the pseudopotential plane-wave method of first principles based on density functional theory, the band structure, density of states, and optical properties of Cu_2ZnSnS_4 under isotropic stress are calculated and analyzed. The CZTS band gap is reduced by stretching the lattice under isotropic stress, and increased by compressing the lattice. The band gap is 0.16 eV when no stress is applied. When the material is stressed, the atomic spacing becomes larger or smaller, which changes the Coulomb force. These changes lead to anincrease or decrease inexciton binding energy, which ultimately affects the electron transport and makes the band gap wider or narrower. The valence band of CZTS is mainly contributed by the 3d electrons of Cu, and the conduction band is mainly contributed by the 5s electrons of Sn. When the stress changes from −6 GPa to 20 GPa, CZTS always presents a direct band gap semiconductor at the high-symmetry point and maintains a tetragonal crystal system. At −6 GPa, the maximum static dielectric constant is 18.88, and the dielectric peak shifts to the high-energy direction with the increase instress. The absorption coefficient of CZTS in the visible light band is greater than 10^4 cm^{-1}, and the reflectivity is lower than 30% in the energy range of 0.5~4.8 eV, showing good light absorption characteristics and low reflectivity. With the increase incompressive stress, the absorption peak of CZTS shifts towards the direction of high energy, and the peak value gradually increases. When the energy is greater than 16 eV, the refractive index is essentially constant, and the extinction coefficient is approximately 0. The maximum value of energy loss function is 57.22 near 15.89 eV, and the peak position of the energy loss function shifts to the direction of high energy with the increase incompressive stress, while the peak value decreases slowly with the increase incompressive stress. In summary, the band gap, optical absorption, dielectric constant, and other properties of CZTS can be adjusted by applying stress; this work provides a strong reference for the experimental preparation of high-quality CZTS films and can help in the development of efficient CZTS solar cells.

Author Contributions: Contributions, X.Y.; experimental design, X.Y.; simulation calculation, X.Y.; writing—review and editing, X.Q.; model building, X.Q.; data analysis, W.Y.; overall planning, W.Y.; review and revision of the thesis, C.Z.; drawing, C.Z.; literature review, D.Z.; software, D.Z.; analysis, B.G.; design of the question, B.G. All authors have read and agreed to the published version of the manuscript.

Funding: This research was funded by the Key Laboratory of Materials Simulation and Computing of Anshun University (Asxyxkpt201803), and the Youth Growth Project of Guizhou Provincial Department of Education, grant number KY (2020) 134.

Institutional Review Board Statement: Not applicable.

Informed Consent Statement: Not applicable.

Data Availability Statement: Not applicable.

Acknowledgments: The Youth Science and Technology Talent Growth Project of the Education Department of Guizhou Province (No.2020138), the Key Supporting Discipline of Materials and Aviation of Anshun College (2020), and the Guizhou Province JMRH Integrated Key Platform Funding Project.

Conflicts of Interest: The authors declare no conflict of interest.

References

1. Yue, Q.; Liu, W.; Zhu, X. N-Type Molecular Photovoltaic Materials: Design Strategies and Device Applications. *J. Am. Chem. Soc.* **2020**, *142*, 11613–11628. [CrossRef] [PubMed]
2. Sun, J.M.; Zhao, E.; Liang, J.; Li, H.; Zhao, S.; Wang, G.; Gu, X.; Tang, B.Z. Diradical-Featured Organic Small-Molecule Photothermal Material with High-Spin State in Dimers for Ultra-Broadband Solar Energy Harvesting. *Adv. Mater.* **2022**, *34*, 2108048. [CrossRef] [PubMed]
3. Harijan, D.; Gupta, S.; Ben, S.K.; Srivastava, A.; Singh, J.; Chandra, V. High photocatalytic efficiency of α-Fe$_2$O$_3$-ZnO composite using solar energy for methylene blue degradation. *Phys. B Condens. Matter* **2022**, *627*, 413567. [CrossRef]
4. Hu, Y.-H.; Li, M.-J.; Zhou, Y.-P.; Xi, H.; Hung, T.-C. Multi-physics investigation of a GaAs solar cell based PV-TE hybrid system with a nanostructured front surface. *Sol. Energy* **2021**, *224*, 102–111. [CrossRef]
5. Celline, A.C.; Subagja, A.Y.; Suryaningsih, S.; Aprilia, A.; Safriani, L. Synthesis of TiO$_2$-rGO Nanocomposite and its Application as Photoanode of Dye-Sensitized Solar Cell (DSSC). *Mater. Sci. Forum* **2021**, *1028*, 151–156. [CrossRef]
6. Guo, W.H.; Zhu, Y.H.; Zhang, M.; Du, J.; Cen, Y.; Liu, S.; He, Y.; Zhong, H.; Wang, X.; Shi, J. The Dion–Jacobson perovskite CsSbCl$_4$: A promising Pb-free solar-cell absorber with optimal bandgap 1.4 eV, strong optical absorption 10^5 cm^{-1}, and large power-conversion efficiency above 20%. *J. Mater. Chem. A* **2021**, *9*, 16436–16446. [CrossRef]
7. Das, B.; Hossain, S.M.; Nandi, A.; Samanta, D.; Pramanick, A.K.; Chapa, S.O.M.; Ray, M. Spectral conversion by silicon nanocrystal dispersed gel glass: Efficiency enhancement of silicon solar cell. *J. Phys. D Appl. Phys.* **2021**, *55*, 025106. [CrossRef]
8. Yan, F.R.; Yang, P.Z.; Li, J.B.; Guo, Q.; Zhang, Q.; Zhang, J.; Duan, Y.; Duan, J.; Tang, Q. Healing soft interface for stable and high-efficiency all-inorganic CsPbIBr$_2$ perovskite solar cells enabled by S-benzylisothiourea hydrochloride. *Chem. Eng. J.* **2021**, *430*, 132781. [CrossRef]

9. Bi, P.; Zhang, S.; Chen, Z.; Xu, Y.; Cui, Y.; Zhang, T.; Ren, J.; Qin, J.; Hong, L.; Hao, X.; et al. Reduced non-radiative charge recombination enables organic photovoltaic cell approaching 19% efficiency. *Joule* **2021**, *5*, 2408–2419. [CrossRef]
10. Khataee, A.; Azevedo, J.; Dias, P.; Ivanou, D.; Draževič, E.; Bentien, A.; Mendes, A. Integrated design of hematite and dye-sensitized solar cell for unbiased solar charging of an organic-inorganic redox flow battery. *Nano Energy* **2019**, *62*, 832–843. [CrossRef]
11. Shen, L.; Li, H.; Meng, X.; Li, F. Transfer printing of fully formed microscale InGaP/GaAs/InGaNAsSb cell on Ge cell in mechanically-stacked quadruple-junction architecture. *Sol. Energy* **2020**, *195*, 6–13. [CrossRef]
12. Lin, S.; Xie, S.; Lei, Y.; Gan, T.; Wu, L.; Zhang, J.; Yang, Y. Betavoltaic battery prepared by using polycrystalline CdTe as absorption layer. *Opt. Mater.* **2022**, *127*, 112265.
13. Yeojun, Y.; Sunghyun, M.; Sangin, K.; Jaejin, L. Flexible fabric-based GaAs thin-film solar cell for wearable energy harvesting applications. *Sol. Energy Mater. Sol. Cells* **2022**, *246*, 111930.
14. Fatemeh, G.Y.; Ali, F. Performance enhancement of CIGS solar cells using ITO as buffer layer. *Micro Nanostruct.* **2022**, *168*, 207289.
15. Minemoto, T.; Matsui, T.; Takakura, H.; Hamakawa, Y.; Negami, T.; Hashimoto, Y.; Uenoyama, T.; Kitagawa, M. Theoretical analysis of the effect of conduction band offset of window/CIS layers on performance of CIS solar cells using device simulation. *Sol. Energy Mater. Sol. Cells* **2001**, *67*, 83–88.
16. Pan, B.; Wei, M.; Liu, W.; Jiang, G.; Zhu, C. Fabrication of Cu_2ZnSnS_4 absorber layers with adjustable Zn/Sn and Cu/Zn+Sn ratios. *J. Mater. Sci. Mater. Electron.* **2014**, *25*, 3344–3352. [CrossRef]
17. Tablero, C. Effect of the oxygen isoelectronic substitution in Cu_2ZnSnS_4 and its photovoltaic application. *Thin Solid Films* **2012**, *520*, 5011–5013. [CrossRef]
18. Su, Z.H.; Liang, G.X.; Fan, P.; Luo, J.; Zheng, Z.; Xie, Z.; Wang, W.; Chen, S.; Hu, J.; Wei, Y.; et al. Device Postannealing Enabling over 12% Efficient Solution-Processed Cu_2ZnSnS_4 Solar Cells with Cd^{2+} Substitution. *Adv. Mater.* **2020**, *32*, 2000121. [CrossRef]
19. Andrea, C.; Ole, H. What is the band alignment of $Cu_2ZnSn(S,Se)_4$ solar cells. *Sol. Energy Mater. Sol. Cells* **2017**, *169*, 177–194.
20. Bao, W.; Ichimura, M. Prediction of the Band Offsets at the CdS/Cu_2ZnSnS_4 Interface Based on the First-Principles Calculation. *Jpn. J. Appl. Phys.* **2012**, *51*, 10NC31. [CrossRef]
21. Su, Z.; Tan, J.; Li, X.; Zeng, X.; Batabyal, S.K.; Wong, L.H. Cation Substitution of Solution-Processed Cu_2ZnSnS_4 Thin Film Solar Cell with over 9% Efficiency. *Adv. Energy Mater.* **2015**, *5*, 1500682. [CrossRef]
22. Kim, S.; Kim, K.M.; Tampo, H.; Shibata, H.; Niki, S. Improvement of voltage deficit of Ge-incorporated kesterite solar cell with 12.3% conversion efficiency. *Appl. Phys. Express* **2016**, *9*, 102301. [CrossRef]
23. Chen, S.; Gong, X.G.; Wei, S.H. Band-structure anomalies of the chalcopyrite semiconductors $CuGaX_2$ versus $AgGaX_2$ (X=S and Se) and their alloys. *Phys. Rev. B* **2007**, *75*, 205209. [CrossRef]
24. Walsh, A.; Chen, S.; Wei, S.H.; Gong, X.G. Kesterite Thin-Film Solar Cells: Advances in Materials Modelling of Cu_2ZnSnS_4. *Adv. Energy Mater.* **2012**, *2*, 400–409. [CrossRef]
25. Tian, Y.; Zheng, Q.; Zhao, J. Tensile Strain-Controlled Photogenerated Carrier Dynamics at the van der Waals Heterostructure Interface. *J. Phys. Chem. Lett.* **2020**, *11*, 586–590. [CrossRef] [PubMed]
26. Fan, L.L.; Chen, S.; Luo, Z.L.; Liu, Q.H.; Wu, Y.F.; Song, L.; Ji, D.X.; Wang, P.; Chu, W.S.; Gao, C.; et al. Strain Dynamics of Ultrathin VO_2 Film Grown on TiO_2 (001) and the Associated Phase Transition Modulation. *Nano Lett.* **2014**, *14*, 4036–4043. [CrossRef]
27. LYU, L.; Yang, Y.Y.; CEN, W.F.; YAO, B.; OU, J.K. First-principles Study on Optical Properties of Cubic Ca_2Ge under Stress Effect. *Bull. Chin. Ceram. Soc.* **2019**, *38*, 3788–3795.
28. Yan, W.J.; Zhang, C.H.; Gui, F.; Zhang, Z. Electronic Structure and Optical Properties of Stressed β-$FeSi_2$. *Acta Opt. Sin.* **2013**, *33*, 243–249.
29. Lazarovits, B.; Kim, K.; Haule, K.; Kotliar, G. Effects of strain on the electronic structure of VO_2. *Phys. Rev. B* **2010**, *81*, 115117. [CrossRef]
30. Manyk, T.; Rutkowski, J.; Kopytko, M.; Martyniuk, P. Theoretical Study of the Effect of Stresses on Effective Masses in the InAs/InAsSb Type-II Superlattice. *Eng. Proc.* **2022**, *21*, 16.
31. Schorr, S.; Hoebler, H.J.; Tovar, M. A neutron diffraction study of the stannite-kesterite solid solution series. *Eur. J. Mineral.* **2007**, *19*, 65–73.
32. Segall, M.D.; Lindan, P.; Probert, M.J.; Pickard, C.J.; Hasnip, P.J.; Clark, S.J.; Payne, M.C. First-principles simulation: Ideas, illustrations and the CASTEP code. *J. Phys. Condens. Matter* **2002**, *14*, 2717–2744. [CrossRef]
33. Perdew, J.P.; Burke, K.; Ernzerhof, M. Generalized Gradient Approximation Made Simple. *Phys. Rev. Lett.* **1996**, *77*, 3865–3868. [CrossRef] [PubMed]
34. Vanderbilt, D. Soft self-consistent pseudopotentials in a generalized eigenvalue formalism. *Phys. Rev. B* **1990**, *41*, 7892–7895. [CrossRef] [PubMed]
35. Liu, J. First-Principles Prediction of Structural, Elastic, Mechanical, and Electronic Properties of Cu_2ZnSnS_4 under Pressure. *ECS J. Solid State Sci. Technol.* **2022**, *11*, 073011. [CrossRef]
36. Kumar, M.; Zhao, H.; Persson, C. Cation vacancies in the alloy compounds of $Cu_2ZnSn(S_{1-x}Se_x)_4$ and $CuIn(S_{1-x}Se_x)_2$. *Thin Solid Films* **2013**, *535*, 318–321. [CrossRef]
37. Kheraj, V.; Patel, K.K.; Patel, S.J.; Shah, D.V. Synthesis and characterisation of Copper Zinc Tin Sulphide (CZTS) compound for absorber material in solar-cells. *J. Cryst. Growth* **2013**, *362*, 174–177.
38. Zhao, H.; Persson, C. Optical properties of $Cu(In,Ga)Se_2$ and $Cu_2ZnSn(S,Se)_4$. *Thin Solid Films* **2011**, *519*, 7508–7512. [CrossRef]

39. Hoo, Q.Y.; Xu, Y. Detection of dielectric screening effect by excitons in two-dimensional semiconductors and its application. *Acta Phys. Sin.* **2022**, *71*, 124–138.
40. Prokopidis, K.; Kalialakis, C. Physical interpretation of a modified Lorentz dielectric function for metals based on the Lorentz–Dirac force. *Appl. Phys. B* **2014**, *117*, 25–32. [CrossRef]
41. Nainaa, F.Z.; Bekkioui, N.; Abbassi, A.; Ez, Z.H. First principle study of structural, electronic optical and electric properties of Ag_2MnSnS_4. *Comput. Condens. Matter* **2019**, *22*, e00443. [CrossRef]
42. Scragg, J.J.; Dale, P.J.; Peter, L.M.; Zoppi, G.; Forbes, I. New routes to sustainable photovoltaics: Evaluation of Cu_2ZnSnS_4 as an alternative absorber material. *Phys. Status Solidi* **2010**, *245*, 1772–1778. [CrossRef]
43. Kahlaoui, S.; Belhorma, B.; Labrim, H.; Boujnah, M.; Regragui, M. Strain effects on the electronic, optical and electrical properties of Cu_2ZnSnS_4: DFT study. *Heliyon* **2020**, *6*, e03713. [CrossRef] [PubMed]

MDPI AG
Grosspeteranlage 5
4052 Basel
Switzerland
Tel.: +41 61 683 77 34

Crystals Editorial Office
E-mail: crystals@mdpi.com
www.mdpi.com/journal/crystals

Disclaimer/Publisher's Note: The title and front matter of this reprint are at the discretion of the Guest Editors. The publisher is not responsible for their content or any associated concerns. The statements, opinions and data contained in all individual articles are solely those of the individual Editors and contributors and not of MDPI. MDPI disclaims responsibility for any injury to people or property resulting from any ideas, methods, instructions or products referred to in the content.

www.ingramcontent.com/pod-product-compliance
Lightning Source LLC
LaVergne TN
LVHW072348090526
838202LV00019B/2501